The Cell Biology of Fertilization

CELL BIOLOGY: A Series of Monographs

EDITORS

D. E. BUETOW
*Department of Physiology
and Biophysics
University of Illinois
Urbana, Illinois*

I. L. CAMERON
*Department of Anatomy
The University of Texas
Health Science Center at San Antonio
San Antonio, Texas*

G. M. PADILLA
*Department of Physiology
Duke University Medical Center
Durham, North Carolina*

A. M. ZIMMERMAN
*Department of Zoology
University of Toronto
Toronto, Ontario, Canada*

Recently published volumes

The Cell Biology of Fertilization

Edited by

Heide Schatten

Integrated Microscopy Resource for Biomedical Research
The University of Wisconsin–Madison
Madison, Wisconsin

Gerald Schatten

Integrated Microscopy Resource for Biomedical Research
The University of Wisconsin–Madison
Madison, Wisconsin

ACADEMIC PRESS, INC.

Harcourt Brace Jovanovich, Publishers

San Diego New York Berkeley Boston
London Sydney Tokyo Toronto

ACADEMIC PRESS, INC.
San Diego, California 92101

United Kingdom Edition published by
ACADEMIC PRESS LIMITED
24-28 Oval Road, London NW1 7DX

Library of Congress Cataloging-in-Publication Data

The Cell biology of fertilization / edited by Heide Schatten, Gerald
  Schatten.
      p.   cm. — (Cell biology)
    Includes bibliographies and index.
    Companion v. to: The Molecular biology of fertilization / edited
  by Heide Schatten, Gerald Schatten.   1989
    ISBN   0-12-622590-7   (alk. paper)
    1. Fertilization.   2. Spermatozoa—Motility.   3. Embryology.
  I. Schatten, Heide.   II. Schatten, Gerald.   III. Molecular biology
  of fertilization.   IV. Series.
    [DNLM: 1. Fertilization.   2. Germ Cells—physiology.   3. Molecular
  Biology.   QH 485 C393]
  QP273.C45   1988
  591.3′2—dc19
  DNLM/DLC
  for Library of Congress   88-10515
                                                                  CIP

PRINTED IN THE UNITED STATES OF AMERICA
89 90 91 92   9 8 7 6 5 4 3 2 1

To our predecessors and successors . . .

Contents

I Sperm Behavior and Motility

1 Ionic Regulation of the Sea Urchin Sperm Acrosome Reaction and Stimulation by Egg-Derived Peptides

Robert W. Schackmann

2 Caltrin and Calcium Regulation of Sperm Activity

Henry Lardy and Jovenal San Agustin

3 Sperm Motility in Nematodes: Crawling Movement without Actin

Thomas M. Roberts, Sol Sepsenwol, and Hans Ris

II Remodeling of Egg Architecture

4 Whole-Mount Analyses of Cytoskeletal Reorganization and Function during Oogenesis and Early Embryogenesis in *Xenopus*

Joseph A. Dent and Michael W. Klymkowsky

5 Egg Cortical Architecture

Frank J. Longo

6 Cytoplasmic Mictrotubule-Associated Motors

J. M. Scholey, M. E. Porter, R. J. Lye, and J. R. McIntosh

7 The Fine Structure of the Formation of Mitotic Poles in Fertilized Eggs

Neidhard Paweletz and Daniel Mazia

8 Intermediate Filaments during Fertilization and Early Embryogenesis

Harald Biessmann and Marika F. Walter

9 Nuclear Architectural Changes during Fertilization and Development

Stephen Stricker, Randall Prather, Calvin Simerly, Heide Schatten, and Gerald Schatten

10   Extracellular Remodeling during Fertilization

Bennett M. Shapiro, Cynthia E. Somers, and Peggy J. Weidman

11   Dispermic Human Fertilization: Violation of Expected Cell Behavior

Ismail Kola and Alan Trounson

III Ionic Regulation and Its Controls

12   G-Proteins and the Regulation of Oocyte Maturation and Fertilization

Paul R. Turner and Laurinda A. Jaffe

13   The Relaxation State of Water in Unfertilized and Fertilized Sea Urchin Eggs

Selma Zimmerman, Ivan L. Cameron, and Arthur M. Zimmerman

14 Calcium and Mitosis: A Mythos?

Christian Petzelt and Mathias Hafner

15 Arousal of Activity in Sea Urchin Eggs at Fertilization

David Epel

Contributors

Numbers in parentheses indicate the pages on which the authors' contributions begin.

Harald Biessmann (189), Developmental Biology Center, University of California, Irvine, Irvine, California 92717

Ivan L. Cameron (319), Department of Cellular and Structural Biology, The University of Texas Health Science Center at San Antonio, San Antonio, Texas 78229

Joseph A. Dent (63), Department of Molecular, Cellular, and Developmental Biology, University of Colorado at Boulder, Boulder, Colorado 80309

David Epel (361), Department of Biological Sciences, Stanford University, Hopkins Marine Station, Pacific Grove, California 93950

Mathias Hafner[1] (341), Institute of Cell and Tumor Biology, German Cancer Research Center, D-6900 Heidelberg 1, Federal Republic of Germany

Laurinda A. Jaffe (297), Department of Physiology, The University of Connecticut Health Center, School of Medicine, Farmington, Connecticut 06032

Michael W. Klymkowsky (63), Department of Molecular, Cellular, and Developmental Biology, University of Colorado at Boulder, Boulder, Colorado 80309

Ismail Kola (277), Centre for Early Human Development, Monash Medical Centre, Monash University, Clayton, Victoria, Australia 3168

Henry Lardy (29), Institute for Enzyme Research, The University of Wisconsin–Madison, Madison, Wisconsin 53705

Frank J. Longo (105), Department of Anatomy, College of Medicine, The University of Iowa, Iowa City, Iowa 52242

R. J. Lye (139), Department of Molecular, Cellular, and Developmental Biology, University of Colorado at Boulder, Boulder, Colorado 80309

[1]Present address: Knoll AG, 6700 Ludwigshafen, Federal Republic of Germany

Daniel Mazia (165), Department of Biological Sciences, Stanford University, Hopkins Marine Station, Pacific Grove, California 93950

J. R. McIntosh (139), Department of Molecular, Cellular, and Developmental Biology, University of Colorado at Boulder, Boulder, Colorado 80309

Neidhard Paweletz (165), Institute of Cell and Tumor Biology, German Cancer Research Center, D-6900 Heidelberg 1, Federal Republic of Germany

Christian Petzelt (341), Institute of Cell and Tumor Biology, German Cancer Research Center, D-6900 Heidelberg 1, Federal Republic of Germany

M. E. Porter (139), Department of Molecular, Cellular, and Developmental Biology, University of Colorado at Boulder, Boulder, Colorado 80309

Randall Prather (225), Integrated Microscopy Resource for Biomedical Research, The University of Wisconsin–Madison, Madison, Wisconsin 53706

Hans Ris (41), Department of Zoology, The University of Wisconsin–Madison, Madison, Wisconsin 53706

Thomas M. Roberts (41), Department of Biological Science, The Florida State University, Tallahassee, Florida 32306

Jovenal San Agustin (29), Institute for Enzyme Research, The University of Wisconsin–Madison, Madison, Wisconsin 53705

Robert W. Schackmann (3), Department of Biology, The University of Utah, Salt Lake City, Utah 84112

Gerald Schatten (225), Integrated Microscopy Resource for Biomedical Research, The University of Wisconsin–Madison, Madison, Wisconsin 53706

Heide Schatten (225), Integrated Microscopy Resource for Biomedical Research, The University of Wisconsin–Madison, Madison, Wisconsin 53706

J. M. Scholey (139), Department of Molecular, Cellular, and Developmental Biology, University of Colorado at Boulder, Boulder, Colorado 80309, and Department of Molecular and Cellular Biology, National Jewish Center for Immunology and Respiratory Medicine, Denver, Colorado 80206

Sol Sepsenwol (41), Department of Biology, The University of Wisconsin–Stevens Point, Stevens Point, Wisconsin 54481

Bennett M. Shapiro (251), Department of Biochemistry, University of Washington, Seattle, Washington 98195

Calvin Simerly (225), Integrated Microscopy Resource for Biomedical Research, The University of Wisconsin–Madison, Madison, Wisconsin 53706

Cynthia E. Somers (251), Department of Biochemistry, University of Washington, Seattle, Washington 98195

Stephen Stricker (225), Integrated Microscopy Resource for Biomedical Research, The University of Wisconsin–Madison, Madison, Wisconsin 53706

Alan Trounson (277), Centre for Early Human Development, Monash Medical Centre, Monash University, Clayton, Victoria, Australia 3168

Paul R. Turner (297), Department of Zoology, University of California, Berkeley, Berkeley, California 94720

Marika F. Walter (189), Developmental Biology Center, University of California, Irvine, Irvine, California 92717

Peggy J. Weidman (251), Department of Biochemistry, Princeton University, Princeton, New Jersey 08544

Arthur M. Zimmerman (319), Department of Zoology, University of Toronto, Toronto, Ontario, Canada M5S 1A1

Selma Zimmerman (319), Division of Natural Sciences, Glendon College, York University, Toronto, Ontario, Canada M4N 3M6

Preface

The origins of cell and molecular biology are rooted firmly in studies on fertilization. Those familiar with the classic monograph of E. B. Wilson (1928), "The Cell in Development and Heredity," will recognize that almost all of the central and still challenging problems in cell and molecular biology were investigated first in a developmental system, often an invertebrate gamete or embryo. Experimental manipulations of eggs from lower vertebrates, especially amphibians, expanded the conclusions derived from these fertilization studies. Moreover, the recent advances in routinely reliable methods for *in vitro* fertilization and embryo culture of mammalian oocytes, including those from humans, coupled with the power of molecular probes are resulting in conclusions with important and often surprising implications for cell and molecular biology.

While the fields of cell and molecular biology have profited from fertilization as a model system for detailed investigations, understanding of the fertilization process has advanced correspondingly owing to this scrutiny as well as to the relative ease of designing experimental approaches. Indeed, the availability of sophisticated methods and probes is generating considerable new knowledge about the mechanisms accounting for gamete formation, recognition and fusion, reinitiation of the egg's metabolism, blocks to polyspermy, and cytoskeletal and motility events, as well as the changes in the pronuclei which permit syngamy and the activation of new gene expression.

The goal of "The Cell Biology of Fertilization" and its companion volume "The Molecular Biology of Fertilization" is to bring together reviews from leading laboratories in which various aspects of the fertilization process are studied. An assortment of experimental approaches is presented, using methods of cell biology, molecular biology, biochemistry, biophysics, enzymology, and immunology. Though our goal was to solicit articles on exciting research areas, a diversity of animal models is considered. Representatives from five invertebrate phyla are presented, including nematodes, clams, insects, ascidians, and the classic sea urchin. Amphibians and mammals are the best understood vertebrates, and it is encouraging that a diversity of mammals are now being explored. The articles consider the familiar mouse, rat, and hamster models, and also inquire about the fertilization process in farm animals, including pigs, sheep, and the Wisconsin favorite, the cow, as well as the animal with consequential clinical and ethical considerations, humans.

The chapters cover various aspects of fertilization as studied from different points of view by various authors. These chapters summarize work at varying levels of organization. In many cases we asked the contributors to restrict themselves to studies of one particular problem or with a specific approach. The authors were asked to include an overview of the field, to review recent and active research in their own laboratories, and to describe the conclusions in a manner which would be readily understood by a broad range of biologists, including those just beginning studies of fertilization as well as those in allied areas. They were encouraged to speculate on the future directions of fertilization research and to contribute new and unpublished material. We anticipate that these volumes will provide background and perspectives into research on fertilization that will be of use to a broad range of scientists, including advanced students interested in fundamental cell and molecular processes, cell biologists, molecular biologists, developmental biologists, geneticists, biochemists, biophysicists, and reproductive biologists.

Each book is subdivided into three sections. "The Cell Biology of Fertilization" first considers sperm behavior and motility. Part II reviews aspects of egg architecture, ranging from extracellular remodeling, through cortical and cytoskeletal structure, to the organization of the nuclei which participate in fertilization and embryogenesis. Part III evaluates the regulatory ions involved in egg activation as well as the manner in which the sperm initiates this cascade of events. Its companion volume "The Molecular Biology of Fertilization" begins with a series of chapters on the molecules involved in sperm–egg recognition and binding. Part II explores pronuclear formation, activation, and the cytoskeletal events resulting in syngamy and cell cycle progression. Part III covers oncogenes, gene expression, and nuclear determination at fertilization and during embryogenesis. It is our hope that these books will provide the reader with a deeper appreciation of the present state of knowledge and the future directions for cellular and molecular investigations on fertilization, which is the critical event bridging our discontinuity in generations.

We are indebted to the Cell Biology series editors, Drs. Dennis Buetow, Ivan Cameron, George Padilla, and Arthur Zimmerman for cheerfully answering a myriad of questions and providing helpful advice. We are grateful for the thoughtful and timely contributions by the authors. Finally, we would like to extend a word of thanks to their funding agencies around the world, without whose support basic biomedical research would be seriously endangered. We thank Ms. Gina Hellenbrand for superb and tireless assistance.

Heide Schatten
Gerald Schatten

I

Sperm Behavior and Motility

1

Ionic Regulation of the Sea Urchin Sperm Acrosome Reaction and Stimulation by Egg-Derived Peptides

ROBERT W. SCHACKMANN

Department of Biology
The University of Utah
Salt Lake City, Utah 84112

I. INTRODUCTION

For successful fertilization, the sperm must locate the egg and be able to fuse with the egg plasma membrane. In many species, these functions are achieved through sperm motility and the acrosome reaction. Sperm motility, the progressive movement and its velocity, is often dependent on the activity of a single flagellum (Gibbons, 1981), and in many invertebrates, chemotaxis occurs in response to material in egg coats (Miller, 1985). In at least one species of sea urchin, a specific egg peptide serves as a chemoattractant and also

3

stimulates sperm respiration (Ward *et al.*, 1985b; Suzuki and Garbers, 1984). The acrosome reaction occurs near the egg surface to prepare the sperm for fusion with the egg plasma membrane. Exocytosis of the acrosomal granule exposes proteins necessary for sperm–egg binding and/or fusion (reviewed in Tilney, 1985; Dan, 1967; Shapiro *et al.*, 1981; Trimmer and Vacquier, 1986) and is also triggered by specific components of the egg coat (SeGall and Lennarz, 1979; Wassarman *et al.*, 1985; Wassarman, 1987).

This chapter reviews studies that characterize the ability of egg factors to initiate the acrosome reaction and to affect sperm motility by altering ion movements across the sperm plasma membrane. Increases in intracellular pH (pH_i) and intracellular $[Ca^{2+}]$ ($[Ca^{2+}]_i$) are key signals involved in the acrosome reaction in echinoids (Schackmann *et al.*, 1981; Lee *et al.*, 1983; Trimmer *et al.*, 1986) and probably in mammals (Yanagimachi and Usui, 1974; Meizel, 1984; Murphy and Yanagimachi, 1984) and with alteration of sperm motility by peptides isolated from egg jelly (Hansbrough and Garbers, 1981a; Repaske and Garbers, 1983). In this review, I summarize existing data and hypotheses about pH_i and $[Ca^{2+}]_i$ changes in sperm and speculate on mechanistic directions anticipated to bring our understanding of sperm behavior to a level that can be interpreted in terms of regulatory molecules in the plasma membrane. Because echinoderms provide a rich source of egg factors important to sperm function, and detailed physiological studies have been performed with these factors, I have focused my attention on studies using these invertebrates. Other reviews have recently been published, which provide more general analyses of the acrosomal reaction (Tilney, 1985), directed sperm movement or chemotaxis (Miller, 1985), mammalian sperm receptors (Wassarman *et al.*, 1985), and the mammalian sperm acrosome reaction (Meizel, 1984; Wassarman, 1987).

II. THE ACROSOME REACTION

A. Background

The modern cell biology and physiology of the acrosome reaction have their origins in a series of morphological studies by Jean Dan (1952, 1954a,b). Dan observed changes in the acrosomal region in both sea urchin and starfish sperm, when they were exposed to "egg water" (seawater containing egg surface materials). A sticky substance appeared to be released from the acrosomal granule in sea urchin sperm and a small rod of ~1 μm extended from the apical tip within a few seconds after exposure to egg water or in the absence of any egg factors if the sperm were placed in seawater at pH 9.2. In other echinoderms, this acrosomal rod was considerably longer (20–30 μm) and was suggested to be the thin filament observed to connect the sperm and the egg during fertilization (Colwin and Colwin, 1956; Dan, 1954a). Extension of the acrosomal rod was shown to result from actin polymerization (Tilney *et al.*, 1973, 1978). Blebbing of the membrane overlying the filament as it extends suggested that alteration of the intracellular volume results from water and ion movements during the reaction (Tilney and Inoué, 1982).

For both invertebrate (Dan, 1954b) and mammalian (Yanagimachi and Usui, 1974) sperm, extracellular Ca^{2+} is required for the acrosome reaction. This requirement reflects a need for Ca^{2+} entry to increase $[Ca^{2+}]_i$. Increases in cytosolic $[Ca^{2+}]$ attend most, if not all, exocytotic changes in cell biology and can result either from enhanced entry across the plasma membrane or by release from an intracellular site. By following $^{45}Ca^{2+}$ uptake into sea urchin sperm, we confirmed that Ca^{2+} entry was enhanced when egg jelly stimulated the acrosome reaction (Schackmann *et al.*, 1978; Kopf and Garbers, 1980). Unlike the morphological changes of the reaction itself, which are complete within seconds, $^{45}Ca^{2+}$ uptake continues for tens of minutes and represents Ca^{2+} accumulation by the mitochondria (Cantino *et al.*, 1983; Schackmann and Shapiro, 1981). Release of Ca^{2+} from an intracellular site is unlikely in sperm on morphological grounds. Intracellular Ca^{2+} release occurs from the endoplasmic reticulum which is absent in sperm. The sperm midpiece does contain mitochondria, the number varies among different species, and though mitochondria can accumulate and release Ca^{2+} (Carafoli, 1982), these organelles are not currently thought to do so under physiological conditions (Somlyo *et al.*, 1985). Using $[Ca^{2+}]_i$ indicators, no evidence for intracellular release has been found in sea urchin sperm (see Section II,C).

Measurement of H^+ efflux accompanying the acrosome reaction in echinoid sperm (Tilney *et al.*, 1978; Schackmann *et al.*, 1978) suggested that an increase in pH_i is important as well. Tilney *et al.* (1978) found that even in the absence of seawater Ca^{2+}, ionophores that increased pH_i (nigericin and X537A) caused actin polymerization, although organized filaments did not form. Membrane fusion between acrosomal and plasma membranes did not occur unless Ca^{2+} was present. When higher concentrations of the weak-base NH_4^+ were used to increase pH_i in starfish sperm, extended filaments were observed, although the acrosomal granule was still present (Schroeder and Christen, 1982). Tilney suggested that the acrosome reaction could be reduced to a two-step process, Ca^{2+}-dependent membrane fusion and pH-dependent actin polymerization. The basic principals of this hypothesis have remained correct. However, subsequent measurements revealed egg jelly increases pH_i by only ~0.2 pH units. In contrast, ionophores and the weak-base NH_4^+ induced substantially larger increases (on the order of 0.5 pH units). Additional regulatory functions for pH_i are likely (see Section II,E).

B. Initiation of the Acrosome Reaction, Ion Requirements, and Inhibitors

In sea urchin sperm, the acrosome reaction is stimulated by high-molecular-weight material in the "jelly" coat surrounding the egg. "Egg jelly" is a mixture of several components including peptides, glycoproteins, and a sulfated fucose polymer (Hotta et al., 1970; SeGall and Lennarz, 1979; Kopf et al., 1979). Purification of a single active component has been achieved, and the fucose sulfate molecule demonstrates species specificity in its ability to initiate the acrosome reaction. However, in its most pure form, the fucose sulfate polymer is reported to initiate the acrosome reaction only at seawater Ca^{2+} concentrations in excess (36 mM) of those required for native egg jelly (half-maximal response at $\sim$3 mM). The data suggest that the fucose sulfate molecule may be altered during isolation. Its ability to bind Ca^{2+} is decreased (SeGall and Lennarz, 1981).

Identification of an egg jelly receptor has yet to be achieved for echinoderm sperm. Species-specific binding has been measured by SeGall and Lennarz (1981) using ^{125}I-labeled egg jelly (presumably labeled on the protein components). Receptor molecule(s) were not identified, and the interaction between the fucose sulfate molecule and specific sperm membrane components remains incompletely defined. In the mouse, a single glycoprotein, ZP3, in the zona pellucida surrounding the egg has been found to initiate the acrosome reaction (Bleil and Wassarman, 1983, 1987; Wassarman et al., 1985) and has recently been sequenced. The oliogosaccharide portion of the molecule retains the ability to bind sperm after proteolysis, but the protein part of the molecule is necessary to initiate the acrosome reaction. It has yet to be determined how ZP3 binding increases $[Ca^{2+}]_i$.

Several monovalent cation movements (in addition to H^+) are important to the acrosome reaction. Egg jelly stimulates not only Ca^{2+} and H^+ efflux, but also Na^+ uptake and efflux of K^+ (Schackmann and Shapiro, 1981; Cantino et al., 1983). The degree of Na^+ uptake and K^+ efflux is extensive. At least half the cellular $[K^+]$ is lost, and the intracellular $[Na^+]$ increases by $>$2-fold, as the ability of the sperm plasma membrane to maintain a normal monovalent cation distribution (high $[K^+]$ inside) is severely compromised. Influx of Na^+ occurs with a time course similar to H^+ efflux (Schackmann and Shapiro, 1981) and suggests that Na^+–H^+ exchange takes place, although the stoichiometry is $>$1 : 1 (Cantino et al., 1983). Removal of Na^+ from the seawater by substitution of choline$^+$, to maintain osmotic and ionic strength, prevents initiation of the acrosome reaction by egg jelly, but also lowers the pH$_i$ from $\sim$7.4 to 6.8 (Christen et al., 1982; Lee et al., 1983; Lee, 1984a,b; Bibring et al., 1984). From these initial observations, it is clear that multiple ionic changes occur in response to egg jelly. Several approaches have been used to arrive at the conclusion that the increase in both pH$_i$ and $[Ca^{2+}]_i$ is essential for the acrosome reaction and that they are the primary ionic parameters.

Alternate mechanisms of initiating the acrosome reaction have been investigated as an approach to understanding the changes induced by egg jelly. Each of these alternate methods of initiating the acrosome reaction is associated with increased $^{45}Ca^{2+}$ uptake and elevation of the pH_i to ~7.6 or greater. The Ca^{2+} ionophore A23187 initiates the acrosome reaction (Decker *et al.*, 1976; Collins and Epel, 1977; Talbot *et al.*, 1976) and increases both Ca^{2+} entry and H^+ efflux (Tilney *et al.*, 1978). Elevating the extracellular pH to ~9 or more initiates the acrosome reaction (Dan, 1952; Decker *et al.*, 1976; Gregg and Metz, 1976; Collins and Epel, 1977), increases the pH_i, initiates $^{45}Ca^{2+}$ uptake independent of seawater [Na^+] (Garcia-Soto and Darszon, 1985), and increases [Ca^{2+}]$_i$ (R. W. Schackmann, unpublished). Additionally, placing sperm in an artificial seawater in which all but 20–30 mM Na^+ has been replaced by choline$^+$ also triggers the acrosome reaction (Shapiro *et al.*, 1980; Schackmann and Shapiro, 1981). Under these conditions, Na^+–H^+ exchange occurs to increase pH_i to 7.6 or higher (Christen *et al.*, 1982, 1983c; Lee *et al.*, 1982, 1983) and to initiate $^{45}Ca^{2+}$ uptake and increased [Ca^{2+}]$_i$ (Schackmann and Shapiro, 1981; Lee *et al.*, 1983; Schackmann and Chock, 1986). The ionophore nigericin at concentrations that initiate the acrosome reaction in sea urchin sperm initially catalyzes K^+–H^+ exchange and subsequently causes a secondary, Na^+-dependent increase in pH_i and [Ca^{2+}]$_i$ (Schackmann *et al.*, 1978; Schackmann and Chock, 1986; Garbers, 1981; Lee *et al.*, 1983; Garcia-Soto *et al.*, 1987). The initial K^+–H^+ exchange collapses the transplasma membrane [K^+] gradient, depolarizes the plasma membrane potential, and leads initially to a decrease in pH_i (Schackmann *et al.*, 1984; Schackmann and Chock, 1986). It is only after the Na^+-dependent increase in pH_i that morphological changes are observed. In contrast, the [Ca^{2+}]$_i$ increase initiated by nigericin begins immediately and demonstrates that Ca^{2+} entry alone does not initiate the acrosome reaction. Further evidence for this point is presented in Section II,D.

Inhibitors known to block some types of either Ca^{2+} or K^+ channel activity prevent the acrosome reaction. Verapamil, D600, and several dihydropyridines inhibit Ca^{2+} channel activity in other cell types and inhibit the acrosome reaction (Schackmann *et al.*, 1978; Kazazoglou *et al.*, 1985; Garcia-Soto and Darszon, 1985). Increasing the seawater [K^+] from 10 to 20 mM prevents egg jelly from initiating the acrosome reaction as does tetraethylammonium, an inhibitor of K^+ channels (Schackmann *et al.*, 1978). These inhibitors all prevent $^{45}Ca^{2+}$ uptake (Schackmann *et al.*, 1978) and part of the increase in pH_i (Christen *et al.*, 1983b; Garcia-Soto *et al.*, 1987). Reducing the seawater pH to 7 or lower effectively blocks the acrosome reaction as well. Low pH can block Ca^{2+} channel activity (Iijima *et al.*, 1986), but also lowers pH_i (Christen *et al.*, 1982, 1983c).

The observations on initiation and inhibition of the acrosome reaction suggest that egg jelly stimulates multiple ionic changes to increase pH_i and [Ca^{2+}]$_i$.

A limitation of this general approach is that inhibitors which block the egg jelly acrosome reaction in many cases do not prevent the acrosome reaction triggered by these alternate methods. For example, the Ca^{2+} channel blockers do not prevent initiation of the acrosome reaction by the 20 mM Na^+– choline$^+$ medium, but do inhibit triggering by high seawater pH (Garcia-Soto and Darszon, 1985; Garcia-Soto et al., 1987). Elevated extracellular $[K^+]$ blocks induction by 20 mM Na^+/choline$^+$, but not by high extracellular pH. Tetraethylammonium inhibits neither alternate method, but effectively blocks egg jelly. These examples serve to demonstrate that each of the alternate methods of initiating the reaction has unique properties that are not identical with egg jelly and, therefore, cannot be assumed to initiate the identical sequence of biochemical events.

Studies with mammalian sperm suggest some similar mechanisms may operate. Besides the requirement for external Ca^{2+}, evidence that increased $[Ca^{2+}]_i$ is necessary for the acrosome reaction is derived from the ability of the Ca^{2+} ionophore A23187 to initiate the acrosome reaction in several species (Yanagimachi and Usui, 1974; Talbot et al., 1976). In hamster sperm, a rise in the acrosomal granule pH$_i$ was found to be associated with the acrosome reaction (Meizel and Deamer, 1978). Guinea pig sperm also do not undergo the acrosome reaction at reduced extracellular pH (Murphy and Yanagimachi, 1984). A requirement for extracellular $[K^+]$ and altered activity of the Na^+, K^+-ATPase has also been implicated in the acrosome reaction (Mrsny and Meizel, 1981), but how or if these observations are linked to changes in $[Ca^{2+}]_i$ or pH$_i$ is not yet understood. The Ca^{2+} channel blockers verapamil and dihydropyridines do not inhibit the acrosome reaction in guinea pig sperm (Roldan et al., 1986), but membrane potential-sensitive Ca^{2+} entry mechanisms have recently been reported in other mammalian sperm (Babcock and Pfeiffer, 1987).

C. $[Ca^{2+}]_i$ Changes and the Acrosome Reaction

$^{45}Ca^{2+}$ measurements allow us to demonstrate that egg jelly enhances Ca^{2+} influx. For further definition of the Ca^{2+} changes important to the acrosome reaction, isotope uptake is inadequate, as mitochondrial accumulation prevents evaluation of the amount of Ca^{2+} entering the sperm during the few seconds of the reaction itself. Additionally, it does not provide information about changes in free $[Ca^{2+}]_i$. Isotope uptake remains useful as a simple means of ascertaining whether Ca^{2+} entry is enhanced (Kopf et al., 1983, 1984).

Incorporation of the Ca^{2+} indicators fura-2, indo-1 (Grynkiewicz et al., 1985), or quin2 (Tsien et al., 1982) into sperm allows direct measurement of $[Ca^{2+}]_i$ changes associated with the acrosome reaction (Trimmer et al., 1986). These dyes are sensitive to $[Ca^{2+}]_i$ in the submicromolar range routinely found in cells. In sperm loaded with either indo-1 or fura-2, egg jelly initiates a rapid rise in $[Ca^{2+}]_i$ from a basal concentration of ~ 100 nM to 1–2 μM within 10 sec. The absolute values are dependent on assumptions made to calibrate the indicators within the sperm. We can directly demonstrate that the rapid increase stimulated by egg jelly arises from Ca^{2+} entry and not intracellular release, as the $[Ca^{2+}]_i$ changes are dependent on extracellular $[Ca^{2+}]$. Figure 1 shows increases in $[Ca^{2+}]_i$ as a function of the seawater $[Ca^{2+}]$. When jelly is added (arrow 1) to the sperm in the absence (<10 μM) of extracellular $[Ca^{2+}]$, no increase occurs within the sperm. This is true even if jelly is added immediately after dilution of sperm into Ca^{2+}-free seawater, following prolonged incubation in a medium containing Ca^{2+} to load any potential intracellular release site. Addition of Ca^{2+} (arrow 2) after jelly results in a rapid, substantial increase in $[Ca^{2+}]_i$. If Ca^{2+} is added to a concentration of 9 mM or more (A and B), the increase in $[Ca^{2+}]_i$ is sustained. This is characteristic of successful initiation of the acrosome reaction. If insufficient extracellular Ca^{2+} exists to initiate the acrosome reaction to a high percentage, the increase becomes transient. In this particular experiment, addition of 4 mM Ca^{2+} after egg jelly allows for a ~ 20-fold increase in $[Ca^{2+}]_i$, followed by a substantial decrease (C). This type of transient increase in $[Ca^{2+}]_i$ is characteristic of changes in $[Ca^{2+}]_i$ stimulated by egg jelly when the acrosome reaction is inhibited by verapamil, tetraethylammonium, or by elevated seawater $[K^+]$ (R. W. Schackmann, unpublished data). That is, a substantial initial increase occurs to a lower peak $[Ca^{2+}]_i$, and this is followed by a secondary decrease in $[Ca^{2+}]_i$.

The transient behavior of $[Ca^{2+}]_i$ when the reaction is inhibited demonstrates that sperm have the capacity to regulate $[Ca^{2+}]_i$. The sperm must contain within their plasma membranes molecules that not only allow for enhanced Ca^{2+} entry, but also molecules that can lower $[Ca^{2+}]_i$. Presumably a Ca^{2+}-translocating ATPase and/or Na^+–Ca^{2+} exchange is responsible for Ca^{2+} removal. Both activities exist in sperm of other species (Rufo et al., 1984; Breitbart et al., 1985), but have yet to be documented in sea urchin sperm. It is unlikely that the decreasing part of the transient results from mitochondrial accumulation, as it is not prevented by uncoupling agents (R. W. Schackmann, unpublished data).

The ability to rapidly measure $[Ca^{2+}]_i$ greatly enhances our knowledge of early changes important to the acrosome reaction, but does not define the type of entry mechanism(s) activated by egg jelly. Activation of Ca^{2+} channels is accepted as the most likely hypothesis, and the following observations support this. The Ca^{2+} channel antagonists verapamil, D600, and several dihydropyridines block the acrosome reaction and part of the initial rise in $[Ca^{2+}]_i$ over a similar concentration range (Schackmann et al., 1978; Kazazoglou et al., 1985). For example, verapamil blocks both the acrosome reaction and the initial increase in $[Ca^{2+}]_i$ with half-maximal inhibition at ~ 10 μM (R. W.

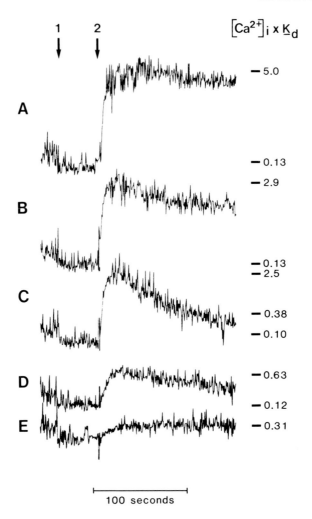

Fig. 1. Sperm were loaded with the Ca^{2+} indicator indo-1 as described in Trimmer *et al.* (1986) and diluted into Ca^{2+}-free artificial seawater to a final concentration of 2×10^8 sperm/ml. At arrow 1, egg jelly (35 μg fucose equivalents/ml) was added. Fluorescence emission was followed at 405 nm with excitation at 355 nm. At arrow 2, Ca^{2+} was added from a 1 *M* stock to give final concentrations of (A) 19 m*M*, (B) 9 mM, (C) 4 m*M*, (D) 1 m*M*, and (E) 0.1 mM. $[Ca^{2+}]_i$ calibrations were performed as described in Trimmer *et al.* (1986).

Schackmann, unpublished). Several dihydropyridines are also effective in the same concentration range. These concentrations are quite high compared to those used to selectively inhibit Ca^{2+} channels in many excitable tissues (Fleckenstein, 1977; Hagiwara and Byerly, 1981). The high concentrations necessary to inhibit the acrosome reaction limit the strength of arguments that the drugs act specifically to block Ca^{2+} channels in the sperm. For example, high (micromolar) concentrations of the dihydropyridines can inhibit phospodiesterase activity as well (Minocherhomjee and Roufogalis, 1984; Norman *et al.*, 1983). Even in mammalian tissues, however, Ca^{2+} channels show considerable variability in responsiveness to these inhibitors (Hagiwara *et al.*, 1981; Cognard *et al.*, 1986), and peptide toxins known to differentially inhibit vertebrate Ca^{2+} channels (Reynolds *et al.*, 1986) are ineffective at blocking Ca^{2+} movements in invertebrate Ca^{2+} channels (McCleskey *et al.*, 1987). A less selective Ca^{2+} channel inhibitor, the inorganic cation Co^{2+} (Hagiwara and Takahashi, 1967), blocks both the acrosome reaction and part of the egg jelly-stimulated increase in $[Ca^{2+}]_i$ (Trimmer *et al.*, 1987). As shown in Fig. 2, egg jelly allows for Ba^{2+} entry into the sperm, and this activity is partially inhibited by verapamil. Ca^{2+} channels pass Ba^{2+} among several other divalent ions (Hagiwara and Ohmori, 1982). Co^{2+} partially blocks Ba^{2+} entry into sperm as well (R. W. Schackmann, unpublished). All of these data are consistent with the Ca^{2+} channel hypothesis, but further analysis requires identification and characterization of single channel activities using electrophysiological

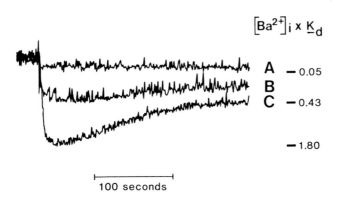

Fig. 2. Sperm were loaded with fura-2 as described in Trimmer *et al.* (1986) and diluted into Ca^{2+}-free artificial seawater to a final concentration of 2×10^8 sperm/ml. Ba^{2+} was added to a total concentration of 10 mM in B and C from a 1 M stock solution. No Ca^{2+} or Ba^{2+} was added in A. In B, 40 μM verapamil was added prior to the recording. Egg jelly was added to 35 μg fucose equivalents/ml to initiate the changes. Fluorescence emission was followed at 500 nm with excitation at 380 nm. Calibration was performed as described (Trimmer *et al.*, 1986).

methods. A preliminary report of Ca^{2+} channel activity from sperm was recently presented (Lievano et al., 1987).

It is likely that egg jelly activates sperm K^+ channels and that this activity contributes to regulation of $[Ca^{2+}]_i$ and pH_i. Inhibition of the acrosome reaction and the sustained increase in $[Ca^{2+}]_i$ by tetraethylammonium or by increasing the seawater $[K^+]$ suggests K^+ channel involvement and possible regulation by the plasma membrane potential (Schackmann et al., 1978, 1981; Garcia-Soto et al., 1987). Tetraethylammonium inhibits a variety of types of $[K^+]$ channels and tetraethylammonium-sensitive channels have been identified in sperm membrane preparations (Lievano et al., 1985; Guerrero et al., 1987). Activation of K^+ channels is expected to hyperpolarize the sperm plasma membrane potential and recent membrane potential measurements using the fluorescent dye, 3,3′-dipropylthiadicarbocyanin [diS-C_3-(5)], reveal a rapid transient hyperpolarization initiated with egg jelly (Garcia-Soto et al., 1987; Gonzalez-Martinez and Darszon, 1987). Such a hyperpolarization possibly activates the membrane potential-sensitive Na^+–H^+ exchange described in Section III,B (Lee, 1984a,b, 1985; Lee and Garbers, 1986) and may give rise to at least part of the increase in pH_i (Schackmann and Shapiro, 1981). Additionally, activation of the Na^+–H^+ exchange is directly associated with a Ca^{2+} entry mechanism that is insensitive to verapamil (see Section III,B; and Schackmann and Chock, 1986), and the acrosome reaction may be associated with activity of multiple Ca^{2+} entry mechanisms. In support of this idea is the observation that verapamil, when added after induction of the acrosome reaction, inhibits only ~30% of the $^{45}Ca^{2+}$ uptake (Schackmann et al., 1978). In contrast, addition of verapamil prior to egg jelly blocks 95% of $^{45}Ca^{2+}$ uptake.

Inhibition of the acrosome reaction by increased seawater $[K^+]$ can also be explained if a hyperpolarizing step occurs and is required for the reaction. The sperm membrane potential is depolarized by increased extracellular $[K^+]$ (Schackmann et al., 1981, 1984); this inhibits Na^+–H^+ exchange (Lee, 1984a,b, 1985; Lee and Garbers, 1986) and may prevent part of the increase in pH_i and activity of the verapamil-insensitive Ca^{2+} entry mechanism. However, not all of the rise in pH_i occurs by means of Na^+–H^+ exchange. Christen et al. (1983b) found that, even in the absence of both Na^+ and Ca^{2+}, egg jelly was capable of increasing pH_i.

Subsequent to the initial hyperpolarization, substantial (~25 mV), sustained depolarization of the plasma membrane potential occurs (Schackmann et al., 1981, 1984; Garcia-Soto et al., 1987). Since depolarization of the plasma membrane potential activates Ca^{2+} channels in other cells, it has been suggested to activate Ca^{2+} channels in sperm (Shapiro et al., 1985). However, because membrane potential measurements rely on equilibrium distribution of diffusible lipophilic ions which takes many seconds or even minutes in sperm, it is not known whether depolarization accompanies or follows the rapid increases in

$[Ca^{2+}]_i$ and pH_i (which are complete within ~ 10 sec). Further experimentation must be performed to determine if rapid, as yet undetected, depolarizing steps occur to directly activate Ca^{2+} channels or if the initial rise in $[Ca^{2+}]_i$ results from a type of Ca^{2+} channel active at hyperpolarized potentials.

D. Membrane Components Regulating Ca^{2+} Entry and the Increase in pH_i

Attempts to identify the membrane components involved in ion movements during the acrosome reaction fall into two categories. A traditional pharmacological approach has characterized binding of a calcium channel antagonist, verapamil. A second approach has been to identify membrane proteins important to the acrosome reaction with antibodies to sperm membrane proteins. Antibodies (Lopo and Vacquier, 1980; Saling et al., 1982; Podell and Vacquier, 1984a; Trimmer et al., 1985, 1987) have been developed that affect the acrosome reaction and fertilization.

Both Na^+ channels and Ca^{2+} channels have been purified from excitable tissue by virtue of their high affinity for ligands, which block or in other ways alter their activity. Using a sperm plasma membrane preparation, binding of the Ca^{2+} channel blocker, [³H]verapamil, has been analyzed (Kazazoglou et al., 1985). Specific binding with a K_d of 11 μM was determined and corresponds closely to concentrations effective at inhibiting the acrosome reaction and blocking the increase in $[Ca^{2+}]_i$. The affinity for the antagonist is low when compared with binding to high-affinity sites in muscle or nerve (Flockerzi et al., 1986) and the total number of specific sites (600 pmol/mg of membrane protein) is quite high. It is unlikely that such a concentration of Ca^{2+} channels exists, and it has been suggested that verapamil may bind to other membrane components as well as to Ca^{2+} channels (Kazazoglou et al., 1985). Competition for [³H]verapamil binding by unlabeled verapamil or by stereoisomers of D600 (methoxyverapamil) shows binding to sperm membranes is highly selective, although inhibition of the acrosome reaction shows no such stereoisomer selectivity. The acrosome reaction is blocked by a number of dihydropyridines at significantly different concentrations, indicating a degree of selectivity among these drugs. Because the affinity for verapamil is low, this approach has yet to identify specific membrane proteins that might serve as Ca^{2+} channels.

Use of antibodies to sperm membrane proteins as a tool to investigate sperm function and the acrosome reaction is well established. Metz et al. (1964) showed that fertilization could be blocked with antisperm antibodies. More recently, Lopo and Vacquier (1980) and Saling et al. (1982) showed that antibodies to specific sperm membrane proteins inhibited the acrosome reaction in *Strongylocentrotus purpuratus* and *Arbacia punctulata*, respectively. Lopo found that antibodies to an 80-kDa protein blocked the acrosome reaction initiated with egg jelly, but that elevated pH- or ionophore-initiated reactions were unaffected. Eckberg and Metz (1982) showed with *A. punctulata* sperm that an antibody that inhibited the acrosome reaction also immunoprecipiated a 68-kDa membrane protein. They suggested this might be an "egg jelly" receptor. Podell and Vacquier (1984a) found with *S. purpuratus* sperm that antibodies binding both 80- and 210-kDa membrane proteins not only blocked the acrosome reaction initiated by egg jelly, but also inhibited $^{45}Ca^{2+}$ uptake and H^+ efflux. The reaction could be initiated with alternate triggering methods, such as increased extracellular pH which bypassed the normal jelly-triggered events. A lectin, wheat germ agglutinin, was also found to bind to the 210-kDa protein and inhibited the acrosome reaction (Podell and Vacquier, 1984b).

The interest in the 210-kDa antigen led Trimmer *et al.* (1985) to develop monoclonal antibodies (mAb) to this and other sperm membrane proteins. Different antibodies to this protein have different physiological effects on the sperm. Two antibodies mAb J10/14 and mAb J4/4 both bind to the 210-kDa antigen and reveal its location on the sperm tail and in the acrosomal region. Each selectively reacts with the antigen by immunoprecipitation of sperm membranes. However, mAb J4/4 does not inhibit the acrosome reaction and does not prevent $^{45}Ca^{2+}$ uptake or the increase in pH_i. In contrast, mAb J10/14 and Fab fragments of J10/14 are both potent inhibitors of the acrosome reaction and block $^{45}Ca^{2+}$ uptake and the increase in pH_i.

Based on its ability to block the ion movements, Trimmer *et al.* (1985) initially suggested that mAb J10/14 could bind to and block a Ca^{2+} entry mechanism in the sperm plasma membrane. Using the intracellular Ca^{2+} indicators fura-2 and indo-1, we were surprised to find that mAb J10/14 actually increased $[Ca^{2+}]_i$ to even higher concentrations than egg jelly does (Trimmer *et al.*, 1986), in apparent contradiction to the isotope uptake data. However, as noted above in Section II,C and by Trimmer *et al.* (1986), $^{45}Ca^{2+}$ uptake assays mitochondrial accumulation after induction of the acrosome reaction. No long-term increase in $^{45}Ca^{2+}$ is stimulated by mAb J10/14, because the acrosome reaction is not initiated. In contrast, fura-2 is distributed throughout the sperm and responds to submicromolar $[Ca^{2+}]_i$ changes. The amount of $^{45}Ca^{2+}$ which enters the sperm to cause these changes is quite limited (<1 nmol/10^8 sperm) and cannot be easily resolved by following $^{45}Ca^{2+}$ uptake. These observations illustrate the limits of $^{45}Ca^{2+}$ uptake as a measure of physiologically significant changes in $[Ca^{2+}]_i$.

Consistent with their ability to inhibit the acrosome reaction, both mAb J10/14 and the Fab fragment to mAb J10/14 are effective at stimulating Ca^{2+} entry, but mAb J4/4 is without effect. In the absence of extracellular Ca^{2+}, mAb J10/14 causes no increase in $[Ca^{2+}]_i$ (Trimmer *et al.*, 1986). mAb J10/14 does not increase pH_i and prevents the increase when egg jelly is subsequently added. The monoclonal antibodies such as J10/14 effectively separate the pH_i and $[Ca^{2+}]_i$ steps (Trimmer et al., 1986). Since mAb J10/14 does not initiate the acrosome reaction, these data also directly demonstrate that increased $[Ca^{2+}]_i$ alone is insufficient to initiate the acrosome reaction. However, if $[Ca^{2+}]_i$ is increased with mAb J10/14 and the pH_i is increased by an alternate method, as with added NH_4^+, the acrosome reaction is initiated. The importance of both the increase of pH_i and $[Ca^{2+}]_i$ to initiation of the acrosome reaction is confirmed, and the data suggest the changes may result from activity of separate biochemical pathways.

The ability to inhibit the egg jelly-stimulated acrosome reaction or to initiate it, if the pH_i is separately increased, suggests that antibody-stimulated Ca^{2+} entry may be related to or identical with Ca^{2+} entry stimulated by egg jelly. Further similarities are evident from examination of inhibitor effects on mAb J10/14-stimulated increases in $[Ca^{2+}]_i$. As shown in Table I, each of the inhibitors of the egg jelly-stimulated changes is also effective at partially inhibiting the increase in $[Ca^{2+}]_i$ stimulated with mAb J10/14. Tetraethylammonium, verapamil, or increased $[K^+]$ (not shown) all partially prevent the increase in $[Ca^{2+}]_i$. Wheat germ agglutinin blocks nearly all of the antibody-stimulated increase, but it is not yet known whether this affects antibody binding. As with egg jelly, Ba^{2+} entry is also stimulated by mAb J10/14 and the inorganic Ca^{2+} channel blocker, Co^{2+}, is effective at inhibiting either Ba^{2+} or Ca^{2+} entry (R. W. Schackmann, unpublished).

An interesting question is whether the 210-kDa antigen is by itself a Ca^{2+}

TABLE I

Inhibition of mAb J10/14 Induced $[Ca^{2+}]_i$ Increase[a]

Conditions	$[Ca^{2+}]_i \times K_d$
Control	0.20
mAb J10/14 (30 µg/ml)	1.3
mAb J10/14 (30 µg/ml) + 30 µM verapamil	0.60
mAb J10/14 (30 µg/ml) + 10 mM TEA[b]	0.95
mAb J10/14 (30 µg/ml) + 40 µg/ml WGA[c]	0.35

[a] *Strongylocentrotus purpuratus* sperm were loaded with indo-1 as described (Schackmann and Chock, 1986; Trimmer et al., 1986). Following dilution into artificial seawater, mAb J10/14 was added in the presence or absence (control) of inhibitors of the acrosome reaction and changes in $[Ca^{2+}]_i$ were followed. Data presented were determined 4 min after antibody was added. The initial increase has partly collapsed. The behavior of the inhibitors is more readily observed during the relaxation phase of the $[Ca^{2+}]_i$ changes.

[b] TEA, Tetraethylammonium.

[c] WGA, Wheat germ agglutinin.

entry mechanism (Trimmer *et al.*, 1985; Trimmer, 1987). On the simple basis of numbers of binding sites for the antigen on the sperm surface ($\sim$150,000/cell), it seems an unlikely candidate for this role. Using a plasma membrane surface area of $\sim$40 μm^2/sperm, this yields a molecular density of >3000/μm^2, a value which exceeds reported values for all but the highest Na^+ channel densities (Hille, 1984). The amount of Ca^{2+} entering sperm stimulated with antibodies does not require such a large channel density. The rate during the first 10 sec of antibody stimulation is estimated from $^{45}Ca^{2+}$ data at approximately <0.2 nmol/10^8 sperm sec or $\sim$$10^6$ ions/sec for a single sperm. Ca^{2+} channels and channels in general pass ions at the rate of $\sim$$10^6$/sec. Even if the channels were only open for an average of 10 msec, the rate of Ca^{2+} entry into sperm can be accounted for by as few as 100 channels, which is three orders of magnitude less than the number of antigen molecules present. The issue then becomes one of how many molecules might be activated by initial antibody binding and how many molecules comprise a single channel. At this time, no information exists on how much binding occurs within the first few seconds following antibody addition, but if the number is sufficiently small and if subsequent binding does not activate additional channels, the numerical arguments presented above may not be valid. Interestingly, recent biochemical experiments with rod outer segments have also found more channel molecules than predicted by physiological measurements (Applebury, 1987). Alternatively, binding of the mAbs to the 210-kDa protein may stimulate Ca^{2+} channel activity indirectly through alteration of a second messenger system within the sperm or perhaps by activating other channel types to alter the plasma membrane potential. The data strongly argue that the 210-kDa antigen is involved in Ca^{2+} entry into sperm, and further experimentation should reveal properties of this important sperm membrane protein and how it functions to regulate Ca^{2+} entry into sperm.

A second antibody, mAb J18/29, stimulates the acrosome reaction by a mechanism that closely approximates egg jelly (Trimmer *et al.*, 1987). Unlike mAb J10/14, mAb J18/29 increases both $[Ca^{2+}]_i$ and pH_i. The inhibitory pattern is identical to that of egg jelly. That is, tetraethylammonium, $[K^+]_e$ and verapamil are all effective at blocking the acrosome reaction stimulated by this antibody (Trimmer *et al.*, 1987) at concentrations close to those that inhibit egg jelly. Increases in $[Ca^{2+}]_i$ by mAb J18/29 are partially blocked by inhibitors of the acrosome reaction. Inhibition is similar to that of egg jelly-initiated changes. mAb J18/29 is currently the only "non-jelly" trigger of the reaction that is sensitive to tetraethylammonium, and it looks quite promising as a defined model method of initiation. In principle, binding of this antibody may reveal the identity of the "egg jelly receptor." Interestingly, mAb J18/29 binds to several membrane proteins on Western blots apparently by recognizing a common feature of each of the proteins. Among these proteins is the 210-kDa glycoprotein. Binding of antibodies to multiple proteins with a common determinant has been reported in other organisms (Bloodgood *et al.*, 1986; S. Ward *et al.*, 1986). The observation that multiple membrane proteins are involved in mAb J18/29 binding is consistent with the idea that multiple ionic changes need to occur for the acrosome reaction to proceed.

E. Relationship of Increased pH_i to Increased $[Ca^{2+}]_i$

Initiation of the acrosome reaction is always accompanied by an increase in pH_i to ~7.6 or greater. Increased $[Ca^{2+}]_i$ without the increase in pH_i does not initiate the acrosome reaction and suggests that the rise in pH_i is not a direct consequence of the $[Ca^{2+}]_i$ increase. Additionally, using the weak base NH_4^+ to increase pH_i, several laboratories have found that NH_4^+ alone is insufficient to cause a complete acrosome reaction (Trimmer *et al.*, 1986; Schroeder and Christen, 1982). NH_4^+ (5–10 mM), which increases pH_i to a greater degree than does egg jelly, does not substantially increase $[Ca^{2+}]_i$ (Trimmer *et al.*, 1986; Schackmann and Chock, 1986). The data suggest that an egg jelly-induced increase in pH_i of ~0.2 U is not the sole regulatory mechanism for activating Ca^{2+} entry. Another laboratory, however, has used higher

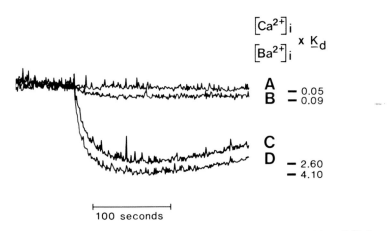

Fig. 3. Sperm were loaded with fura-2 as described in Fig. 2 and diluted into Ca^{2+}-free artificial seawater at pH 7.8. In A, no Ca^{2+} or Ba^{2+} was added. In B, 10 mM Ca^{2+} was added prior to recording. In C and D, 10 mM Ba^{2+} was added prior to recording. In C, 40 μM verapamil was also added 10 sec after recording was begun. NH_4^+ (10 mM) was added from a 1 M stock solution, pH adjusted to 8.0, to initiate the changes in intracellular Ca^{2+} or Ba^{2+}. Fluorescence emission was at 500 nm with excitation at 380 nm. The Ba^{2+}–fura-2 dissociation constant has been determined to be a ~5-fold higher than the Ca^{2+}–fura-2 dissociation constant; hence the increase in the $[Ba^{2+}]_i$ is ~200 times that for $[Ca^{2-}]_i$.

concentrations of NH_4^+ to initiate the acrosome reaction and has reported some initiation at 10 mM NH_4^+ (Garcia-Soto and Darszon, 1985; Garcia-Soto et al., 1987). The reason for these different observations has not been resolved. The differences may reflect aspects of $[Ca^{2+}]_i$ regulation by pH_i, which we have yet to adequately address. For example, Fig. 3 shows that 10 mM NH_4^+ increases $[Ca^{2+}]_i$ only slightly. In contrast, 10 mM NH_4^+ is very effective at stimulating increased Ba^{2+} entry into the sperm. Since Ba^{2+} activates the Ca^{2+}-ATPase much less effectively than Ca^{2+} (Pfleger and Wolf, 1975) and since Ba^{2+} entry through Ca^{2+} channels can prevent inactivation of the channel (Eckert and Tillotson, 1981), it is possible that increased pH_i stimulates not only Ca^{2+} entry, but also Ca^{2+} removal, and that entry of Ba^{2+} is stimulated to a greater degree than Ba^{2+} removal. Ba^{2+} entry stimulated with NH_4^+ is not blocked as efficiently by verapamil as is that stimulated by egg jelly (Fig. 2); it remains to be seen if the entry mechanisms stimulated are the same. However this current problem is resolved, it is likely that manipulation of pH_i affects several biochemical regulatory systems within the sperm.

III. INCREASES IN pH_i AND $[Ca^{2+}]_i$ BY PEPTIDES FROM EGG JELLY

A. Background

Early observations (Ohtake, 1976a,b) demonstrated the existence in egg jelly of material that stimulated sperm respiration. The material, which could be dialyzed away from the acrosome reaction-triggering substance in egg jelly, was highly effective at reduced seawater pH (6.6), but was almost ineffective on sperm in normal seawater at pH 8. The low-molecular-weight material was sensitive to proteases and had other properties, suggesting that it was a peptide. Subsequent purification of this respiratory-activating material revealed the early predictions to be correct, and several peptides have been isolated and sequenced (Kopf et al., 1979; Hansbrough and Garbers, 1981a; Garbers et al., 1982; Suzuki et al., 1981, 1984). The peptides "speract" isolated from S. purpuratus and H. pulcherrimus and "resact" from A. punctulata do not cross-react. Synthetic peptides mimic all biological effects of the native substance, and structural analogs have been used to study the biochemistry and physiology resulting from peptide binding (Garbers et al., 1982; Dangott and Garbers, 1984; Shimomura and Garbers, 1986; Shimomura et al., 1986). The peptides stimulate sperm motility, as well as respiration in seawater at pH 6.6 (Kopf et al., 1979; Hansbrough and Garbers, 1981b; Shimomura and Garbers, 1986) and can, under appropriate conditions, stimulate respiration at alkaline pH (Suzuki and Garbers, 1984). They stimulate rapid, transient increases in cyclic nucleotide concentrations (Hansbrough and Garbers, 1981a; Garbers *et al.*, 1982; Suzuki *et al.*, 1984; Shimomura and Garbers, 1986). At pH 8, the peptide resact acts as a chemoattractant for *Arbacia* sperm (Ward *et al.*, 1985b). Chemotaxis requires Ca^{2+} (Ward *et al.*, 1985b), whereas stimulation of respiration requires Na^+, but not Ca^{2+} (Hansbrough and Garbers, 1981b).

B. Mechanism of Peptide Activity

The basis of the respiratory stimulation at low-seawater pH can be understood in terms of the regulation of sea urchin sperm motility and respiration by pH_i (Christen *et al.*, 1982, 1983a,c; Repaske and Garbers, 1983; Lee *et al.*, 1983). Binding of the peptides speract or resact initiates a rapid Na^+-dependent efflux of H^+ (Hansbrough and Garbers, 1981b; Repaske and Garbers, 1983). Speract stimulates Na^+ uptake, and this Na^+–H^+ exchange increases pH_i (Hansbrough and Garbers, 1981b; Repaske and Garbers, 1983). Motility and respiration are strongly coupled through the dynein ATPase activity in sea urchin sperm, since dynein activity is regulated by pH_i (Christen *et al.*, 1983a). Dynein activity is absent *in vivo*, when the pH is ~7 or below and respiration is minimal. Increased pH_i initiates ATP hydrolysis which in turn stimulates oxidative phosphorylation. When sperm are in normal seawater, the pH_i in sperm is ~7.4 and respiration is nearly maximal. Further increases in pH_i do not further stimulate respiration. By placing sperm at reduced seawater pH to lower pH_i, basal respiration rates are decreased so that peptide-stimulated increases in pH_i can stimulate respiration.

Peptide-stimulated Na^+–H^+ exchange has properties similar to the Na^+–H^+ exchange activated by Na^+ addition to sperm in Na^+-free seawater (Lee, 1984a,b, 1985; Lee and Garbers, 1986; Schackmann and Chock, 1986). The exchange is inhibited by increasing seawater $[K^+]$ from 10 to 20 mM or more. It is unique among Na^+–H^+ exchanges in that it is insensitive to amiloride and can be inhibited by depolarization of the plasma membrane potential. It has been characterized in intact sperm, in intact flagella, and in flagellar membrane vesicles (Lee, 1984a,b, 1985). Recent work by Lee and Garbers (1986) suggests that the exchange is not active when sperm are in normal seawater, but is stimulated by peptide binding. The membrane potential of sperm in seawater is depolarized compared to Na^+-free conditions (Schackmann *et al.*, 1984). The initial ionic step following peptide binding is thought to be activation of K^+ channels to hyperpolarize the plasma membrane potential and to thereby activate the Na^+–H^+ exchange (Lee and Garbers, 1986).

Though Ca^{2+} is not required for respiratory- or motility-stimulating activity, the peptide resact will not stimulate chemotaxis in the absence of seawater $[Ca^{2+}]$ (Ward *et al.*, 1985b). If Ca^{2+} is present in seawater, the peptide speract stimulates a rapid, transient increase in $[Ca^{2+}]_i$, which results from Ca^{2+} entry across the plasma membrane (Schackmann and Chock, 1986). This response is more limited in size (3-fold increase) and duration (relaxes within 60 sec) than the increase stimulated by intact egg jelly or the monoclonal antibodies mentioned in the previous section. The $[Ca^{2+}]_i$ increase is secondary to the Na^+–H^+ exchange, as the exchange occurs without any extracellular Ca^{2+}, and peptide stimulation of Ca^{2+} entry is not observed without extracellular Na^+. If the Na^+–H^+ exchange is blocked with extracellular $[K^+]$, Ca^{2+} entry is also prevented. While a precise mechanism for Ca^{2+} entry cannot yet be presented, the data suggest that entry results either from Na^+–Ca^{2+} exchange following Na^+ entry or else from activation of a cation channel insensitive to verapamil (Schackmann and Chock, 1986). Several features of the Ca^{2+} entry stimulated by speract suggest the mechanism is substantially different from that initiated during the acrosome reaction or by the antibodies. Although excess seawater $[K^+]$ inhibits both egg jelly and speract increases in $[Ca^{2+}]_i$, the peptide-stimulated increase in $[Ca^{2+}]_i$ is completely blocked, whereas only the sustained part of the egg jelly change is blocked. The initial rise in $[Ca^{2+}]_i$ by egg jelly is only slightly decreased (R. W. Schackmann, unpublished data). The speract-initiated increase is also unaffected by the Ca^{2+} channel blocker verapamil or by Co^{2+} (R. W. Schackmann, unpublished data). The data suggest that at least two types of Ca^{2+} entry mechanisms may be stimulated by different egg factors.

Unlike the egg jelly "receptor," speract and resact receptors have been identified by covalently cross-linking radiolabeled receptor analogs to sperm membranes (Dangott and Garbers, 1984; Shimomura et al., 1986). In S. purpuratus, speract binding is to a 77-kDa protein. In A. punctulata, the resact receptor appears to be the plasma membrane guanylate cyclase. In both species, peptide binding is associated with a rapid, transient activation of the guanylate cyclase and a rapid (within 3 sec) increase in intracellular cGMP (Bentley et al., 1986a,b; Hansbrough and Garbers, 1981b; Ramarao and Garbers, 1985; Shimomura and Garbers, 1986; Suzuki et al., 1984; Ward et al., 1985a). Subsequently, the guanylate cyclase is inactivated, and in A. punctulata, this is associated with dephosphorylation and an apparent molecular-weight shift of this enzyme on SDS gels (Ward and Vacquier, 1983; Ward et al., 1985a; Suzuki et al., 1984; Bentley et al., 1986a,b). Dephosphorylation can be stimulated without peptides by increasing the pH_i with NH_4^+ or by elevating the extracellular pH (G. E. Ward et al., 1986; Shimomura et al., 1986). Phosphorylation of the guanylate cyclase can be achieved by lowering pH_i. These findings suggest that one target for biochemical regulation by pH_i within the sperm is the phosphorylation state and activity of guanylate cyclase.

The similar time course for increases in cGMP and cAMP, $[Ca^{2+}]_i$ and pH_i, suggests relationships exist between the ion movements and cyclic nucleotide concentrations (Hansbrough and Garbers, 1981b; Shimomura *et al.*, 1986; Schackmann and Chock, 1986). At this point, relationships between cGMP and ion changes are only beginning to be explored. For example, it has yet to be determined if increased extracellular K^+ blocks cyclic nucleotide increases or just the ion changes. By analogy with other systems such as the rod outer segment, we anticipate the finding of cGMP-activated channels in sperm membranes. If such channels are activated by increased cGMP and if the channel activity is subsequently blocked by a depolarizing concentration of K^+, changes in cGMP may be less affected by the increased extracellular $[K^+]$ than are $[Ca^{2+}]_i$ and pH_i changes.

Additional data, which support a role for cyclic nucleotides, are derived from use of the cyclic nucleotide phosphodiesterase inhibitor methylisobutylxanthine (MIX) to stimulate peptide-mediated ion changes in sperm. MIX enhances the $[Ca^{2+}]_i$ increase and Na^+ uptake stimulated by the peptide speract (Schackmann and Chock, 1986). The resulting increases are sustained similar to those initiated with egg jelly during the acrosome reaction. If sufficient MIX is present, speract can initiate the acrosome reaction in *S. purpuratus* sperm. However, unlike the egg jelly increase, speract-plus-MIX changes are insensitive to verapamil, and this combination reflects an alternative method of initiating the acrosome reaction. MIX also can enhance $^{45}Ca^{2+}$ uptake in abalone sperm (Kopf *et al.*, 1983, 1984).

IV. SUMMARY

In this chapter, I have briefly reviewed studies on ion movements important to the sea urchin acrosome reaction and to the stimulation of sperm by peptides. Rapid increases of both pH_i and $[Ca^{2+}]_i$ are early events initiated by binding of the egg-derived materials to receptors on the sperm plasma membrane. Both peptides and egg jelly stimulate rapid Na^+ uptake, H^+ efflux, and Ca^{2+} entry and probably activate channels that pass K^+. Increased $[Ca^{2+}]_i$ occurs from Ca^{2+} entry across the plasma membrane. If initiated by egg jelly or by antibodies that bind the 210-kDa antigen, increases are partially blocked by either Ca^{2+} antagonists such as verapamil or by alterations expected to affect K^+ channel activity (increased extracellular $[K^+]$ or tetraethylammonium). If initiated with egg-derived peptides such as speract, increases in $[Ca^{2+}]_i$ are completely blocked by increased extracellular $[K^+]$ and are insensitive to verapamil. Activity of a membrane potential-sensitive Na^+–H^+ exchange is involved in regulating motility and the acrosome reaction, and this exchange can stimulate Ca^{2+} entry by a mechanism similar to that stimulated by speract. It is not yet known whether this mechanism of Ca^{2+} entry is active during the acrosome reaction and constitutes the verapamil-insensitive entry mechanism mentioned in Section II,C, but the ability of extracellular $[K^+]$ to block the reaction is suggestive. Sea urchin sperm, therefore, contain at least two different Ca^{2+} entry mechanisms. One of these is functionally linked to the 210-kDa membrane protein. Evidence also suggests multiple types of K^+ channels may function during sperm activation. Increased $[K^+]$ blocks both peptide- and egg jelly-stimulated changes, but tetraethylammonium only affects the acrosome reaction. Whether these various membrane functions are delineated along the length of the sperm remains an important consideration for sperm cell biology.

Other general and important biological questions remain unanswered. For example, how do the transient increases in pH_i, $[Ca^{2+}]_i$, and cGMP relate to chemotaxis or other alterations in flagellar bending? Brokaw and Nagayama (1985) demonstrated that alteration of Ca^{2+}–calmodulin concentrations affects flagellar bending in permeant sperm models, but it seems unlikely that Ca^{2+} changes, even with increased pH_i, are sufficient by themselves for the complex requirements of directed movement. We have only to look at the complexity of bacterial chemotaxis to realize that studies with sperm have just begun to examine the biochemistry involved. Likewise, although the acrosome reaction is intimately connected to increased pH_i and $[Ca^{2+}]_i$, multiple sites of action are expected for each of these intracellular signals. For example, actin polymerization requires the pH_i increase, but the pH_i increases may be directly involved in regulating $[Ca^{2+}]_i$ as well (Section II,E).

Finally, changes in pH_i and $[Ca^{2+}]_i$ are by no means the only important signaling processes known to occur during sperm activation. cAMP changes resulting from or possibly involved in initiating the acrosome reaction have been reviewed elsewhere (Garbers and Kopf, 1980), and studies are now being performed that implicate roles for GTP-binding proteins (Kopf *et al.*, 1986) and phosphatidylinositol turnover (Domino and Garbers, 1988) in sperm biochemistry and physiology. The sperm, as do other cells, makes use of common regulatory pathways to prepare itself for fertilization, but many of these regulatory systems are only beginning to be explored in the male gamete.

REFERENCES

Applebury, M. L. (1987). Biochemical puzzles about the cyclic-GMP-dependent channel. *Nature (London)* **326**, 546–547.

Babcock, D. F., and Pfeiffer, D. R. (1987). Independent elevation of cytosolic $[Ca^{2+}]$ and pH of mammalian sperm by voltage-dependent and pH sensitive mechanisms. *J. Biol. Chem.* **262**, 15041–15047.

Babcock, D. F., Rufo, G. A., and Lardy, H. A. (1983). Potassium dependent increases in cytosolic pH stimulate metabolism and motility of mammalian sperm. *Proc. Natl. Acad. Sci. U.S.A.* **80**, 1327–1331.

Bentley, J. K., Tubb, D. J., and Garbers, D. L. (1986a). Receptor-mediated activation of spermatozoan gyanylate cyclase. *J. Biol. Chem.* **261**, 14859–14862.

Bentley, J. K., Shimomura, H., and Garbers, D. L. (1986b). Retention of a functional resact receptor in isolated sperm membranes. *Cell (Cambridge, Mass.)* **45**, 281–288.

Bibring, T., Baxandall, J., and Harter, C. C. (1984). Sodium-dependent pH regulation in active sea urchin sperm. *Dev. Biol.* **101**, 425–435.

Bleil, J. D., and Wassarman, P. M. (1983). Sperm–egg interactions in the mouse: Sequence of events and induction of the acrosome reaction by zona pellucida glycoprotein. *Dev. Biol.* **95**, 317–324.

Bloodgood, R. A., Woodward, M. P., and Salomonsky, N. L. (1986). Redistribution and shedding of flagellar membrane glycoproteins visualized using an anti-carbohydrate monoclonal antibody and concanavalin A. *J. Cell Biol.* **102**, 1797–1812.

Breitbart, H., Rubinstein, S., and Nass-Arden, L. (1985). The role of calcium and Ca^{2+}-ATPase in maintaining motility in ram spermatozoa. *J. Biol. Chem.* **260**, 11548–11553.

Brokaw, C. J., and Nagayama, S. M. (1985). Modulation of the asymmetry of sea urchin sperm flagellar bending by calmodulin. *J. Cell Biol.* **100**, 1875–1883.

Cantino, M. E., Schackmann, R. W., and Johnson, D. E. (1983). Changes in subcellular elemental distribution accompanying the acrosome reaction in sea urchin sperm. *J. Exp. Zool.* **226**, 255–268.

Carafoli, E. (1982). The transport of calcium across the inner membrane of mitochondria. *In* "Membrane Transport of Calcium" (E. Carafoli, ed.), pp. 109–139. Academic Press, London.

Christen, R., Schackmann, R. W., and Shapiro, B. M. (1982). Elevation of intracellular pH activates respiration and motility of sperm of the sea urchin, *Strongylocentrotus purpuratus. J. Biol. Chem.* **257**, 14881–14890.

Christen, R., Schackmann, R. W., and Shapiro, B. M. (1983a). Metabolism of sea urchin sperm. Interrelationships between intracellular pH, ATPase activity and mitochondrial respiration. *J. Biol. Chem.* **258**, 5392–5399.

Christen, R., Schackmann, R. W., and Shapiro, B. M. (1983b). Interactions between sperm and sea urchin egg jelly. *Dev. Biol.* **98**, 1–14.

Christen, R., Schackmann, R. W., Dahlquist, F. W., and Shapiro, B. M. (1983c). ^{31}P-energy phosphate compounds by changes in intracellular pH. *Exp. Cell Res.* **149**, 289–294.

Cognard, C., Lazdunski, M., and Romey, G. (1986). Different types of Ca^{2+} channels in mammalian skeletal muscle cells in culture. *Proc. Natl. Acad. Sci. U.S.A.* **83**, 517–521.

Collins, F., and Epel, D. (1977). The role of calcium ions in the acrosome reaction of sea urchin sperm. *Exp. Cell Res.* **106**, 211–222.

Colwin, L. H., and Colwin, A. L. (1956). The acrosome filaments and sperm entry in *Thyone briareus* and *Asterias. Biol. Bull. (Woods Hole, Mass.)* **113**, 243–257.

Dan, J. C. (1952). Studies on the acrosome. I. Reaction to egg water and other stimuli. *Biol. Bull. (Woods Hole, Mass.)* **103**, 54–66.

Dan, J. C. (1954a). Studies on the acrosome. II. Acrosome reaction in starfish spermatozoa. *Biol. Bull. (Woods Hole, Mass.)* **107**, 203–218.

Dan, J. C. (1954b). Studies on the acrosome. III. Effect of calcium deficiency. *Biol. Bull. (Woods Hole, Mass.)* **107**, 335–349.

Dan, J. C. (1967). Acrosome reaction and lysins. *In* "Fertilization" (C. Metz and A. Monroy, eds.), Vol. 1, pp. 237–293. Macmillan, New York.

Dangott, L. J., and Garbers, D. L. (1984). Identification and partial characterization of the receptor for speract. *J. Biol. Chem.* **259**, 13712–13716.

Decker, G. L., Joseph, D. B., and Lennarz, W. J. (1976). A study of factors involved in induction of the acrosomal reaction in sperm of the sea urchin, *Arbacia punctulata*. *Dev. Biol.* **53**, 115–125.

Domino, S. E., and Garbers, D. L. (1988). The fucose sulfate glycoconjugate that induces the acrosome reaction in spermatozoa stimulates inositol 1,4,5-trisphosphate accumulation. *J. Biol. Chem.* **263**, 690–695.

Eckberg, W. R., and Metz, C. B. (1982). Isolation of an *Arbacia* sperm fertilization antibody. *J. Exp. Zool.* **221**, 101–105.

Eckert, R., and Tillotson, D. L. (1981). Calcium-mediated inactivation of the calcium conductance in caesium-loaded giant neurons of *Aplysia californica*. *J. Physiol. (London)* **314**, 265–280.

Fleckenstein, A. (1977). Specific pharmacology of calcium in myocardium, cardiac pacemakers, and vascular smooth muscle. *Annu. Rev. Pharmacol. Toxicol.* **17**, 149–166.

Flockerzi, V., Oeken, H.-J., Hofmann, F., Pelzer, D., Cavalie, A., and Trautwein, W. (1986). Purified dihydropyridine-binding site from skeletal muscle t-tubules is a functional calcium channel. *Nature (London)* **323**, 66–68.

Garbers, D. L. (1981). The elevation of cyclic AMP concentrations in flagella-less sea urchin sperm heads. *J. Biol. Chem.* **256**, 620–624.

Garbers, D. L., and Kopf, G. S. (1980). The regulation of spermatozoa by calcium and cyclic nucleotides. *Adv. Cyclic Nucleotide Res.* **13**, 251–306.

Garbers, D. L., Watkins, H. D., Hansbrough, J. R., Smith, A., and Misono, K. S. (1982). The amino acid sequence and chemical synthesis of speract and of speract analogues. *J. Biol. Chem.* **257**, 2734–2737.

Garcia-Soto, J., and Darszon, A. (1985). High pH-induced acrosome reaction and Ca^{2+} uptake in sea urchin sperm suspended in Na^+-free seawater. *Dev. Biol.* **110**, 338–345.

Garcia-Soto, J., Gonzalez-Martinez, M., De La Torre, L., and Darszon, A. (1987). Internal pH can regulate Ca^{2+} uptake and the acrosome reaction in sea urchin sperm. *Dev. Biol.* **120**, 112–120.

Gibbons, I. (1981). Cilia and flagella of eukaryotes. *J. Cell Biol.* **91**, 107–124.

Gonzalez-Martinez, M., and Darszon, A. (1987). A fast transient hyperpolarization occurs during the sea urchin sperm acrosome reaction induced by egg jelly. *FEBS Lett.* **212**, 247–250.

Gregg, K. W., and Metz, C. B. (1976). Physiological parameters of the sea urchin acrosome reaction. *Biol. Reprod.* **14**, 405–411.

Grynkiewicz, G., Poenie, M., and Tsien, R. Y. (1985). A new generation of Ca^{2+} indicators with greatly improved fluorescence properties. *J. Biol. Chem.* **260**, 3440–3450.

Guerrero, A., Sanchez, J. A., and Darszon, A. (1987). Single-channel activity in sea urchin sperm revealed by the patch–clamp technique. *FEBS Lett.* **220**, 295–298.

Hagiwara, S., and Byerly, L. (1981). Calcium channel. *Annu. Rev. Neurosci.* **4**, 69–125.

Hagiwara, S., and Ohmori, H. (1982). Studies of calcium channels in rat clonal pituitary cells with patch electrode voltage clamp. *J. Physiol. (London)* **331**, 231–252.

Hagiwara, S., and Takahashi, K. (1967). Surface density of calcium ion and calcium spikes in the barnacle muscle fiber membrane. *J. Gen. Physiol.* **50**, 583–601.

Hansbrough, J. R., and Garbers, D. L. (1981a). Speract. Purification and characterization of a peptide associated with eggs that activates spermatozoa. *J. Biol. Chem.* **256**, 1447–1452.

Hansbrough, J. R., and Garbers, D. L. (1981b). Sodium-dependent activation of sea urchin spermatozoa by speract and monensin. *J. Biol. Chem.* **256**, 2235–2241.

Hille, B. (1984). "Ionic Channels of Excitable Membranes." Sinauer Associates, Sunderland, Massachusetts.

Hotta, K., Hamazaki, H., and Kurokawa, M. (1970). Isolation and properties of a new type of sialopolysaccharide–protein complex from jelly coat of sea urchin eggs. *J. Biol. Chem.* **245,** 5434–5440.

Iijima, T., Ciani, S., and Hagiwara, S. (1986). Effects of the external pH on Ca channels: Experimental studies and theoretical considerations using a two-site, two-ion model. *Proc. Natl. Acad. Sci. U.S.A.* **83,** 654–658.

Kazazoglou, T., Schackmann, R. W., Fosset, M., and Shapiro, B. M. (1985). Calcium channel antagonists inhibit the acrosome reaction and bind to plasma membranes of sea urchin sperm. *Proc. Natl. Acad. Sci. U.S.A.* **82,** 1460–1464.

Kopf, G. S., and Garbers, D. L. (1980). Calcium and a fucose-sulfate-rich polymer regulate sperm cyclic nucleotide metabolism and the acrosome reaction. *Biol. Reprod.* **22,** 1118–1126.

Kopf, G. S., Tubb, D. J., and Garbers, D. L. (1979). Activation of sperm respiration by a low molecular weight egg factor and by 8-bromoguanosine 3′,5′-monophosphate. *J. Biol. Chem.* **254,** 8554–8560.

Kopf, G. S., Lewis, C. A., and Vacquier, V. D. (1983). Methylxanthines stimulate calcium transport and inhibit cyclic nucleotide phosphodiesterases in abalone sperm. *Dev. Biol.* **99,** 115–120.

Kopf, G. S., Lewis, C. A., and Vacquier, V. D. (1984). Characterization of basal and methylxanthine stimulated Ca^{2+} transport in abalone spermatozoa. *J. Biol. Chem.* **259,** 5514–5520.

Kopf, G. S., Woolkalis, M. J., and Garton, G. L. (1986). Evidence for a guanine nucleotide-binding regulatory protein in invertebrate and mammalian sperm. *J. Biol. Chem.* **261,** 7327–7331.

Lee, H. C. (1984a). Sodium and proton transport in flagella isolated from sea urchin spermatozoa. *J. Biol. Chem.* **259,** 4957–4963.

Lee, H. C. (1984b). A membrane potential-sensitive Na^+/H^+ exchange system in flagella isolated from sea urchin spermatozoa. *J. Biol. Chem.* **259,** 15315–15319.

Lee, H. C. (1985). The voltage sensitive Na^+/H^+ exchange in sea urchin spermatozoa flagellar membrane vesicles studied with an entrapped pH probe. *J. Biol. Chem.* **260,** 10794–10799.

Lee, H. C., and Garbers, D. L. (1986). Modulation of the voltage-sensitive Na^+/H^+ exchange in sea urchin spermatozoa through membrane potential changes induced by the egg peptide speract. *J. Biol. Chem.* **261,** 16026–16032.

Lee, H. C., Forte, J. G., and Epel, D. (1982). The use of fluorescent amines for the measurement of pH_i: Applications in liposomes, gastric microsomes, and sea urchin gametes. *In* "Intracellular pH: Its Measurement, Regulation, and Utilization in Cellular Functions" (R. Nuccitelli and D. W. Deamer, eds.), pp. 135–160. Alan R. Liss, New York.

Lee, H. C., Johnson, C., and Epel, D. (1983). Changes in internal pH associated with the initiation of motility and acrosome reaction of sea urchin sperm. *Dev. Biol.* **95,** 31–45.

Lievano, A., Sanchez, J., and Darszon, A. (1985). Single-channel activity of bilayers derived from sea urchin sperm plasma membranes at the tip of a patch-clamp electrode. *Dev. Biol.* **112,** 253–257.

Lievano, A., Sanchez, J., and Darszon, A. (1987). Ca^{2+} channels in the plasma membrane of *Strongylocentrotus purpurtus* sea urchin sperm. *Biophys. J.* **51,** 433a.

Lopo, A. C., and Vacquier, V. D. (1980). Antibody to a specific sperm surface glycoprotein inhibits the egg-jelly-induced acrosome reaction. *Dev. Biol.* **79,** 325–333.

McCleskey, E. W., Fox, A. P., Feldman, D. H., Cruz, L. J., Olivera, B. M., Tsien, R. W., and Yoshikami, D. (1987). ω-Conotoxin: Direct and persistent block of specific types of calcium channels in neurons but not muscle. *Proc. Natl. Acad. Sci. U.S.A.* **84**, 4327–4331.

Meizel, S. (1984). The importance of hydrolytic enzymes to an exocytotic event, the mammalian sperm acrosome reaction. *Biol. Rev. Cambridge Philos. Soc.* **59**, 125–157.

Meizel, S., and Deamer, D. W. (1978). The pH of the hamster sperm acrosome. *J. Histochem. Cytochem.* **26**, 98–110.

Metz, C. B., Schuel, H., and Bischoff, E. R. (1964). Inhibition of the fertilizing capacity of sea urchin sperm by papain digested nonagglutinating antibody. *J. Exp. Zool.* **155**, 261–272.

Miller, R. L. (1985). Sperm chemo-orientation in the Metazoa. *In* "Biology of Fertilization" (C. B. Metz and A. Monroy, eds.), Vol. 2, pp. 275–337. Academic Press, Orlando, Florida.

Minocherhomjee, A.-E.-V. M., and Roufogalis, B. D. (1984). Antagonism of calmodulin and phosphodiesterase by nifedipine and related calcium entry blockers. *Cell Calcium* **5**, 57–63.

Mrsny, R. J., and Meizel, S. (1981). Potassium ion influx and $Na^+ : K^+$ ATPase activity are required for the hamster sperm acrosome reaction. *J. Cell Biol.* **91**, 77–82.

Murphy, S. J., and Yanagimachi, R. (1984). The pH dependence of motility and the acrosome reaction of guinea pig spermatozoa. *Gamete Res.* **10**, 1–8.

Norman, J. A., Ansell, J., and Phillipps, M. A. (1983). Dihydropyridine Ca^{2+} entry blockers selectively inhibit peak I cAMP phosphodiesterase. *Eur. J. Pharmacol.* **93**, 107–112.

Ohtake, H. (1976a). Respiratory behavior of sea urchin spermatozoa. I. Effect of pH and egg water on the respiratory rate. *J. Exp. Zool.* **198**, 303–312.

Ohtake, H. (1976b). Respiratory behavior of sea urchin spermatozoa. II. Sperm activating substance obtained from jelly coat of sea urchin eggs. *J. Exp. Zool.* **198**, 313–322.

Pfleger, H., and Wolf, H. U. (1975). Activation of membrane-bound high-affinity calcium ion-sensitive adenosine triphosphatase of human erythrocytes by bivalent metal ions. *Biochem. J.* **147**, 359–361.

Podell, S. B., and Vacquier, V. D. (1984a). Inhibition of sea urchin sperm acrosome reaction by antibodies directed against two sperm membrane proteins. *Exp. Cell Res.* **155**, 467–476.

Podell, S. B., and Vacquier, V. D. (1984b). Wheat germ agglutinin blocks the acrosome reaction in *Strongylocentrotus purpuratus* sperm by binding a 210,000-mol-wt. membrane protein. *J. Cell Biol.* **99**, 1598–1604.

Ramarao, C. S., and Garbers, D. L. (1985). Receptor mediated regulation of guanylate cyclase activity in spermatozoa. *J. Biol. Chem.* **260**, 8390–8396.

Repaske, D. R., and Garbers, D. L. (1983). A hydrogen ion flux mediates stimulation of respiratory activity by speract in sea urchin spermatozoa. *J. Biol. Chem.* **258**, 6025–6029.

Reynolds, I. J., Wagner, J. A., Snyder, S. H., Thayer, S. A., Olivera, B. M., and Miller, R. J. (1986). Brain voltage-sensitive calcium channel subtypes differentiated by ω-conotoxin fraction GVIA. *Proc. Natl. Acad. Sci. U.S.A.* **83**, 8804–8807.

Roldan, E. R. S., Shibata, S., and Yanagimachi, R. (1986). Effect of Ca^{2+} channel antagonists on the acrosome reaction of guinea pig and golden hamster spermatozoa. *Gamete Res.* **13**, 281–292.

Rufo, G. A., Schoff, P., and Lardy, H. A. (1984). Regulation of calcium content in bovine spermatozoa. *J. Biol. Chem.* **259**, 2547–2552.

Saling, P. M., Eckberg, W. R., and Metz, C. B. (1982). Mechanism of univalent antisperm antibody inhibition of fertilization in the sea urchin *Arbacia punctulata*. *J. Exp. Zool.* **221**, 93–99.

Schackmann, R. W., and Chock, P. B. (1986). Alteration of intracellular $[Ca^{2+}]$ in sea urchin sperm by the egg peptide speract: Evidence that increased intracellular Ca^{2+} is coupled to Na^+ entry and increased intracellular pH. *J. Biol. Chem.* **261**, 8719–8728.

Schackmann, R. W., and Shapiro, B. M. (1981). A partial sequence of ionic changes associated with the acrosome reaction of *Strongylocentrotus purpuratus*. *Dev. Biol.* **81**, 145–154.

Schackmann, R. W., Eddy, E. M., and Shapiro, B. M. (1978). The acrosome reaction of *Strongylocentrotus purpuratus* sperm. Ion requirements and movements. *Dev. Biol.* **65**, 483–495.

Schackmann, R. W., Christen, R., and Shapiro, B. M. (1981). Membrane potential depolarization and increased pH_i accompany the acrosome reaction of sea urchin sperm. *Proc. Natl. Acad. Sci. U.S.A.* **78**, 6066–6070.

Schackmann, R. W., Christen, R., and Shapiro, B. M. (1984). Measurement of plasma membrane and mitochondrial potentials in sea urchin sperm. Changes upon activation and induction of the acrosome reaction. *J. Biol. Chem.* **259**, 13914–13922.

Schroeder, T. E., and Christen, R. (1982). Polymerization of actin without acrosomal exocytosis in starfish sperm. Visualization with NBD-phallicidin. *Exp. Cell Res.* **140**, 363–371.

SeGall, G. K., and Lennarz, W. J. (1979). Chemical characterization of the component of the egg jelly coat from sea urchin eggs responsible for induction of the acrosome reaction. *Dev. Biol.* **71**, 33–48.

SeGall, G. K., and Lennarz, W. J. (1981). Jelly coat and induction of the acrosome reaction in echinoid sperm. *Dev. Biol.* **86**, 87–93.

Shapiro, B. M., Schackmann, R. W., Gabel, C. A., Foerder, C. A., and Farrance, M. L. (1980). Molecular alterations in gamete surfaces during fertilization and early development. *Symp. Soc. Dev. Biol.* **38**, 127–149.

Shapiro, B. M., Schackmann, R. W., and Gabel, C. A. (1981). Molecular approaches to the study of fertilization. *Annu. Rev. Biochem.* **50**, 815–843.

Shapiro, B. M., Schackmann, R. W., Tombes, R. M., and Kazazoglou, T. (1985). Coupled ionic and enzymatic regulation of sperm behavior. *Curr. Top. Cell. Regul.* **26**, 97–113.

Shimomura, H., and Garbers, D. L. (1986). Differential effects of resact analogues on sperm respiration rates and cyclic nucleotide concentrations. *Biochemistry* **25**, 3405–3410.

Shimomura, H., Dangott, L. J., and Garbers, D. L. (1986). Covalent coupling of a resact analogue to guanylate cyclase. *J. Biol. Chem.* **261**, 15778–15782.

Somlyo, A. P., Bond, M., and Somlyo, A. V. (1985). Calcium content of mitochondria and endoplasmic reticulum in liver frozen rapidly *in vivo*. *Nature (London)* **314**, 622–625.

Suzuki, N., and Garbers, D. L. (1984). Stimulation of sperm respiration rates by speract and resact at alkaline extracellular pH. *Biol. Reprod.* **30**, 1167–1174.

Suzuki, N., Nomura, K., Ohtake, H., and Isaka, S. (1981). Purification and primary structure of sperm-activating peptides from the jelly coat of sea urchin eggs. *Biochem. Biophys. Res. Commun.* **99**, 1238–1244.

Suzuki, N., Shimomura, H., Radany, E. W., Ramarao, C. S., Ward, G. E., Bentley, J. K., and Garbers, D. L. (1984). A peptide associated with eggs causes a mobility shift in a major plasma membrane protein of sea urchin spermatozoa. *J. Biol. Chem.* **259**, 14874–14879.

Talbot, P., Summers, R. G., Hylander, B. L., Keogh, E. M., and Franklin, L. E. (1976). The role of calcium in the acrosome reaction: An analysis using ionophore A23187. *J. Exp. Zool.* **198**, 383–392.

Tilney, L. G. (1985). The acrosomal reaction. *In* "Biology of Fertilization" (C. B. Metz and A. Monroy, eds.), Vol. 2, pp. 157–213. Academic Press, Orlando, Florida.

Tilney, L. G., and Inoué, S. (1982). The acrosomal reaction of *Thyone* sperm. II. The kinetics and possible mechanism of acrosomal process elongation. *J. Cell Biol.* **93**, 820–827.

Tilney, L. G., Hatano, S., Ishikawa, H., and Mooseker, M. (1973). The polymerization of actin: Its role in the generation of the acrosomal process of certain echinoderm sperm. *J. Cell Biol.* **59**, 109–126.

Tilney, L. G., Kiehart, D., Sardet, C., and Tilney, M. (1978). The polymerization of actin. IV. The role of Ca^{++} and H^+ in the assembly of actin and in membrane fusion in the acrosomal reaction of echinoderm sperm. *J. Cell Biol.* **77**, 536–550.

Trimmer, J. S., and Vacquier, V. D. (1986). Activation of sea urchin gametes. *Annu. Rev. Cell Biol.* **2**, 1–26.

Trimmer, J. S., Trowbridge, I. S., and Vacquier, V. D. (1985). Monoclonal antibody to a membrane glycoprotein inhibits the acrosome reaction and associated Ca^{2+} and H^+ fluxes of sea urchin sperm. *Cell (Cambridge, Mass.)* **40**, 697–703.

Trimmer, J. S., Schackmann, R. W., and Vacquier, V. D. (1986). Monoclonal antibodies increase intracellular Ca^{2+} in sea urchin spermatozoa. *Proc. Natl. Acad. Sci. U.S.A.* **83**, 9055–9059.

Trimmer, J. S., Ebina, Y., Schackmann, R. W., Meinhof, C.-G., and Vacquier, V. D. (1987). Characterization of a monoclonal antibody which induces the acrosome reaction of sea urchin sperm. *J. Cell Biol.* **105**, 1120–1128.

Tsien, R. Y., Pozzan, T., and Rink, T. J. (1982). Calcium homeostatis in intact lymphocytes: Cytoplasmic free calcium monitored with a new intracellularly trapped fluorescent indicator. *J. Cell Biol.* **94**, 325–334.

Ward, G. E., and Vacquier, V. D. (1983). Dephosphorylation of a major sperm membrane protein is induced by egg jelly during sea urchin fertilization. *Proc. Natl. Acad. Sci. U.S.A.* **80**, 5578–5582.

Ward, G. E., Garbers, D. L., and Vacquier, V. D. (1985a). Effects of extracellular egg factors on sperm guanylate cyclase. *Science* **227**, 768–770.

Ward, G. E., Brokaw, C. J., Garbers, D. L., and Vacquier, V. D. (1985b). Chemotaxis of *Arbacia punctulata* spermatozoa to resact, a peptide from the egg jelly layer. *J. Cell Biol.* **101**, 2324–2329.

Ward, G. E., Moy, G. W., and Vacquier, V. D. (1986). Phosphorylation of membrane-bound guanylate cyclase of sea urchin spermatozoa. *J. Cell Biol.* **103**, 95–101.

Ward, S., Roberts, T. M., Strome, S., Pavalko, F. M., and Hogan, E. (1986). Monoclonal antibodies that recognize a polypeptide antigenic determinant shared by multiple *Caenorhabditis elegans* sperm-specific proteins. *J. Cell Biol.* **102**, 1778–1786.

Wassarman, P. M. (1987). The biology and chemistry of fertilization. *Science* **235**, 553–560.

Wassarman, P. M., Florman, H. M., and Greve, J. M. (1985). Receptor-mediated sperm–egg interactions in mammals. *In* "Biology of Fertilization" (C. B. Metz and A. Monroy, eds.), Vol. 2, pp. 341–360. Academic Press, Orlando, Florida.

Yanagimachi, R., and Usui, N. (1974). Calcium dependence of the acrosome reaction and activation of guinea pig spermatozoa. *Exp. Cell Res.* **89**, 161–174.

2

Caltrin and Calcium Regulation of Sperm Activity

HENRY LARDY AND JOVENAL SAN AGUSTIN

Institute for Enzyme Research
The University of Wisconsin–Madison
Madison, Wisconsin 53705

Calcium is more universally involved in regulation and control of biological processes than any other single agent (Heilbrunn, 1937, 1952; Rasmussen, 1981). Small wonder then, that Jacques Loeb (1913) found that sea urchin eggs failed to fertilize in media devoid of calcium. We know now that these eggs contain in their endoplasmic reticulum (Eisen and Reynolds, 1985) all the calcium they need for development and that the calcium is required to make the sea urchin sperm competent for fertilization (Dan, 1954).

I. THE ROLE OF CALCIUM IN CAPACITATION

Basic investigations aimed at *in vitro* fertilization led to the discovery of the phenomenon called sperm capacitation (Austin, 1951; Chang, 1951), and, in turn, studies of *in vitro* capacitation led to the finding that extracellular

29

THE CELL BIOLOGY OF
FERTILIZATION

calcium is an essential participant in causing the acrosome reaction (Dan, 1954; Dan et al., 1964; Yanagimachi and Usui, 1974).

The necessity of calcium in extracellular fluid provides no insight as to its mode of action, but the use of radioactive ^{45}Ca soon yielded useful information. This review of calcium function in spermatozoa will be restricted predominantly to the cells of the bull and the guinea pig.

Bovine epididymal spermatozoa contain a pool of 4–10 (6.4 ± 0.8) nmol of calcium/10^8 cells that can be mobilized by the calcium ionophore A23187 (Babcock et al., 1979), but these cells can take up 25–50 nmol when incubated with 0.1–0.2 mM CaCl$_2$ for 20 min (Babcock et al., 1976). The use of a variety of inhibitors, of fluorescent monitors, and of electron microanalysis yielded results consistent with the majority of the calcium going to the sperm mitochondria (Babcock et al., 1975, 1976, 1978). Low concentrations of A23187 (0.01–0.5 nmol/mg of sperm protein) move the accumulated calcium from the mitochondria to a compartment (cytosol?) from which it can apparently be pumped out of the cell to the suspending medium. Increasing the ionophore concentration (0.5–5.0 nmol/mg protein) overcomes the ability of the mitochondrial and plasma membrane transport mechanisms to cope with the ionophore's influence: calcium is equilibrated between mitochondria, cytosol, and extracellular medium whether or not the mitochondria are generating energy by oxidation of metabolites (Babcock et al., 1976). The presence of phosphate enhances calcium retention, and calcium also increases phosphate uptake (Babcock et al., 1975, 1976).

Experiments demonstrating that A23187, in the presence of calcium, would induce the acrosome reaction (Summers et al., 1976; Reyes et al., 1977; Green, 1978) lead to the postulate that "the immediate cause of the acrosome reaction is an increase in the cytoplasmic free calcium concentration" (Green, 1978).

Evidence that calcium *uptake* is indeed associated with the acrosome reaction was first provided by Singh et al. (1978). In contrast to the rapid uptake of calcium by bull epididymal spermatozoa, initial calcium uptake by guinea pig epididymal spermatozoa is restricted and is not affected by carbonyl-cyanide-*p*-trifluoromethoxyphenylhydrazone, an uncoupler of oxidative phosphorylation. This indicates that the calcium is bound to the outer surface of the sperm, not transported across the plasma membrane. After 1 hr of incubation at 37°C in minimal capacitation medium (Rogers and Yanagimachi, 1975) (110 mM NaCl, 25 mM NaHCO$_3$, 1 mM sodium pyruvate, 1 mM CaCl$_2$) energy-dependent calcium uptake began and simultaneously the number of acrosome-reacted cells increased. In the presence of 70 nM A23187, calcium uptake began at once and reached a maximum at 60 min. The 1-hr lag in occurrence of acrosome reaction was cut to less than 30 min in the presence of the ionophore (Singh et al., 1978).

Incubating guinea pig epididymal spermatozoa for 30–60 min in an isotonic solution containing lactate, pyruvate, and bicarbonate renders them permeable to Ca^{2+}. The presence of metabolizable sugars or of 2-deoxyglucose delays the acrosome reaction (Rogers and Yanagimachi, 1975), because they inhibit the enhancement of permeability (Coronel and Lardy, 1987). The biochemical processes involved in altering permeability to calcium under physiological conditions have not been elucidated, but are being studied in several laboratories. A great variety of agents and conditions has been found to induce the acrosome reaction; the common basis for their effect in all species studied is an increased membrane permeability to Ca^{2+}.

II. THE DISCOVERY OF CALTRIN

The rapid uptake of calcium by bovine epididymal spermatozoa is abolished when they are ejaculated. Despite the high concentration of calcium in bovine seminal fluid (Drevius, 1972), ejaculated sperm contain no more mobilizable calcium (7 ± 1 nmol/10^8 cells) than epididymal cells, and the ejaculated cells do not take up calcium either from the seminal fluid or from isotonic media after they are washed free of seminal fluid (Babcock *et al.*, 1979). The inhibitory material of the seminal fluid is retained on the plasma membrane of the sperm, as is evidenced by relatively slow uptake of calcium by membrane vesicles from ejaculated sperm as compared with the uptake by vesicles from epididymal cells (Rufo *et al.*, 1984). The addition of sperm-free seminal fluid to freshly collected epididymal sperm inhibits their normal calcium uptake. Approximately 1 mg of seminal plasma protein in 1 ml exerts 50% inhibition of Ca^{2+} uptake by 5×10^7 bovine epididymal sperm (Babcock *et al.*, 1979). These findings prompted a search for the inhibitory factor in seminal plasma (Singh, 1980; Rufo *et al.*, 1982).

The calcium transport inhibitor (caltrin) was found to be a protein that behaved on Sephadex G-50 and on SDS–polyacrylamide gel electrophoresis in glass tubes, as if it had a molecular weight of 10,000 (Rufo *et al.*, 1982). On slab gels, it migrates as a protein of 5000–6000 MW. Sequencing the protein (Scheme 1) disclosed its molecular weight to be 5411 (Lewis *et al.*, 1985).

12

Ser·Asp·Glu·Lys·Ala·Ser·Pro·Asp·Lys·His·His·Arg·Phe·Ser·Leu·Ser·Arg·Tyr·

24

Ala·Lys·Leu·Ala·Asn·Arg·Leu·Ala·Asn·Pro·Lys·Leu·Leu·Glu·Thr·Phe·Leu·

36   47

Ser·Lys·Trp·Ile·Gly·Asp·Arg·Gly·Asn·Arg·Ser·Val

Scheme 1.   The sequence of amino acids in caltrin.

When the sequence was established, it became apparent that caltrin is identical with bovine seminal plasmin, a protein that had been isolated by Reddy and Bhargava (1979) and shown to possess antibacterial activity. The sequence of the first 24 amino acids in caltrin was identical with that reported by Theil and Scheit (1983) for seminal plasmin. However, the segments 24–34 and 35–45 were transposed in the sequence of Theil and Scheit as compared to that of Lewis *et al.* Theil and Scheit also included an extra lysine at position 46, which was not present in caltrin. We assumed that the proteins were identical and that an error had been made in the Gottingen laboratory in positioning the tryptic peptides. An additional anomaly was the reporting of the molecular mass of the protein from the amino acid composition without correction for the water loss in peptide bond formation. Sedimentation velocity and equilibrium runs yielded an apparent molecular weight in agreement with the erroneous, calculated value. A new, partial sequence determination of seminal plasmin reported from Bhargava's laboratory (Sitaram *et al.*, 1986) is in agreement with the sequence established by Lewis *et al.* and confirms the identity of caltrin and seminal plasmin.

III. PROPERTIES AND FUNCTION OF CALTRIN

The anomalous behavior of this protein during studies of its molecular weight by different procedures has been observed in each of the laboratories investigating it (Reddy and Bhargava, 1979; Singh, 1980; Rufo *et al.*, 1982; Theil and Scheit, 1983). Significant amounts of cysteine were found in the amino acid analyses (Reddy and Bhargava, 1979; Rufo *et al.*, 1982), but not in the peptide sequence (Theil and Scheit, 1983; Lewis *et al.*, 1985). The protein contains no carbohydrate. It does not bind calcium.

Caltrin originates in the seminal vesicles of the bull (Singh, 1980; Rufo *et al.*, 1982), but also has been found in the ampullae and the corpus prostate (Shivaji *et al.*, 1984).

The isoelectric point of caltrin is 8.3 (Rufo *et al.*, 1982). The higher isoelectric points (>9) for a partially effective preparation (Singh, 1980) and for seminal plasmin (pI = 9.8) (Reddy and Bhargava, 1979) can be explained by the partial or total removal of anions from this basic protein during the purification procedures (San Agustin *et al.*, 1987).

Both crude seminal plasma (Babcock *et al.*, 1979) and purified caltrin (Rufo *et al.*, 1982) inhibit calcium uptake by bovine epididymal sperm from 80 to nearly 100%. During purification by Singh (1980), maximal inhibition by the homogeneous protein decreased to 25–40%. It was postulated that "a limited proteolysis or modification of (caltrin) may result in loss of inhibitory activity without impairing its ability to bind to the sperm surface receptors.'' It was assumed that the denatured molecules competed with the inhibitory form at the binding sites on the sperm, thus yielding a plateau in the extent of inhibition. Publication of Singh's purification procedure was delayed until a more effective preparation of the purified caltrin was obtained (Rufo et al., 1982).

It has now been found that the highly inhibitory pure protein can be converted to a noninhibitory form by retention of the protein on a cation exchange column and elution with 0.5 M NaCl. The noninhibitory form of the protein binds to epididymal sperm at the same locations that bind inhibitory caltrin (San Agustin et al., 1987) (see below), which supports Singh's hypothesis.

The procedure of Reddy and Bhargava (1979) yields a protein that does not inhibit calcium transport by epididymal sperm, because the inhibitory property is lost at the CM–Sephadex column purification step (San Agustin et al., 1987). Both native (inhibitory) and noninhibitory forms of caltrin inhibit the growth of Escherichia coli (unpublished data) as was originally described for seminal plasmin (Reddy and Bhargava, 1979).

IV. ENHANCEMENT OF CALCIUM TRANSPORT BY CALTRIN

Caltrin that has been subjected to retention on the cation exchange resin not only lacks the ability to block calcium transport into epididymal spermatozoa, but, as shown in Fig. 1, it is a powerful enhancer of calcium uptake by bull sperm (San Agustin et al., 1987). The significance of this finding will be discussed in a later section of this chapter. The conformation of caltrin changes in the transition from inhibitory to stimulatory form. The latter form combines with tetracycline to enhance fluorescence, whereas the former does not.

Because treatment of the protein with cation exchangers caused a loss of calcium transport inhibitory action, the effect of anions from seminal fluid was examined. Seminal plasma was deproteinized, was acidified to pH 1.5, was heated to 90°C, and after centrifuging to remove denatured protein, was extracted repeatedly with ether. The material remaining after evaporation of the ether was fractionated on a BioGel P-2 column, and two components were found to restore inhibitory activity to deionized caltrin. One of these has been identified as citrate; the other has been partially purified, is apparently quantitatively more effective than citrate in restoring inhibitory activity, but has not yet been characterized. Treating caltrin, from which anions have been removed, with either citrate or the as yet uncharacterized fraction, followed by gel filtration to remove excess anions makes the protein fully as effective in blocking calcium as when originally isolated.

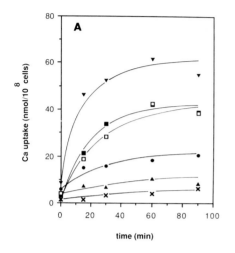

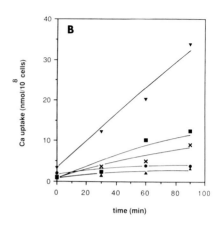

Fig. 1. (A) Calcium uptake in bovine epididymal spermatozoa. Epididymal spermatozoa (8×10^8 cells/ml) previously incubated for 15 min in NKM plus 10 mM DL-β-hydroxybutyrate were transferred to the assay medium without $CaCl_2$ and contained the indicated additions: ●, none; ▼, 30 μM A23187; ▲, 5μM CICCP; ■, 0.400 mg/ml caltrin (inhibitory activity lost during prolonged storage); ×, 0.400 mg/ml fresh caltrin; □, 0.400 mg/ml fresh caltrin passed through CM Sephadex G-25 cation exchanger. The final cell count was 4×10^7 cells/ml, and the final volume was 1 ml. Ten minutes after addition of cells to the assay medium, $CaCl_2$ containing ^{45}Ca was added to a final concentration of 0.2 mM ($t = 0$). At indicated times during the assay, 0.200-ml aliquots were taken out for determination of calcium uptake. A representative result of several experiments is shown. (B) Calcium uptake in bovine ejaculated spermatozoa. The assay was carried out as described in A with the following additions: ●, none; ▼, 30 μM A23187; ▲, 5 μM CICCP; ■, 350 μg/ml anticaltrin IgG; ×, 225 μg/ml anticaltrin IgG. Values shown are the mean of triplicate determination. [From San Agustin *et al.* (1987).]

Bull semen contains 350–1150 mg of citric acid/100 ml (Mann, 1964) with an average concentration of 35 mM, and it is likely that caltrin is maintained in its inhibitory form by this high concentration of citrate together with the second anion not yet identified.

V. EXPERIMENTS WITH ANTIBODIES TO CALTRIN

Homogeneous bovine caltrin was used to induce the formation of antibodies in rabbits. The antibodies reacted with bovine seminal plasma to yield a single precipitin line in the Ouchterlony double diffusion test. The antibody recognized also a protein from the seminal vesicles of the guinea pig, but the precipitin complex differs slightly from that formed with bovine caltrin. The

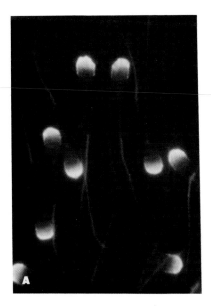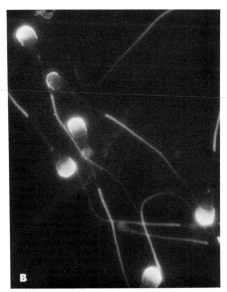

Fig. 2. Site of caltrin binding to (A) ejaculated or (B) epididymal bovine spermatozoa. Washed, ejaculated sperm were treated with anticaltrin rabbit serum and subsequently with goat anti-rabbit IgG labeled with fluorescein isothiocyanate. The epididymal cells (B) were first exposed to 0.4 mg caltrin/10^8 cells in 1 ml, then washed and treated as above. The photos were taken with a Zeiss fluorescence microscope. Epididymal cells not exposed to caltrin, but subjected to the antibodies in sequence, did not fluoresce significantly. [From San Agustin *et al.* (1987).]

antibodies to bovine caltrin did not react with extracts of several other tissues including prostate gland, coagulating gland, and epididymis of the guinea pig, rat, mouse, and hamster.

The site of caltrin binding to bovine sperm has been determined by the standard double antibody technique using goat anti-rabbit IgG labeled with fluorescein isothiocyanate to locate the caltrin–rabbit IgG complex. Bovine epididymal spermatozoa show no regions of intense fluorescence when exposed to the two antibodies in sequence. The fluorescence exhibited by ejaculated cells so treated is shown in Fig. 2A. Washed, ejaculated sperm retain caltrin over the acrosome and on the tail. A small spot of fluorescence is detected at the juncture of the head and midpiece. Caltrin neither binds to the distal portion of the head below the periacrosomal band, nor to the midpiece where the mitochondria are located. Bovine epididymal sperm that had been treated with caltrin retained antibodies on the same areas as ejaculated cells (Fig. 2B).

VI. POSTULATED ROLE FOR CALTRIN

The dual behavior of bovine caltrin, depending on its conformational state, invites the following hypothesis concerning its function. At the time of ejaculation, caltrin serves to inhibit calcium transport by the Na^+-Ca^{2+} exchanger (Rufo et al., 1984) from seminal fluid into the acrosome. It thus prevents premature development of the acrosome reaction, which would result in the loss of acrosomal enzymes as the sperm move up the female reproductive tract. Once sperm have moved away from the seminal fluid deposited in the female, anions may slowly diffuse from the caltrin bound to the sperm surface. It is also possible that the bound organic anions are used as an energy source by either sperm or the tissues of the female tract. Once sufficient amounts of the anions have been lost, the conformation of caltrin alters so as to make it an inducer of calcium uptake. Calcium taken into the acrosome activates phospholipase A_2; the resulting discomposition of the phospholipid bilayer causes the fusion of acrosomal and plasma membrane components, their vesiculation, and rupture so as to release the acrosomal contents.

In several other species, the epididymal spermatozoa do not take up calcium as do bovine epididymal sperm. These include the guinea pig, rabbit, and dog. The guinea pig has a caltrinlike protein in its seminal vesicles, and the seminal vesicle contents inhibit calcium uptake by epididymal cells that have been incubated at 37°C for 30 min under conditions that activate calcium uptake (Coronel et al., 1988). Whether the guinea pig caltrin eventually becomes an inducer of calcium uptake has not been determined, but the deionized and stimulatory form of bovine caltrin has been found to induce rapid calcium uptake by dog epididymal sperm (unpublished data).

If caltrin bound over the acrosomal region regulates the development of the acrosome reaction, it is logical to inquire as to the function of the caltrin bound over the tail. Many years ago Yanagimachi found that sperm acquire a "hyperactivation" state just prior to fertilizing ova either in vitro (Yanagimachi, 1969) or in vivo (Yanagimachi and Mahi, 1976). Hyperactivated motility involves a greater degree of bending of the sperm tail, greater tail excursion, and moving in circles rather than linearly. This hyperactivity is dependent on the presence of calcium in sperm of the guinea pig (Yanagimachi and Usui, 1974), bull (Babcock et al., 1976; Singh et al., 1983), rabbit (Suarez et al., 1983), mouse (Morton et al., 1978; Cooper, 1984), and hamster (Morton et al., 1974; Katz et al., 1978; Suarez et al., 1984). It seems likely that caltrin could be regulating the availability of Ca^{2+} to the dynein ATPase and thus influencing the character of the motility. Immediately after mating, the exclusion of Ca^{2+} by the inhibitory form of caltrin would permit moderately vigorous, linearly progressive motility thus facilitating movement through the uterus and into the fallopian tubes. When caltrin changes conformation so as to enhance calcium uptake, the more vigorous motility and circular path that ensues might facilitate moving spermatozoa out of mucosal folds in the reproductive tract and into the central lumen (Suarez and Osman, 1987) and then aid in driving the spermatozoon through the cumulus layer and the zona pellucida and thus permit fusion with the egg plasma membrane.

This small protein exhibits a variety of properties that suggest it to have important biological functions. We must remember, however, that bovine epididymal sperm are fully capable of impregnating cows, despite the absence of factors from seminal fluid (Lardy and Ghosh, 1952).

REFERENCES

Austin, C. R. (1951). Observations on the penetration of the sperm into the mammalian egg. *Aust. J. Sci. Res., Ser. B* **4**, 581–596.

Babcock, D. F., First, N. L., and Lardy, H. A. (1975). Transport mechanism for succinate and phosphate localized in the plasma membrane of bovine spermatozoa. *J. Biol. Chem.* **250**, 6488–6495.

Babcock, D. F., First, N. L., and Lardy, H. A. (1976). Action of ionophore A23187 at the cellular level. Separation of effects at the plasma and mitochondrial membranes. *J. Biol. Chem.* **251**, 3881–3886.

Babcock, D. F., Stamerjohn, D. M., and Hutchinson, T. (1978). Calcium redistribution in individual cells correlated with ionophore action on motility. *J. Exp. Zool.* **204**, 391–399.

Babcock, D. F., Singh, J. P., and Lardy, H. A. (1979). Alteration of membrane permeability to calcium ions during maturation of bovine spermatozoa. *Dev. Biol.* **69**, 85–93.

Chang, M. C. (1951). Fertilizing capacity of spermatozoa deposited into the fallopian tubes. *Nature (London)* **168**, 697–698.

Cooper, T. (1984). The onset of hyperactivated motility of spermatozoa from the mouse. *Gamete Res.* **9**, 55–74.

Coronel, C., and Lardy, H. A. (1987). Characterization of Ca^{2+}-uptake by guinea pig epididymal spermatozoa. *Biol. Reprod.* **37**, 1097–1107.

Coronel, C., San Agustin, J., and Lardy, H. A. (1988). Identification and partial characterization of caltrin-like proteins in the reproductive tract of the guinea pig. *Biol. Reprod.* **38**, 713–722.

Dan, J. C. (1954). Studies on the acrosome III. Effect of calcium deficiency. *Biol. Bull. (Woods Hole, Mass.)* **107**, 335–349.

Dan, J. C., Ohori, Y., and Kushida, H. (1964). Studies on the acrosome. VII. Formation of the acrosomal process in sea urchin spermatozoa. *J. Ultrastruct. Res.* **11**, 508–524.

Drevius, L. O. (1972). Water content, specific gravity and concentrations of electrolytes in bull spermatozoa. *J. Reprod. Fertil.* **28**, 15–28.

Eisen, A., and Reynolds, G. T. (1985). Source and sinks for the calcium released during fertilization of single sea urchin eggs. *J. Cell Biol.* **100**, 1522–1527.

Green, D. P. L. (1978). The induction of the acrosome reaction in guinea-pig sperm by the divalent metal cation ionophore A23187. *J. Cell Sci.* **32**, 137–151.

Heilbrunn, L. V. (1937). "An Outline of General Physiology," 1st ed. Saunders, Philadelphia, Pennsylvania.

Heilbrunn, L. V. (1952). "An Outline of General Physiology," 3rd ed. Saunders, Philadelphia, Pennsylvania.

Katz, D. F., Yanagimachi, R., and Dresdner, R. D. (1978). Movement characteristics and power output of guinea pig and hamster spermatozoa in relation to activation. J. Reprod. Fertil. 52, 167–172.

Lardy, H. A., and Ghosh, D. (1952). Comparative metabolic behavior of epididymal and ejaculated mammalian spermatozoa. Ann. N.Y. Acad. Sci. 55, 594–596.

Lewis, R. V., San Agustin, J., Kruggel, W., and Lardy, H. A. (1985). The structure of caltrin, the calcium-transport inhibitor of bovine seminal plasma. Proc. Natl. Acad. Sci. U.S.A. 82, 6490–6491.

Loeb, J. (1913). "Artificial Parthenogenesis and Fertilization." Univ. of Chicago Press, Chicago, Illinois.

Mann, T. (1964). "The Biochemistry of Semen and of the Male Reproductive Tract." Wiley, New York.

Morton, B., Harrigan-Lum, J., Albagli, L., and Jooss, T. (1974). Biochem. Biophys. Res. Commun. 56, 372–379.

Morton, B., Sagadraca, R., and Fraser, C. (1978). Fert. Steril. 29, 695–698.

Rasmussen, H. (1981). "Calcium and cAMP as Synarchic Messengers." Wiley (Interscience), New York.

Reddy, E. S. P., and Bhargava, P. M. (1979). Seminal plasmin—an antimicrobial protein from bovine seminal plasma which acts in E. coli by specific inhibition of rRNA synthesis. Nature (London) 279, 725–728.

Reyes, A., Goicoechea, B., and Rosado, A. (1977). In vitro capacitation of mammalian spermatozoa by the calcium ionophore A23187. Fertil. Steril. 28, 356.

Rogers, B. J., and Yanagimachi, R. (1975). Retardation of guinea pig sperm acrosome reaction by glucose: The possible importance of pyruvate and lactate metabolism in capacitation and the acrosome reaction. Biol. Reprod. 13, 568–575.

Rufo, G. A., Jr., Singh, J. P., Babcock, D. F., and Lardy, H. A. (1982). Purification and characterization of a calcium transport inhibitor protein from bovine seminal plasma. J. Biol. Chem. 257, 4627–4632.

Rufo, G. A., Jr., Schoff, P. K., and Lardy, H. A. (1984). Regulation of calcium content in bovine spermatozoa. J. Biol. Chem. 259, 2547–2552.

San Agustin, J. T., Hughes, P., and Lardy, H. A. (1987). Properties and function of caltrin, the calcium-transport inhibitor of bull seminal plasma. FASEB J. 1, 60–66.

Shivaji, S., Bhargava, P. M., and Scheit, K. H. (1984). Immunological identification of seminalplasmin in tissue extracts of sex glands of bull. Biol. Reprod. 30, 1237–1241.

Singh, J. P. (1980). Calcium in acrosome reaction and flagellar contractility of mammalian spermatozoa. Ph.D. Dissertation, University of Wisconsin-Madison, Madison.

Singh, J. P., Babcock, D. F., and Lardy, H. A. (1978). Increased calcium-ion influx is a component of capacitation of spermatozoa. Biochem. J. 172, 549–556.

Singh, J. P., Babcock, D. F., and Lardy, H. A. (1983). Motility activation, respiratory stimulation, and alteration of Ca^{2+} transport in bovine sperm treated with amine local anesthetics and calcium transport antagonists. Arch. Biochem. Biophys. 221, 291–303.

Sitaram, N., Kumari, V. K., and Bhargava, P. M. (1986). Seminalplasmin and caltrin are the same protein. FEBS Lett. 201, 233–236.

Suarez, S. S., and Osman, R. A. (1987). Initiation of hyperactivated flagellar bending in mouse sperm within the female reproductive tract. Biol. Reprod. 36, 1191–1198.

Suarez, S. S., Katz, D. F., and Overstreet, J. W. (1983). Movement characteristics and acrosomal status of rabbit spermatozoa recovered at the site and time of fertilization. *Biol. Reprod.* **29**, 1277–1287.

Suarez, S. S., Katz, D. F., and Meizel, S. (1984). Changes in motility that accompany the acrosome reaction in hyperactivated hamster spermatozoa. *Gamete Res.* **10**, 253–265.

Summers, R. G., Talbot, P., Keough, E. M., Hylander, B. L., and Franklin, L. E. (1976). Ionophore A23187 induces acrosome reactions in sea urchin and guinea pig spermatozoa (1). *J. Exp. Zool.* **196**, 381–385.

Theil, R., and Scheit, K. H. (1983). Amino acid sequence of seminalplasmin, an antimicrobial protein from bull semen. *EMBO J.* **2**, 1159–1163.

Yanagimachi, R. (1969). In vitro capacitation of hamster spermatozoa by follicular fluid. *J. Reprod. Fertil.* **18**, 275–286.

Yanagimachi, R., and Mahi, C. (1976). The sperm acrosome reaction and fertilization in the guinea pig: A study *in vivo*. *J. Reprod. Fertil.* **46**, 49–54.

Yanagimachi, R., and Usui, N. (1974). Ca^{++} dependence of sperm acrosome reaction and activation of guinea pig spermatozoa. *Exp. Cell Res.* **89**, 161–174.

3

Sperm Motility in Nematodes: Crawling Movement without Actin

THOMAS M. ROBERTS,[*] SOL SEPSENWOL,[†] AND HANS RIS[‡]

[*]Department of Biological Science
The Florida State University
Tallahassee, Florida 32306

[†]Department of Biology
The University of Wisconsin–Stevens Point
Stevens Point, Wisconsin 54481

[‡]Department of Zoology
The University of Wisconsin–Madison
Madison, Wisconsin 53706

I. INTRODUCTION

Sperm motility usually refers to swimming motion propelled by a beating flagellum. In fact, the abundance and availability of flagellated sperm, particularly from sea urchins, have made these cells valuable models for studying all aspects of microtubule-based motility (see reviews by Gibbons, 1981; Linck, 1982; Brokaw, 1986). There are, however, other types of sperm that lack fla-

41

gella and must use alternative methods to reach oocytes (reviewed in Roosen-Runge, 1977). Among these are the amoeboid sperm of nematodes.

Interest in nematode sperm dates back over a century to van Beneden's study of spermatogenesis in *Ascaris megalocephala* (van Beneden and Julian, 1884). These cells have continued to fascinate biologists because they lack two of the cardinal features of other sperm—the flagellum and the acrosome. In 20 years of extensive examination of nematode sperm with the electron microscope, the equivalent of an acrosome or a motile axoneme have never been found (reviewed by Foor, 1970; Anya, 1976). These studies did reveal that when sperm complete development they take on an amoeboid shape by extending a pseudopod. As a result, nematode sperm have been frequently referred to as amoeboid cells. It was not until 1977, however, that the first report of motility *in vitro* confirmed that the pseudopod is responsible for locomotion (Wright and Somerville, 1977).

Nematode sperm motility has now become the subject of intensive analysis. As predicted by their morphology, these cells exhibit many of the features of other amoeboid cells, such as their patterns of locomotion, substrate attachment, and plasma membrane mobility. Nematode sperm, however, lack the actin–myosin contractile system typically associated with crawling movement in eukaryotic cells. In this review, we describe progress toward understanding how these sperm move without using either the microtubule-based motor expected for a male gamete or the actin filament-based motor expected for an amoeboid cell.

II. PREPARATION FOR MOTILITY : ACTIVATION OF NEMATODE SPERM

Like many flagellated sperm, it appears that all nematode sperm are stored in the male in an immotile form and go through a rapid transformation to an actively motile state, a process we call *sperm activation*. Unlike their flagellated counterparts, however, activation of nematode sperm involves a morphological transformation after the completion of spermatogenesis. Specialized membranous organelles fuse with the plasma membrane, and the cell extends a prominent pseudopod. This process has been studied in detail in *Ascaris lumbricoides* (Foor and McMahon, 1973; Abbas and Cain, 1979; Sepsenwol, 1982; Sepsenwol *et al.*, 1986; Sepsenwol and Taft, 1988) and in *Caenorhabditis elegans* (Nelson and Ward, 1980; Ward *et al.*, 1983). Sperm from both species form their pseudopods in stages (Fig. 1): (1) elaboration of elongated filopodia over one hemisphere of the cell, (2) consolidation of the filopodia into a single bleb, (3) construction of a small spatulate pseudopod, (4)

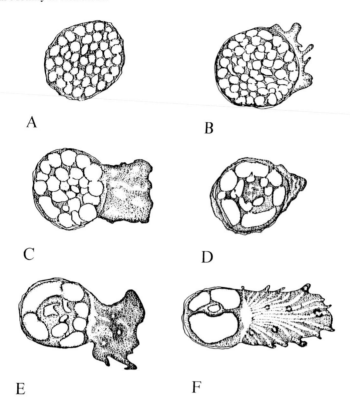

Fig. 1. Stages of *in vitro* activation of *Ascaris* sperm from time-lapse films. Drawings are of inactive spermatids removed from the seminal vesicle into HEPES–NaCl buffer, 7.4, and activated with a vas deferens supernatant. Suspension is sealed in a well slide, is imaged with phase-contrast optics, and is recorded by time-lapse cinematography. (A) Stage 0. Inactive cells from seminal vesicle are spherical to ovoid cells containing numerous small vesicles; no evidence of blebbing is indicated. (B) Stage 1 (2–7 min after addition of activator). Cell attaches to the substrate. Pseudopod first appears as a bleb or ruffle to one side of the sperm cell. (C) Stage 2 (4–8 min). A pulsatile, spatulate pseudopod extends and rotates overhead. Cytoplasmic vesicles begin to coalesce. Phase-dark villar projections become visible, originating at the leading edge of the pseudopod and migrating to cell body. (D) Stage 3 (6–8 min). Pulsatile pseudopod extends and makes rapid sweeps overhead; coalescence of vesicles into a refractile ring of material occurs. (E) Stage 4 (16–32 min). Marked increase in pseudopodial activity occurs: the pseudopod extends, flattens on the substrate, retracts, rotates to a new position and again extends, flattens, retracts, etc. Vesicles have fused into one or two large refractile bodies. (F) Stage 5 (19–42 min). Onset of translational motility (crawling). Pseudopod extends, flattens out and attaches to the substrate, and begins to move rapidly along a relatively straight track. Migration of villar projections over the pseudopod and the major branches of the radial fiber complexes are best observed at this stage. As cells move, they break free of cell-body attachment.

establishment of unidirectional membrane flow, and (5) extension of the pseudopod and the onset of membrane ruffling. During the last stage, the pseudopod is elevated and waves back and forth with the cell body anchored to the substrate. The pseudopod eventually attaches to the substrate and the cell begins to crawl forward (see Section III,C). The membrane movement that begins early in activation persists through full motility.

In *Ascaris*, the native agent that triggers activation is secreted from the glandular epithelium of the vas deferens, the terminal structure in the male reproductive tract (Foor and McMahon, 1973). The vas is sealed off from the seminal vesicle by a muscular valve (Foor, 1976). Other nematodes contain similar glandular cells in their vas deferens, but there is no direct evidence that they produce a sperm activator. In general, however, nematode males accumulate inactive spermatids in their seminal vesicles that activate only after passing through the vas to the female reproductive tract during copulation. The presence of activated sperm in the seminal vesicle, noted in some reports (e.g., Wright and Somerville, 1984), may be misleading. During copulation or even in handling and dissection, activator may regurgitate through the muscular valve into the seminal vesicle (Nelson and Ward, 1980). A clear exception to this anatomical arrangement for activation occurs in *C. elegans* hermaphrodites. They contain no homologs to the vas deferens and muscular valve, but their spermatids activate and move to the sperm storage point, the spermatheca, to await their oocyte successors from the gonad.

In the two species in which activation has been studied *in vitro*, *Ascaris* and *C. elegans*, there are several common requirements. The process is dependent on monovalent cations, usually Na^+, but K^+ and Li^+ are also effective. The cells become irreversibly inactive when incubated briefly in media lacking these cations. Low concentrations of proteases can initiate activation *in vitro*, although neither species exhibits detectable endogenous sperm protease activity, nor do protease inhibitors block *in vitro* activation by other agents (Abbas and Cain, 1979; Ward *et al.*, 1983). In flagellated sperm, the sudden onset of rapid flagellar beating is accompanied by a sharp increase in adenylate cyclase activity and 3',5'-cyclic AMP (cAMP) levels (reviewed by Hoskins and Casillas, 1975). Agents which increase endogenous cAMP as well as cell-permeating analogs of cAMP can induce activation *in vitro*. In *Ascaris* sperm, however, there is no detectable endogenous adenylate cyclase enzyme activity, nor does the lipophilic analog, dibutyryl cAMP, induce activation or increase motility in activated sperm (Sepsenwol *et al.*, 1986).

The initiation of sperm activation in *C. elegans* is reminiscent of the current acid-efflux model for activation of sea urchin sperm. The onset of flagellar motility is preceded by a sharp rise in internal pH, monitored externally as a sudden efflux of H^+. This acid efflux has been linked to a Na^+–H^+ exchanger in the sperm plasma membrane (Hansbrough and Garbers, 1981; Lee, 1984).

The monovalent cationophore, monensin, thought to favor Na^+-H^+ exchange, activates *C. elegans* sperm *in vitro*. Weak bases such as triethanolamine, which enter the cell and raise internal pH, are also effective sperm activators. Proteases, however, activate sperm without raising intracellular pH; their mode of action is still unclear (Ward *et al.*, 1983).

Ascaris sperm are anaerobic and activate poorly in oxygenated media. Except for proteases, synthetic agents that activate *C. elegans* sperm are not effective on *Ascaris* sperm *in vitro*. There has, however, been some progress toward identifying the native activator in *Ascaris*. Foor and McMahon (1973) first demonstrated that *Ascaris* sperm could be activated by extracts of the vas deferens. The active agent is both heat and protease sensitive (Abbas and Cain, 1979). Using radiolabeled vas deferens proteins, Abbas and Cain (1981) demonstrated that two species, 56,000 and 9,000 Da by SDS–gel electrophoresis, bind to isolated sperm membranes. Monoclonal antibodies generated against activator in vas deferens homogenates bind to biologically active components in the range of 50,000–70,000 Da (S. Sepsenwol and C. Johnson, unpublished observations). Recent studies, using rapid ion-exchange fractionation of vas deferens supernatants, show a single, pepsin-sensitive 58,000-Da band on silver-stained SDS gels associated with sperm activation activity (S. Sepsenwol, unpublished observations). Other fractionation procedures are underway to further characterize this protein.

Even though there are obvious differences in the mechanism underlying sperm activation among various species of nematodes, in each case, the end result is the formation of a pseudopod and the initiation of surface motility. Together, these events prepare nematode sperm for a type of locomotion that depends on an interplay between the pseudopod cytoplasm and the mobile plasma membrane.

III. MEMBRANE DYNAMICS AND LOCOMOTION

A. The Pattern of Locomotion

In vitro motility has now been examined in sperm from four species, *C. elegans* (Nelson *et al.*, 1982), *A. lumbricoides* (Nelson and Ward, 1981; Sepsenwol and Taft, 1988), *Nematospiroides dubius* (Wright and Sommerville, 1977, 1985), and *Nippostrongylus brasiliensis* (Wright and Sommerville, 1984). These cells differ in size and shape, but exhibit the same morphological asymmetry (Fig. 2a and b) and pattern of movement (Fig. 2c) by extending a pseudopod to pull themselves over the substrate. Usually when the pseudopod is formed, it persists so that changes in the direction of movement are accom-

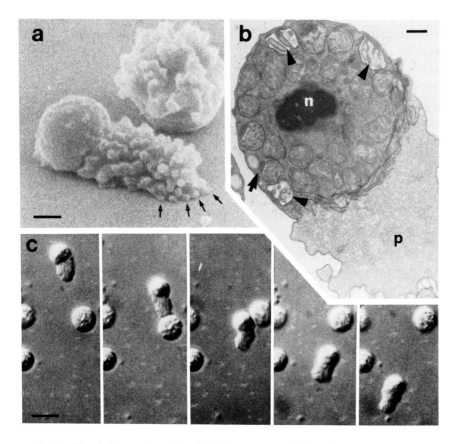

Fig. 2. Morphology and motility of *C. elegans* sperm. (a) Scanning electron micrograph of a crawling cell. The pseudopod is studded with short projections several of which (arrows) extend forward to contact the substrate in front of the rest of the cell. Photographed at 45° stage tilt. Bar, 1 μm. (b) Thin section electron micrograph showing internal asymmetry. The organelles are packed into the rounded cell body. The arrow indicates an intact, bilobed membranous organelle; arrowheads indicate several membranous organelles that have fused with the plasma membrane leaving a permanent saclike invagination (no fusion pores are visible in this section). The pseudopod (p) is organelle free. n, Nucleus. Bar, 500 nm. (c) A sequence of Nomarsky DIC micrographs showing translocation of a spermatozoon over a glass slide. In the second panel, the cell collides with a rounded spermatid, but continues movement. The interval between panels equals 20 sec. Bar, 5 μm. [Reproduced, with copyright permission of the publisher, from Roberts and Streitmatter (1984).]

plished by altering the pseudopodial contour; only occasionally is the pseudopod retracted and reformed in a new direction (Wright and Somerville, 1977; Nelson *et al.*, 1982). Sperm from *C. elegans, N. brasiliensis,* and *N. dubius* are about 7–10 μm long and crawl at 10–20 μm/min. *Ascaris* sperm are much larger. When their pseudopods are fully extended, these cells are about 25 μm in length and move at more than 70 μm/min (Sepsenwol, 1982).

Locomotion of each of these cells is accompanied by vigorous surface mobility. This is often manifest as the movement of knoblike projections that form at the front of the cell, sweep rearward over the surface, and disappear at the base of the pseudopod (Wright and Somerville, 1977; Nelson *et al.*, 1982; Sepsenwol and Taft, 1988; see also Fig. 6). *Nippostrongylus brasiliensis* sperm form constriction rings around their pseudopods that exhibit a similar tip-to-base movement (Wright and Somerville, 1984). As existing constrictions migrate toward the cell body, new ones are formed, so that as many as three such rings may be visible at one time. In all cases, the membrane articulations move rearward at about the same rate as the cell crawls forward.

B. Mobility of Surface Membrane Components

The use of several types of membrane markers to study membrane dynamics on *C. elegans* sperm has shown that the apparent movement of pseudopodial projections reflects actual movement of membrane components. For example, latex beads attached to the pseudopod surface exhibit the same behavior as the pseudopodial projections (Roberts and Ward, 1982a). As soon as these beads bind to the pseudopod surface, they are swept rearward to the base where they stop. They are not internalized or released, nor do they migrate onto the cell body or return to the pseudopod. Beads that bind to the cell body do not move at all. Parallel experiments using fluorescent-tagged lectins (Roberts and Ward, 1982b) confirmed the asymmetry of membrane movement, by showing that sperm can clear the fluorescent probe from the pseudopod, but not the cell body.

Recently, colloidal gold conjugates of a monoclonal antibody to membrane proteins (gold-ABY) have been used to examine the pattern of membrane dynamics in greater detail (Pavalko and Roberts, 1987). Sperm incubated in gold-ABY exhibit an initial uniform distribution of labeled antigen on their pseudopodial surface (Fig. 3a). Movement of antigen over the surface begins with clearance of gold particles from the tips of the pseudopodial projections (Fig. 3b) and proceeds until most of the gold-ABY has been removed from the pseudopod and accumulated on the surface of the cell body–pseudopod junction (Fig. 3c). The removal of antigen from the pseudopod is completed in about 2 min.

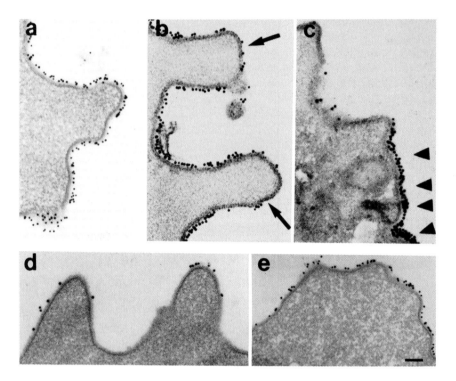

Fig. 3. (a–c) Clearance of gold-ABY-labeled membrane proteins from the pseudopod surface on *C. elegans* sperm. Cells fixed immediately after a 15-min incubation in gold-ABY exhibit a uniform distribution of surface-bound gold particles (a). A 5-sec wash before fixation results in clearance of gold particles from the tips of the pseudopodial projections (arrows in b). After a 2-min wash, most of the label has been removed from the pseudopod and the gold particles have accumulated on the surface at the cell body–pseudopod junction (arrowheads in c). (d and e) Insertion of new membrane protein onto the pseudopod surface. Cells were incubated first in unlabeled antibody then pulse labeled with gold-ABY. A 5-sec pulse results in labeling only at the tips of the pseudopodial projections (d). After a 2-min pulse, the entire pseudopod surface is labeled (e). Bar, 100 nm. [Reproduced, with copyright permission of the publisher, from Pavalko and Roberts (1987).]

A variation of this experiment showed that the tip-to-base movement of pseudopodial membrane proteins is accompanied by continuous, localized assembly of new antigen onto the pseudopod surface. In this case, sperm were incubated first in unlabeled antibody to saturate their existing surface antigen. Brief exposure of these cells to gold-ABY revealed insertion of new antigen onto the surface at the tips of the pseudopodial projections (Fig. 3d). The new antigen then moves rearward, and localized assembly continues so that within

2 min the population of pseudopodial surface antigen is completely renewed (Fig. 3e).

C. Correlation between Membrane Movement and Substrate Attachment

The pattern of centripetal membrane movement accompanied by polarized membrane assembly exhibited by *C. elegans* sperm is shared by a number of other types of crawling cells (reviewed in Singer and Kupfer, 1986). Observations of the pattern of cell–substrate attachment (see, for example, Kolega *et al.*, 1982) suggest that the mobility of the plasma membrane may be essential in order for crawling cells to gain the traction necessary for translocation. The pattern of substrate attachment under *C. elegans* sperm supports this hypothesis.

When *C. elegans* spermatozoa crawl, only the pseudopod contacts the substrate (Roberts, 1983). At the front of the cell, several of the pseudopodial projections extend forward and attach to the substrate ahead of the main mass of the pseudopod (Fig. 2a). The cell body is dragged along at the rear, either elevated slightly off the substrate or carried atop the trailing portion of the pseudopod. Interference reflection microscopy of live sperm has shown that only rarely is the entire underside of the pseudopod in contact with the substrate (Roberts and Streitmatter, 1984). Instead, the cell continuously forms discrete contact zones along the leading edge. These sites remain stationary relative to the substrate as the cell progresses over them. When the contacts reach the rear of the pseudopod, they disappear. Thus, the contact sites, like pseudopodial projections and surface-attached beads, move from the front to the rear of the cell at the same speed that the cell crawls forward.

Taken together, the results of these experiments show that the precise choreography of membrane movement allows *C. elegans* sperm to establish and maintain the traction that they need for locomotion. The assembly of membrane at the tips of the pseudopodial projections provides a continuous supply of components for constructing contacts at the forwardmost point on the cell. Thus, as old attachments move to the rear of the cell, new ones are formed along the advancing front, allowing locomotion to continue.

The relationship of pseudopodial membrane mobility to substrate attachment and translocation leave open the question of the machinery generating the force for motility. Early on, Roberts and Ward (1982b) speculated that centripetal membrane flow alone could propel sperm locomotion. In other cells that crawl, propulsion is a function usually associated with contraction of cytoplasmic actin–myosin complexes (reviewed in Pollard, 1981), but neither actin filaments nor myosin are present in *C. elegans* spermatozoa (see Section IV). Nematode sperm do, however, contain a new type of nonactin filaments that may be actively involved in locomotion. The attention of several investigators is now focused on the structure, composition, and function of these unusual fibers.

IV. CYTOSKELETAL ELEMENTS IN SPERM

Caenorhabditis elegans sperm undergo a complete change in cytoskeletal organization during the spermatogenesis. Before meiosis, spermatocytes contain numerous actin filaments (Nelson *et al.*, 1982; Roberts, 1987a) and microtubules (Ward, 1986). After meiosis, each secondary spermatocyte gives rise to four spermatids, which bud from the poles of the parent cell (Wolf *et al.*, 1978). The microtubules and actin filaments fail to segregate to the developing spermatids and, thus, are left behind in an anucleate residual body. As a result, the juxtanuclear centriole is the only tubulin-containing structure in spermatids (Ward, 1986). Postmeiotic loss of microtubules occurs in sperm from most other species of nematodes as well. There are only a few reports of microtubule-like structures in nematode spermatozoa, such as the bundles of tubules in the nonmotile taillike extension from the cell body in *Aspicularis tetraptera* sperm (Lee and Anya, 1967) and the single layer of regularly spaced tubules that underlies the entire plasmalemma of *Heterodera* sperm (Shepherd *et al.*, 1973; Shepherd and Clark, 1976, 1983). Even in these cases, the identification of tubulin in these structures has not been confirmed.

Because nematode sperm locomotion is so similar to the actin-based crawling movement of other metazoan cells, the loss of actin filaments during spermatid formation is surprising. Nonetheless, several independent approaches have shown that actin is only a minor sperm protein and is not involved in motility. Cytochalasins, which are potent inhibitors of actin-based movements (Tanenbaum, 1978), have no effect on the motility of *Ascaris, C. elegans,* or *N. brasiliensis* sperm (Nelson and Ward, 1981; Nelson *et al.*, 1982; Wright and Somerville, 1985). Parallel biochemical analyses have shown that actin comprises only 0.5% of the total cellular protein in *Ascaris* sperm (Nelson and Ward, 1981) and even less, 0.02%, in sperm from *C. elegans* (Nelson *et al.*, 1982). This small amount of actin can be detected by indirect immunofluorescence, but corresponding filaments cannot be found using probes specific for F-actin, such as phalloidin or heavy meromyosin (Nelson *et al.*, 1982; Roberts, 1987a). The actin filaments present in the prespermatid stages of spermatogenesis provide the positive control for these assays. Spermatocyte actin filaments decorate with both phalloidin and heavy meromyosin, and meiotic cytokinesis is blocked by cytochalasins (Nelson *et al.*, 1982; Roberts, 1987a).

Even though nematode sperm lack actin filaments, they are not devoid of cytoplasmic fibers. In fact, filaments have been reported in the pseudopods of several species of nematodes (Foor, 1970; Beams and Sekhon, 1972; Clark et al., 1972; McLaren, 1973; Burghardt and Foor, 1975; Shepherd and Clark, 1976, 1983; Ugwunna and Foor, 1982; Wright and Somerville, 1985). The organization and composition of these filaments have been examined only in C. elegans and Ascaris sperm. The pattern of locomotion exhibited by these two types of sperm is nearly identical (Section III,A), but the organization of their filaments is not.

The pseudopods of C. elegans sperm contain an amorphous granular cytoplasm that is so dense that the filaments embedded within are difficult to distinguish (Roberts, 1983). Fixation in the presence of detergents, such as Triton X-100 or saponin, removes enough of the cytoplasm to allow the filaments to be examined by both conventional and high-voltage electron microscopy (HVEM) (T. M. Roberts, unpublished observations). The filaments, which are 2–3 nm wide, are linked together to form a loose meshwork that extends throughout the pseudopod. Many of the cortical filaments abut the inner face of the plasma membrane so that the filament network appears to be linked at multiple, but apparently random, sites to the cell surface. Both sectioned specimens and critical-point dried whole mounts have been examined by HVEM without revealing more highly ordered arrays of filaments in these cells.

Ascaris sperm, in contrast, contain filaments organized into long, frequently branched *fiber complexes* that course for several micrometers through the pseudopod to terminate in each of the villar projections that move along the surface (Figs. 4 and 5) (Pawley et al., 1986; Sepsenwol and Ris, 1988). The complexes are about 150 nm in diameter. The individual filaments range in thickness from 2 to 10 nm and, thus, are more variable than the thin filaments in C. elegans sperm. Within each complex, the filaments are intermeshed and radiate outward in all directions in a "bottle-brush" configuration. At their free ends, these radiating fibers often associate either with similar filaments from adjacent complexes or with the plasma membrane (Fig. 5). The inner face of the plasma membrane is thick and electron opaque and, in some regions, appears to be composed of a carpet of short fibers. Filaments from the fiber complexes that approach the plasma membrane, particularly at the ends of the complexes in the villar projections, but also elsewhere along the surface, insert into this thickened, inner face of the plasmalemma.

The fiber complexes are so prominent that they are visible under differential interference (DIC) and phase contrast optics and have been examined in live sperm by enhanced video DIC microscopy. Surprisingly, the entire assemblage of fiber complexes moves rearward as a unit in concert with the villar projections on the cell surface (Fig. 6; Sepsenwol and Ris, 1988). The anatomical

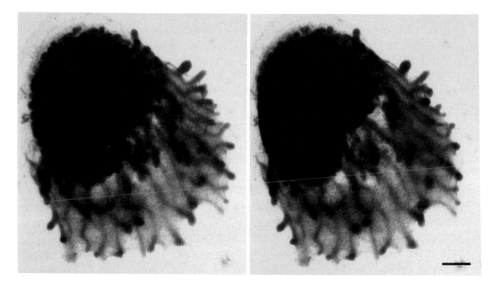

Fig. 4. High-voltage electron microscopy stereomicrograph of a whole mount of an *Ascaris* sperm activated on a gold grid and fixed in 2% glutaraldehyde in 0.1 *M* HEPES buffer (pH 7.4) containing 0.1% Triton X-100. Postfixation in 0.1% OsO_4. Stained with 1% uranyl acetate and critical-point dried. Bar, 2 μm; tilt angle, 12°.

association of fiber complexes and the plasma membrane together with the observation of comigration of complexes and villar projections in live sperm suggest that the sperm cytoskeleton plays a role in moving membrane components over the cell surface.

The nonactin filaments in nematode sperm may be composed of major sperm protein (MSP), a family of 14,200-Da polypeptides that comprises 10–15% of the total cellular protein in both *C. elegans* (Klass and Hirsh, 1981) and *Ascaris* (Nelson and Ward, 1981) sperm. In *C. elegans,* MSP is synthesized in spermatocytes from a multigene family (Burke and Ward, 1983; Klass *et al.,* 1984) and accumulates exclusively in the fibrous bodies (Ward and Klass, 1982; Roberts *et al.,* 1986). These organelles, which are characteristic of nematode sperm, are assembled as paracrystalline arrays of fibers (Fig. 7a) with reported diameters that range from 4.5 nm in *C. elegans* (Ward *et al.,* 1981) to 7 nm in *Ascaris* (Favard, 1961) and *Rhabditis pellio* (Pasternak and Samoiloff, 1972). Following meiosis, the fibrous bodies segregate to the cytoplasm of the developing spermatids (Fig. 7b) and, with the loss of actin filaments and microtubules to the residual body, become the only filamentous component in the cell. Ward and Klass (1982) used an antibody to MSP in immunofluorescence assays to demonstrate that, when the fibrous bodies segregate to the

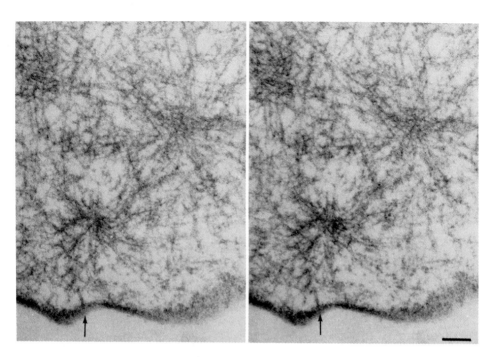

Fig. 5. High-voltage electron microscopy stereomicrograph of a section 0.13 μm thick through the pseudopod of an activated sperm. The cells were fixed in 2.5% glutaraldehyde in 0.1% HEPES buffer (pH 7.4) containing 0.05% saponin and 0.2% tannic acid, followed by postfixation in 0.1% OsO_4, and staining with 1% uranyl acetate. Sections were stained with 7.5% uranyl magnesium acetate and lead citrate. Three fiber complexes are seen in cross section. Fibers radiating from the axes of the fiber complexes intermesh with those of adjacent complexes and insert into the dense layer on the inner side of the plasma membrane (arrow). Bar, 200 nm; tilt angle, 20°.

spermatids and disassemble, MSP spreads throughout the cytoplasm. After sperm activation, MSP concentrates in the pseudopod. Thus, MSP follows the route from fibrous body to pseudopod predicted for a filament polypeptide.

Preliminary results from several experiments suggest that the pseudopodial filaments in *Ascaris* sperm do, in fact, contain MSP. For example, a monoclonal antibody to MSP labels the filament complexes in *Ascaris* sperm by immunofluorescence and decorates many of the individual filaments in immunogold assays. In addition, when cell-free extracts of *Ascaris* sperm are warmed to 39°C under anaerobic conditions (to approximate the environment in the porcine intestine) filaments 4–5 nm in diameter assemble *in vitro*. These filaments decorate with gold conjugates of the same antibody that labels the filaments in the pseudopod (T. Roberts, S. Sepsenwol, and H. Ris, unpublished

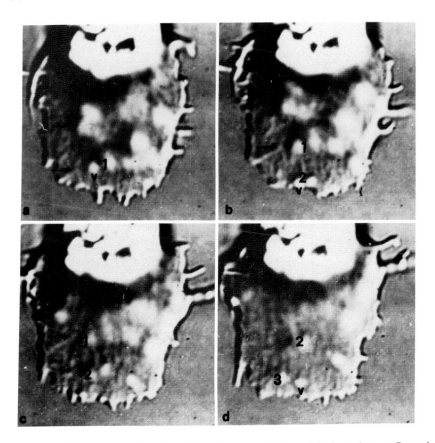

Fig. 6. Light micrographs of an activated sperm with the cell body at the top. Several villi (v) and fiber complexes are visible (compare with Fig. 5). Photographed from videotape; interval between frames equals 5 sec. Villi originate at the forward edge of the pseudopod and move toward the cell body at a uniform rate (see villi numbered 1–3). A branched fiber complex (arrow) moves at the same rate toward the cell body without much change in shape. Nomarski DIC, Axiomat with × 100 oil immersion objective.

observations). Efforts are underway to determine if filaments will assemble from purified fractions of *Ascaris* MSP, an approach that should identify the filament polypeptide decisively.

Major sperm protein is not simply a fragment of actin. The MSP genes from both *C. elegans* (Burke and Ward, 1983; Klass *et al.*, 1984; Ward *et al.*, 1988) and *Ascaris* (Bennett and Ward, 1986) have been cloned; they exhibit no homology to the nematode actin genes. In addition to their relatively low molecular weight, most members of the MSP family are basic ($pI \sim 8.5$), although

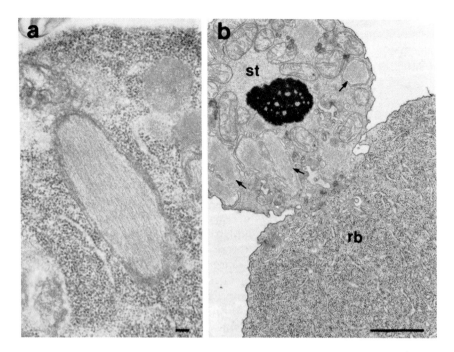

Fig. 7. Fibrous bodies in *C. elegans* sperm. (a) Thin section of a primary spermatocyte showing the parallel array of fibrous body filaments enveloped by membrane derived from a membranous organelle, the remainder of which is not shown in this section. Bar, 100 nm. (b) A spermatid (st) budding from the parent cell leaving behind a residual body (rb). All of the fibrous bodies (arrows) have segregated to the spermatid cytoplasm. Bar, 1 μm. (b) Reproduced from *The Journal of Cell Biology*, 1981, Vol. 91, pp. 26–44 by copyright permission of The Rockefeller University Press.]

one form in *C. elegans* is neutral (pI = 7.1) (Klass and Hirsh, 1981; Burke and Ward, 1983). These properties of MSP are distinct from five other fine, nonactin filament polypeptides, spasmin, tektin, titin, echinonematin, and giardin (reviewed in Roberts, 1987b). Thus, MSP may represent a new type of filament protein.

Are all of the filaments in the fibrous bodies and pseudopods of *C. elegans* and *Ascaris* sperm polymers of MSP? Certainly the variability of filament diameter (e.g., 4.5 nm in the fibrous bodies versus 2–3 nm in the pseudopod in *C. elegans*; 2–10 nm in *Ascaris* pseudopods) requires explanation. Some variability has also been observed among the filaments that assemble *in vitro* from cell-free extracts; close examination has revealed that a few 2- to 3-nm filaments coexist with the 4- to 5-nm fibers. The wider filaments are not uniform and frequently appear stranded and frayed at their ends, as if composed of two tightly twisted subfibrils. We cannot rule out the possibility of multiple types of filaments in these cells, but the available structural and biochemical data suggest that some of the variation in fiber diameter could be the result of different conformations of the same 2- to 3-nm subfibril.

V. TOWARD A MODEL FOR NEMATODE SPERM MOTILITY

It is remarkable that the phenomenology of actin-poor nematode sperm has so many similarities to the actin-based (i.e., cytochalasin-sensitive) locomotion of other crawling metazoan cells. There are two features of actin-based motility that are especially relevant to nematode sperm motility. (1) *Unidirectional membrane flow*—polarized membrane assembly results in directed membrane movement by displacing existing surface components rearward and provides the surface components needed to construct substrate attachments at the leading edge of the cell. (2) *Formation of oriented filament assemblies*—(actin) filaments attach or are nucleated at membrane regions in contact with the substrate and assemble into meshworks or bundles. These assemblies presumably anchor to stable elements of the cytoskeleton toward the passive side of the cell, although in nematode sperm, it is not clear that the antimembrane ends of the MSP filament arrays are attached.

A central question is shared by research on actin-rich crawling cells and MSP-rich nematode sperm: how is force generated for crawling? At present, this question has not been explicitly answered for actin-based amoeboid cells (Harris, 1976; Abercrombie, 1980; Small *et al.*, 1982; Geiger, 1982; Singer and Kupfer, 1986). Construction of a model for nematode sperm locomotion should start with the clear-cut relationship between unidirectional membrane flow over the pseudopod and the specialized cytoskeleton. The ability to correlate cytoskeletal anatomy with its behavior during crawling is a distinct advantage, but many questions about nematode sperm remain. Do the filaments exert tension on the membrane to drive its centripetal flow or vice versa? Are the processes of assembly of filaments at the tip and disassembly of filaments at the base a machine for pseudopod traction? Which components anchor the membrane to the substrate? Are there molecules other than MSP that regulate filament assembly and architecture? These questions are sufficient to propel research on nematode sperm motility for some time. The basic tools for this research already exist: *Ascaris* sperm provide abundant material for biochemical, structural, and behavioral studies. *Caenorhabditis elegans* has become a rich source of mutants that affect sperm development, morphology, and motility (Ward *et al.*, 1981; Ward, 1986). From every aspect, these cells have yielded unexpected results; we expect to be surprised again.

ACKNOWLEDGMENTS

Work from our laboratories described herein was supported by grants from NIH GM29994 (to T. M. R.), UWSP-UPDC, NSF DCB8610475, and NIH GM37435 (to S. S.). Integrated Microscopy Resource for Biomedical Research is supported by a grant from the Biotechnology Resources Program, DRR-NIH (P41RR00570).

REFERENCES

Abbas, M. K., and Cain, G. D. (1979). In vitro activation and behavior of the amoeboid sperm of *Ascaris suum* (Nematoda). *Cell Tissue Res.* **200**, 273–284.

Abbas, M. K., and Cain, G. D. (1981). Surface receptors: Are they involved in the transformation of spermatozoa of *Ascaris? Cell Tissue Res.* **214**, 553–567.

Abercrombie, M. (1980). The crawling movement of metazoan cells. *Proc. R. Soc. Lond.* B **207**, 129–147.

Anya, A. O. (1976). Physiological aspects of reproduction in nematodes. *In Adv. Parasitol.* (B. Dawes, ed.), **14**, 268–351.

Beams, H. W., and Sekhon, S. S. (1972). Cytodifferentiation during spermiogenesis in *Rhabditis pellio. J. Ultrastruct. Res.* **38**, 511–527.

Bennett, K. L., and Ward, S. (1986). Neither a germ-line specific nor several somatically-expressed genes are lost or rearranged during embryonic chromatin diminution in the nematode *Ascaris lumbricoides. Dev. Biol.* **118**, 141–147.

Brokaw, C. J. (1986). Future directions for studies of mechanisms for generating flagellar bending waves. *J. Cell Sci., Suppl.* **4**, 103–113.

Burghardt, R. C., and Foor, W. E. (1975). Rapid morphological transformations of spermatozoa in the uterus of *Brugia pahangi* (Nematoda, Filarioidea). *J. Parasitol.* **61**, 343–350.

Burke, D. J., and Ward, S. (1983). Identification of a large multigene family encoding the major sperm protein of *Caenorhabditis elegans. J. Mol. Biol.* **171**, 1–29.

Clark, W. H., Jr., Moretti, R. L., and Thompson, W. W. (1972). Histochemical and ultrastructural studies of the spermatids and sperm of *Ascaris lumbricoides* var. *suum. Biol. Reprod.* **7**, 145–159.

Favard, P. (1961). Evolution des ultrastructures cellulaires au cours de la spermatogénèse de l'*Ascaris. Ann. Sci. Nat., Zool. Biol. Anim.* [12] **3**, 53–152.

Foor, W. E. (1970). Spermatozoan morphology and zygote formation in nematodes. *Biol. Reprod.* **2**, Suppl., 177–202.

Foor, W. E. (1976). Structure and function of the glandular vas deferens in *Ascaris suum* (Nematoda). *J. Parasitol.* **51**, 849–864.

Foor, W. E., and McMahon, J. T. (1973). Role of the glandular vas deferens in the development of *Ascaris* spermatozoa. *J. Parasitol.* **59**, 753–758.

Geiger, B. (1982). Involvement of vinculin in contact-induced cytoskeletal interactions. *Cold Spring Harbor Symp. Quant. Biol.* **46**, 671–682.

Gibbons, I. R. (1981). Cilia and flagella of eukaryotes. *J. Cell Biol.* **91**, 107s–124s.

Hansbrough, J. R., and Garbers, D. L. (1981). Sodium dependent activation of sea urchin spermatozoa by speract and monensin. *J. Biol. Chem.* **256**, 2235–2241.

Harris, A. K. (1976). Recycling of dissolved plasma membrane components as an explanation of the capping phenomenon. *Nature (London)* **263**, 781–783.

Hoskins, D. D., and Casillas, E. R. (1975). Function of cyclic nucleotides in mammalian spermatozoa. *In* "Handbook of Physiology" (D. W. Hamilton and R. O. Greep, eds.), Sect. 7, Vol. V, pp. 453–460. Am. Physiol. Soc., Washington, D.C.

Klass, M. R., and Hirsh, D. (1981). Sperm isolation and biochemical analysis of the major sperm protein from *Caenorhabditis elegans*. *Dev. Biol.* **84**, 299–312.

Klass, M. R., Kinsley, S., and Lopez, L. C. (1984). Isolation and characterization of a sperm-specific gene family in the nematode *Caenorhabditis elegans*. *Mol. Cell. Biol.* **4**, 529–537.

Kolega, J., Shure, M. S., Chen, W.-T., and Young, N. D. (1982). Rapid cellular translocation is related to close contacts formed between various cultured cells and their substrata. *J. Cell Sci.* **54**, 23–34.

Lee, D. L., and Anya, A. O. (1967). The structure and development of the spermatozoon of *Aspicularis tetraptera* (Nematoda). *J. Cell Sci.* **2**, 537–544.

Lee, H. C. (1984). A membrane potential-sensitive Na^+–H^+ exchange system in flagella isolated from sea urchin spermatozoa. *J. Biol. Chem.* **259**, 15315–15319.

Linck, R. W. (1982). The structure of microtubules. *Ann. N.Y. Acad. Sci.* **383**, 98–121.

McLaren, D. J. (1973). The structure and development of the spermatozoon of *Dipetalonema viteae* (Nematoda: Filarioidea). *Parasitology* **66**, 447–463.

Nelson, G. A., and Ward, S. (1980). Vesicle fusion, pseudopod extension, and amoeboid motility are induced in nematode spermatids by the ionophore monensin. *Cell (Cambridge, Mass.)* **19**, 457–464.

Nelson, G. A., and Ward, S. (1981). Amoeboid motility and actin in *Ascaris lumbricoides* sperm. *Exp. Cell Res.* **131**, 149–160.

Nelson, G. A., Roberts, T. M., and Ward, S. (1982). *Caenorhabditis elegans* spermatozoan locomotion: Amoeboid movement with almost no actin. *J. Cell Biol.* **92**, 121–131.

Pasternak, J., and Samoiloff, M. R. (1972). Cytoplasmic organelles present during spermatogenesis in the free-living nematode *Panagrellus silusiae*. *Can. J. Zool.* **50**, 147–151.

Pavalko, F. M., and Roberts, T. M. (1987). *Caenorhabditis elegans* spermatozoa assemble membrane proteins onto the surface at the tips of pseudopodial projections. *Cell Motil. Cytoskel.* **7**, 169–177.

Pawley, J. B., Sepsenwol, S., and Ris, H. (1986). Four-dimensional microscopy of *Ascaris* sperm motility. *Ann. N.Y. Acad. Sci.* **433**, 171–178.

Pollard, T. D. (1981). Cytoplasmic contractile proteins. *J. Cell Biol.* **91**, 156s–165s.

Roberts, T. M. (1983). Crawling *Caenorhabditis elegans* spermatozoa contact the substrate only by their pseudopods and contain cytoplasmic 2-nm filaments. *Cell Motil.* **3**, 333–347.

Roberts, T. M. (1987a). Nematode sperm as a model for research on cell motility. *In* "Vistas in Nematology" (J. Veech and D. Dickson, eds.). pp. 440–447. Society of Nematologists, Hyattsville, Maryland.

Roberts, T. M. (1987b). Fine (2–5 nm) filaments: New types of cytoskeletal structures. *Cell Motil. Cytoskel.* **8**, 130–142.

Roberts, T. M., and Streitmatter, G. (1984). Membrane-substrate contact under the spermatozoon of *Caenorhabditis elegans*, a crawling cell that lacks filamentous actin. *J. Cell Sci.* **69**, 117–126.

Roberts, T. M., and Ward, S. (1982a). Membrane flow during nematode spermiogenesis. *J. Cell Biol.* **92**, 113–120.

Roberts, T. M., and Ward, S. (1982b). Centripetal flow of pseudopodial surface components could propel the amoeboid movement of *Caenorhabditis elegans* spermatozoa. *J. Cell Biol.* **92**, 132–138.

Roberts, T. M., Pavalko, F. M., and Ward, S. (1986). Membrane and cytoplasmic proteins are transported in the same organelle complex during nematode spermatogenesis. *J. Cell Biol.* **102**, 1787–1796.

Roosen-Runge, E. C. (1977). "The Process of Spermatogenesis in Animals." Cambridge Univ. Press, London and New York.

Sepsenwol, S. (1982). Sperm-motility inducing properties of *A. suum* vas deferens fractions. *Mol. Biochem. Parasitol., Suppl.* **1**, 61.

Sepsenwol, S., and Ris, H. (1988). A unique cytoskeleton associated with crawling in the "amoeboid" sperm of the nematode *Ascaris suum. J. Cell Biol.* (submitted for publication).

Sepsenwol, S., and Taft, S. (1988). *In vitro* induction of crawling in the amoeboid sperm of the nematode parasite, *Ascaris suum. Cell Motil. Cytoskel.* (with videotape Suppl.) (submitted for publication).

Sepsenwol, S., Nguyen, M., and Braun, T. (1986). Adenylate cyclase activity is absent in inactive and motile sperm in the nematode parasite, *Ascaris suum. J. Parasitol.* **72**, 962–964.

Shepherd, A. M., and Clark, S. A. (1976). Spermatogenesis and the ultrastructure of sperm and of the male reproductive tract of *Aphelenchoides blasthophorus* (Nematoda: Tylenchida: Aphelenchina). *Nematologica* **22**, 1–9.

Shepherd, A. M., and Clark, S. A. (1983). Spermatogenesis and sperm structure in some *Meloidogyne* species (Heteroderoidea, Meloidogynidae) and a comparison with those in some cyst nematodes (Heteroderoidea, Heteroderidae). *Rev. Nematol.* **6**, 17–32.

Shepherd, A. M., Clark, S. A., and Kempton, A. (1973). Spermatogenesis and sperm ultrastructure in some cyst nematodes, *Heterodera* spp. *Nematologica* **19**, 551–560.

Singer, S. J., and Kupfer, A. (1986). The directed migration of eukaryotic cells. *Annu. Rev. Cell Biol.* **2**, 337–365.

Small, J. V., Rinnerthaler, G., and Hinssen, H. (1982). Organization of actin meshworks in cultured cells: The leading edge. *Cold Spring Harbor Symp. Quant. Biol.* **46**, 599–611.

Tanenbaum, S. E., ed. (1978). "Cytochalasin: Biochemical and Cell Biological Aspects." Elsevier/North-Holland Biomedical Press, Amsterdam.

Ugwanna, S. C., and Foor, W. E. (1982). Development and fate of the membranous organelles in spermatozoa of *Ancyostoma caninum. J. Parasitol.* **68**, 834–844.

van Beneden, E., and Julian, C. (1884). La spermatogénèse chez l'ascaride megalocéphale. *Bull. Acad. Belg.* **7**, 312–342.

Ward, S. (1986). Asymmetric localization of gene products during the development of *Caenorhabditis elegans* spermatozoa. *In* "Gametogenesis and the Early Embryo" (J. G. Gall, ed.), pp. 55–75. Alan R. Liss, New York.

Ward, S., and Klass, M. R. (1982). The location of the major protein in *Caenorhabditis elegans* sperm and spermatocytes. *Dev. Biol.* **92**, 203–208.

Ward, S., Argon, Y., and Nelson, G. A. (1981). Sperm morphogenesis in wild-type and fertilization-defective mutants of *Caenorhabditis elegans. J. Cell Biol.* **91**, 26–44.

Ward, S., Nelson, G. A., and Hogan, E. (1983). The initiation of *Caenorhabditis elegans* spermiogenesis in vivo and in vitro. *Dev. Biol.* **98**, 70–79.

Ward, S., Burke, D. J., Sulston, J. E., Coulson, A. R., Albertson, D. G., Ammons, D., Klass, M., and Hogan, E. (1988). Genomic organization of major sperm protein genes and pseudogenes in the nematode *Caenorhabditis elegans. J. Mol. Biol.* **199**, 1–13.

Wolf, N., Hirsh, D., and McIntosh, J. R. (1978). Spermatogenesis in males of the free-living nematode, *Caenorhabditis elegans. J. Ultrastruct. Res.* **63**, 155–169.

Wright, E. J., and Somerville, R. I. (1977). Movement of a non-flagellate spermatozoon: A study of the male gamete of *Nematospiroides dubius* (Nematoda). *Int. J. Parasitol.* **7**, 353–359.

Wright, E. J., and Somerville, R. I. (1984). Post-insemination changes in the amoeboid sperm of a nematode *Nippostrongylus brasiliensis*. *Gamete Res.* **10**, 397–413.

Wright, E. J., and Somerville, R. I. (1985). Structure and development of the spermatozoon of the parasitic nematode, *Nematospiroides dubius*. *Parasitology* **90**, 179–192.

II

Remodeling of Egg Architecture

4

Whole-Mount Analyses of Cytoskeletal Reorganization and Function during Oogenesis and Early Embryogenesis in *Xenopus*[1]

JOSEPH A. DENT AND MICHAEL W. KLYMKOWSKY

Department of Molecular, Cellular, and Developmental Biology
University of Colorado at Boulder
Boulder, Colorado 80309

[1]Dedicated to the memories of Andrew M. Browne and Richard C. Parker.

THE CELL BIOLOGY OF
FERTILIZATION

I. INTRODUCTION

During the past century, a relatively small number of organisms have come, through various paths and for various reasons, to be viewed as appropriate model systems for the study of development (see Slack, 1983). Among the most important of these is the amphibian embryo. Nowhere has the amphibian embryo been more important than in the study of the role of the cytoskeleton in development. Cytoskeletal elements have been implicated in the determination of embryonic axes; during gastrulation, changes in the cytoskeleton are presumed to underlie the formation of bottle cells, the involution of the presumptive mesoderm, and the formation of the neural folds. The cytoskeleton also plays a clear role in cellular migration and cellular morphology within later-stage embryos. We were originally drawn to the developing *Xenopus* embryo because it represents an *in vivo* system in which to study the function of intermediate filaments, a major component of the cytoskeleton in higher eukaryotic cells.

Traditionally, the role of the cytoskeleton in morphogenesis has been studied by a combination of light and electron microscopy together with the use of drugs that affect cytoskeleton organization. This approach, while undeniably useful, is also inherently limited by the difficulty in creating three-dimensional images of an object the size and opacity of the amphibian oocyte and early embryo. Conventional means of obtaining detailed three-dimensional information from such a specimen require the preparation and analysis of serial sections. The time required for such an analysis is substantial and the information obtained may not always reveal the true complexity of the structure under study. It was for this reason that we first developed a cortical whole-mount immunocytochemical method that enabled us to resolve details of cytokeratin organization, reorganization, and animal–vegetal asymmetry in *Xenopus* oocytes and embryos (Klymkowsky *et al.,* 1987) that had been completely overlooked by conventional section analysis.

Our original whole-mount method was not able to visualize structures within the interior of the embryo. Andrew Murray (University of California, San Francisco) solved the problem of oocyte–embryo opacity by developing an improved solution for matching the refractive index of yolk platelets, rendering the embryo transparent (Fig. 1A–C). Subsequently, we developed a fixative that preserves structure, while allowing antibodies the size of IgMs (900,000 Da) to penetrate the 1.2-mm-diameter oocyte. We also developed a method to bleach the cortical pigment granules, while leaving the antigenicity of many proteins unaffected. Used together (Table I), these methods make possible high-resolution mapping of proteins and cytoskeletal organization in three dimensions (Fig. 1D and E).

Since our results are for the most part unpublished and the field of the

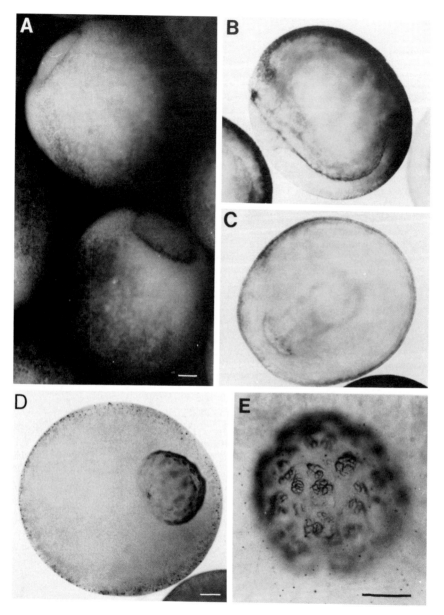

Fig. 1. Whole-mount immunocytochemistry. Stage XI *Xenopus* embryos are shown before clearing (A) and after clearing (B) and after bleaching and clearing (C). B and C are optical sections. Whole-mount immunocytochemistry of a stage V oocyte with RL1, an IgM antibody directed against nuclear pore components (Snow *et al.*, 1987), reveals both the domain structure of the nuclear envelope and the cortical localization of annulate lamellae (D); higher magnification of a RL1-stained stage III oocyte reveals details of structural domains in nuclear envelope (E). Bar in A, 100 μm for A–C; bars in D and E, 100 μm.

TABLE I

Whole-Mount Immunocytochemical Staining of *Xenopus*

1. Place oocytes, eggs, and embryos into a microfuge tube and remove the excess liquid. Fix by adding 1 ml of one-part DMSO : four -parts methanol (Dent's Fixative). (Eggs and embryos should be dejellied prior to fixation with 2% cysteine pH 8.0.) Fix overnight at −20°C.
2. Bleach embryos for 48 hr at room temperature in 20% DMSO : 10% hydrogen peroxide : methanol. Bleached embryos may be stored in 100% methanol at −20°C for future use.
3. Wash embryos 2 times in Tris-buffered saline (TBS) at room temperature for 20 min.
4. Add primary antibody plus 20% newborn calf serum; incubate overnight at 4°C. Keep volume to a minimum to avoid damaging the cortex of the cell(s).
5. Wash 3 times in TBS at room temperature; each wash should be 2-hr long.
6. Incubate with appropriately diluted anti-mouse Ig–peroxidase antibody. Dilute antibody into TBS plus 20% serum. Incubate overnight at 4°C with rocking.
7. Wash 3 times with TBS as in Step 5.
8. Incubate in 0.5 mg/ml diaminobenizidine (DAB) in TBS plus 0.02% hydrogen peroxide for 6 hr at room temperature with rocking—DAB should be filtered before use. Rocking is crucial at this stage!
9. Wash 2 times in 100% methanol for 5 min each time.
10. Soak in one-part benzyl alcohol : two -parts benzyl benzoate (Murray's Clear) for 15 min; mount in Murray's for microscopy. All steps (up to mounting for microscopy) can be done in microfuge tubes. They are not dissolved by Murray's solution.

cytoskeleton in *Xenopus* development is rather eclectic, this review will be somewhat unconventional. After a brief introduction of the cytoskeleton, we will present an exigesis of the role of the cytoskeleton during oogenesis, oocyte maturation, and the early stages of development, up to neurulation. Where appropriate, we will illustrate how whole-mount immunocytochemistry can confirm and extend previous observations. Readers interested in a more complete overview of *Xenopus* development are referred to Gerhart (1980), Slack (1983), Keller (1985), Gerhart and Keller (1986), and Nieuwkoop *et al.* (1985). An overview of a number of topics in amphibian development can be found in Slack (1985).

II. CYTOSKELETAL SYSTEMS

There are three major cytoskeletal systems in eukaryotic cells: microtubules (MTs), microfilaments (MFs), and intermediate filaments (IFs). Microtubules are composed of α- and β-tubulin together with a number of associated proteins

(Dustin, 1984; Olmstead, 1986). They have a distinct polarity, revealed by the different polymerization kinetics of their two ends. Microtubules form the structural basis of the mitotic and meiotic spindles and an extensive network throughout the interphase cell. Within the typical interphase cell, MTs radiate from one or more MT center(s); the MT center is generally associated with a cloud of ill-defined material that often, but not always, surrounds a pair of centrioles (for review, see Peterson and Berns, 1980; Brinkley, 1985). In many cell types, the Golgi apparatus is closely associated with and perhaps organized by the MT center (Thyberg and Moskalewski, 1985). The focal organization of the Golgi apparatus appears to play a major role in directed cell movement and perhaps cellular asymmetry (see Singer and Kupfer, 1986, for review). The MTs themselves act as tracks for the movement of intracellular organelles (Scholey *et al.*, Chapter 6, this volume). The mechanism by which the MT center is positioned within the cell remains unclear (see Kirschner and Mitchison, 1986, for speculation).

Microfilaments are composed of actin and associated proteins (Pollard and Cooper, 1986; Stossel *et al.*, 1985). Microfilaments are concentrated in the cortex of the cell, but also can be found within the cellular interior (Schliwa and Van Blerkom, 1981). They are intimately involved in cell movement, which is mediated by MF-associated cell–substrate adherence junctions (Burridge, 1986; Trinkaus, 1984). The structural integrity of tissues is maintained in large measure by MF-associated cell–extracellular matrix and cell–cell adhesion junctions. In addition, actin appears to be a major component of the amphibian and mammalian oocyte nucleus (Scheer and Dabauvalle, 1985). The results of intranuclear injection of antiactin antibodies and actin-binding proteins suggest that actin or an actinlike protein may be involved in transcription (Scheer *et al.*, 1984; Scheer, 1986).

In most of the higher organisms examined to date, actin and the tubulins are encoded by a family of genes, and specific genes are often expressed in a tissue-specific manner. The physiological significance of this differential gene expression remains obscure (Cowan *et al.*, 1987; Lopata and Cleveland, 1987; Schatz *et al.*, 1986).

Intermediate filaments are more diverse in subunit protein structure than either actin or tubulin. These proteins share a conserved structural motif and form ultrastructurally similar filaments that are insoluble under physiological conditions (see Traub, 1985; Steinert and Perry, 1985; Biessmann and Walter, Chapter 8, this volume). About 30 IF proteins are currently recognized in mammals, including the nuclear lamins and five types of cytoplasmic-IF proteins: vimentin, desmin, glial fibrillary acidic protein, the neurofilament proteins, and the cytokeratins (Franke, 1987). The recent identification of a nerve growth factor-induced IF protein in rat distinct from the previously defined classes (see Leonard *et al.*, 1988 and references therein) illustrates that new intermediate filament proteins remain to be discovered.

Both cytoplasmic and nuclear IFs are expressed in a cell-type-specific manner. Although present in most cells of higher vertebrates and in some cells of invertebrates (see Bartnik *et al.*, 1985, 1986), the function of IFs remains enigmatic. The experimental disruption of IFs by the injection of anti-IF antibodies into cultured cells has no apparent effect on cellular behavior or morphology (Klymkowsky, 1981; Lin and Feramisco, 1981; Gawlitta *et al.*, 1981; Lane and Klymkowsky, 1982; Klymkowsky *et al.*, 1983). Essentially identical results were obtained when cytokeratin filament organization was disrupted by the expression of a truncated cytokeratin protein (Albers and Fuchs, 1987). The inappropriate expression of IF proteins in cultured cells has also provided little clue as to their function (Kreis *et al.*, 1983; Guidice and Fuchs, 1987). To date, the only experimentally demonstrated function of cytoplasmic IFs is in the formation of frog virus 3 cytoplasmic assembly sites (Murti *et al.*, 1988). Whether intermediate filaments actually have a significant function in the normal cell out of its appropriate organismic context remains to be seen (see Lane and Klymkowsky, 1982; Klymkowsky *et al.*, 1983).

III. THE CYTOPLASMIC ASYMMETRY OF THE OOCYTE

Perhaps the most striking feature of the *Xenopus* oocyte, aside from its large size, is its visible asymmetry. The mature oocyte has a pigmented "animal" hemisphere, a lightly pigmented "vegetal" hemisphere, and an unpigmented equatorial band. The development of oocyte asymmetry appears to begin early, arguably with the asymmetry of the parental oocyte (Gerhart, 1980). In *Xenopus*, the cells that give rise to the germ line originate from the very vegetal-most cytoplasmic region of the egg. This region contains cortical material of ill-defined nature referred to as "germ plasm" (for reviews, see Nieuwkoop and Sutasurya, 1979; Heasman *et al.*, 1984). Early in development, primordial germ cells, which are derived from blastomeres that contain germ plasm, can be easily recognized and already possess a distinctive axis of asymmetry defined by a juxtanuclear mass of mitochondria (Al-Muktar and Webb, 1971). In oogenesis, there is a clear indication of nuclear asymmetry as well, since during the early phases of meiosis, the condensed chromosomes become associated with the inner surface of the nuclear envelope at a point where the outer envelope is juxtaposed with the mitochondrial mass. The nuclear–mitochondrial mass axis persists largely undisturbed until vitellogenesis (Al-Muktar and Webb, 1971; Dumont, 1972; Wylie *et al.*, 1986). The mitochondrial mass fragments during stage II of oogenesis (according to the classification scheme of Dumont, 1972)[2] and material from the mitochondrial mass accumulates at what appears to be the future vegetal pole (see Section III,C). Fragmentation of the mitochondrial mass precedes the asymmetric accumulation of specific mRNAs (Capco and Jeffery, 1982; Rebagliati *et al.*, 1985; Melton, 1987) and the asymmetric deposition of pigment (Dumont, 1972) and yolk (Danilchik and Gerhart, 1987).

Whether the cytoskeleton plays a direct role in the establishment of cytoplasmic asymmetry within the oocyte remains unclear. An alternative possibility is that the electrical field associated with the *Xenopus* oocyte is the prime mover. This field, first measured by Robinson (1979) with an extracellular vibrating probe electrode, is present in stage III oocytes, and perhaps earlier. It could provide both the directionality and motive force underlying oocyte asymmetry by electrophoresing proteins and protein–nucleic acid complexes within the oocyte. Even if the oocytes' electrical field is the primary effector of cytoplasmic asymmetry, the cytoskeleton is likely to play an important role in establishing and/or maintaining the asymmetrical distribution of ion channels and pumps that presumably produce this field. The availability of long-term culture methods that support the *in vitro* development of *Xenopus* oocytes (Wallace and Misulovin, 1978; Danilchik and Gerhart, 1987), together with methods for their subsequent maturation and fertilization (J. Roberts and J. C. Gerhart, personal communication), opens the possibility of direct experimental studies of the molecular mechanisms underlying the development of oocyte asymmetry.

[2]Stage I oocytes have not yet begun to accumulate yolk, are between 50 and 300 μm in diameter, and have a transparent cytoplasm. Their nucleus is centrally located and their nuclear envelope is smooth. The mitochondrial mass, also known as the Balbiani body, is spherical. The transition between stage I and stage II oocytes is characterized by the appearance of multiple extrachromosonal nucleoli, the appearance of characteristic folding of the nuclear envelope, an increase in overall diameter to 300–450 μm, and the beginning of yolk deposition (vitellogenesis). In stage II, the mitochondrial mass begins to fragment in a characteristic manner. Stage III oocytes continue in diplotene and grow in diameter to between 450 and 600 μm. They are now opaque due to the accumulation of yolk; pigment begins to appear uniformly. By stage IV, the oocyte has increased in diameter to 0.6–1 mm and has entered late diplotene. Lampbrush chromosomes have now begun to retract; the nucleus has moved toward the animal pole, and some nucleoli have begun to migrate to the center of the nucleus. Animal and vegetal hemispheres are now distinguishable due to their differential pigmentation. In stage V oocytes, large yolk platelets have become localized in the vegetal hemisphere, the chromosomes have retracted into the center of the nucleus, and the cell has reached a diameter of 1–1.2 mm. Yolk accumulation has ended in the stage VI oocyte, and an equatorial band of unpigmented cortex appears. Nucleoli become less prominent and nuclei becomes highly infolded at their vegetal pole.

A. Microtubules and Microtubule Centers

Within the mitochondrial mass of the early oocyte are Golgi elements and a pair of centrioles (Al-Muktar and Webb, 1971; Billet and Adam, 1976). The presence of centrioles suggests that the mitochondrial mass may also be the site of a MT center and that the mitochondrial mass itself may be an elaboration of the association between the Golgi apparatus and the MT center found in many types of cells (see Section II). While it is tempting to speculate that the mitochondrial mass–nuclear axis defines the future asymmetry of the oocyte–embryo, it is worth remembering that not all amphibia have a mitochondrial mass (as pointed out, but not illustrated, by Malacinski in Slack, 1985, p. 15). Nevertheless, they may all have a preestablished asymmetry axis based on the less conspicuous MT center. A simple model for oocyte asymmetry is that the nucleus–mitochondrial mass axis is a residue of the final mitotic division that produced the oocyte and that oocyte asymmetry reflects an elaboration of this original asymmetry. At present, there is precious little direct evidence to support this hypothesis (see Gerhart, 1980).

Using section-based immunocytochemistry, Palacek *et al.* (1985) found that the bulk of the tubulin immunoreactivity in early oocytes was associated with the mitochondrial mass; very little was found in the cytoplasm (see their Fig. 5). According to both Palacek *et al.* (1985) and Wylie *et al.* (1985), tubulin associates with the fragmenting mitochondrial mass in stage II oocytes. In later-stage oocytes, tubulin immunoreactivity is found in a radial pattern (Fig. 2F and G). This radial distribution seems to reflect tubulin-rich cytoplasmic domains rather than the distribution of MTs, since cytoplasmic proteins (Smith *et al.*, 1986), nucleolar antigens, and ribosomes (M. W. Klymkowsky, unpublished observation) display a similar radial distribution.

We have found somewhat different results using whole-mount immunocytochemistry with a monoclonal antibody directed against β-tubulin and rabbit

-->

Fig. 2. Whole-mount labeling of oocytes with antivimentin and antitubulin antibodies. Stage I oocytes were labeled with the antivimentin antibodies 14h7 (A and E) and RV202 (B–D). 14h7 labels the mitochondrial mass (arrow in A) and the vimentin-positive thecal cells of the follicular layer; a through-focus series (B–D) of a single oocyte reveals that RV202 labels only the thecal cells and not the mitochondrial mass. The fragmentation of the mitochondrial mass can be visualized using 14h7 labeling (E). Arabic numerals refer to stages in the process of fragmentation. Note the "basketwork" staining in oocytes indicated by the number 3 and the arrows in E. Stage I oocytes (arrow) are not stained by a monoclonal anti-β-tubulin antibody (F), whereas larger oocytes are stained in a characteristic radial manner. In contrast, antidetyrosylated α-tubulin antibody (G) labels all stage oocytes intensively. In later stages, this antibody reacts with discrete structures in the cortex of the oocyte (white arrows in H). Bar in A, 50 μm for A–D; bar in E, 100 μm; bar in F, 100 μm for F and G; bar in H, 25 μm.

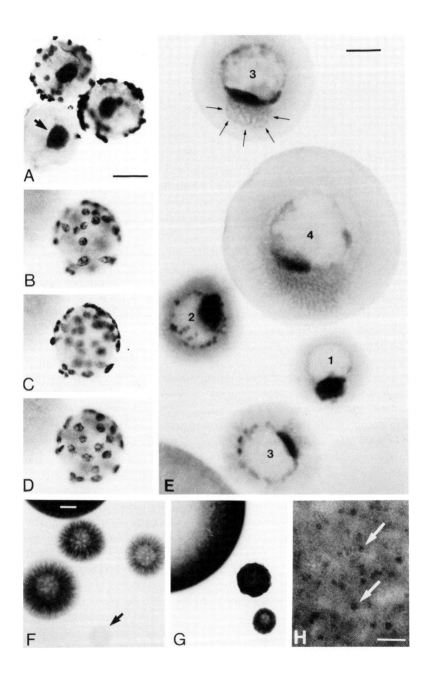

antibodies specific for tyrosylated or detyrosylated α-tubulin (Gundersen *et al.*, 1984). First, none of the antibodies we used labeled the mitochondrial mass (Fig. 2F and G). The anti-β-tubulin antibody did not stain small (stage I) oocytes at all, but labeled larger oocytes in the radial manner described by Palacek *et al.* (1985) (Fig. 2F). Antibodies against both tyrosylated and detyrosylated α-tubulin stained all stage oocytes in a radial manner (Fig. 2G). This result suggests that the epitope recognized by anti-β-tubulin is blocked in early oocytes. The nature of the discrepancy between our results and those of Palacek *et al.* (1985) and Wylie *et al.* (1985) with regards to tubulin and the mitochondrial mass is unclear; it is possible that their antitubulin antibodies reacted with a nontubulin component of the mitochondrial mass (see Section III,C), or alternatively, that the form of tubulin in the mitochondrial mass may be unreactive with the antibodies we used.

Whether there is an association between tubulin and the mitochondrial mass, it is fairly clear that the bulk of the tubulin within the oocyte is not in microtubular form. Electron microscopic studies find few MTs (Dumont and Wallace, 1972; Franke *et al.*, 1976; Heidemann *et al.*, 1985). The reason for the relative absence of MTs within the oocyte is not a lack of tubulin, since colchicine-binding studies indicate that tubulin amounts to 1% of the total soluble protein of stage II through stage VI oocytes and throughout early embryogenesis (Pestell, 1975). Treatment of oocytes with vinblastine induces the formation of large tubulin crystals that are located primarily in the oocyte cortex (Dumont and Wallace, 1972) and in association with the nuclear envelope. *In vitro* experiments indicate that oocyte tubulin will not polymerize and that it will actively inhibit the polymerization of brain and egg tubulin (Gard and Kirschner, 1987a). With oocyte maturation (Section IV), the tubulin becomes readily polymerizable into MTs, as demonstrated by the massive assembly of MTs onto injected MT centers (Gurdon, 1968; Heidemann and Kirschner, 1975, 1978; Karsenti *et al.*, 1984) and the ability of egg tubulin to polymerize *in vitro* (Gard and Kirschner, 1987a). This suggests that there is a positive block to MT polymerization in the oocyte.

Evidence for tubulin complexes comes from studies of oocyte tubulin and tubulin–tyrosine ligase. Tubulin detyrosylation appears to occur primarily on polymeric, but not necessarily microtubular, forms of tubulin (Webster *et al.*, 1987 and references therein). Preston *et al.* (1981) found no tyrosylatable tubulin in stage II–IV oocytes, even after treatment with carboxypeptidase, which in other systems renders tubulin tyrosylatable (Gundersen *et al.*, 1984). This result suggests that much of the tubulin in the early oocyte is held in a form that cannot react with the tubulin–tyrosine ligase or carboxypeptidase. Tyrosylatable tubulin appears at stages V and VI of oogenesis, even though there are still few MTs present (Preston *et al.*, 1981), suggesting a change in the form of tubulin sequestration during oogenesis, perhaps in preparation for oocyte maturation and early development. The nature of this putative aggregated form of tubulin is unknown. Whole-mount immunocytochemistry with antidetyrosylated tubulin antibody reveals discrete tubulin-containing structures in the cortex of later-stage oocytes (Fig. 2H). Whether these structures are in fact involved in tubulin sequestration remains to be demonstrated.

While the bulk of oocyte tubulin appears to be in a nonpolymerizable, non-microtubular form (see above), studies on the effects of drugs that depolymerize conventional microtubules suggest that there is a functionally significant population of microtubules within both *Xenopus* (Colman *et al.*, 1981) and *Rana* (Lessman *et al.*, 1986; Lessman, 1987) oocytes. In *Xenopus,* treatment of late-stage oocytes with either nocodazole or vinblastine results in the movement of the nucleus toward the cortex; this movement is in the antigravity direction, and the nucleus is often found flattened against the cortex (M. W. Klymkowsky, unpublished observations). The observation suggests that microtubules play a role in maintaining the position of the nucleus within the oocyte. Whether this is a direct effect due to interactions between microtubules and the nucleus (see Palachek *et al.*, 1985) or is an indirect effect due to the influence of microtubules on the overall consistency of the oocyte cytoplasm remains to be determined. In any case, these results suggest that microtubules play a role in determining the overall organization of the oocyte.

B. Cortical Organization and Microfilaments

The oocyte, like all eukaryotic cells, has a distinct plasma membrane-associated MF-rich cortex (Bray *et al.*, 1985; Vacquier, 1981; Longo, Chapter 5, this volume). In addition to MFs, the cortical region of the oocyte contains the contractile protein myosin, coated vesicles, cortical granules, and elements of the endoplasmic reticulum (Campanella and Andreuccetti, 1977). During oogenesis, cytokeratin-type IFs (Section III,C) and annulated lamellae (Fig. 1D) (Balinsky and Devis, 1963; M. W. Klymkowsky, unpublished observations) also become localized in the cortical region. The cortex of the later-stage oocyte has a distinct animal–vegetal polarity reflected by the distribution of pigment granules, by the organization of cytokeratin filaments (Section III,C), by the thickness of the cortex, by the distribution of intramembraneous particles, and in the fluidity of the membrane (Dictus *et al.*, 1984; Nieuwkoop *et al.*, 1985).

The presence of MFs in the cortex of the *Xenopus* oocyte was studied in detail by Franke *et al.* (1976). Examination of isolated cortices by immunofluorescence microscopy indicates that the actin is organized in small "whorls" that are interconnected by finer bundles of MFs. As yet, there is little information as to whether this type of MF organization is characteristic of all oocyte stages, or whether there is an animal–vegetal asymmetry in MF organization.

In contrast to tubulin, the bulk of the actin in the stage VI oocyte, while unpolymerized, is clearly polymerizable (Clark and Merriam, 1978; Merriam and Clark, 1978). Treatment of late-stage oocytes with the antimicrofilament drug cytochalasin B causes the mottling of pigment in the animal hemisphere and the dispersion of cortical structures (Colman *et al.*, 1981); interestingly, the mottling of pigment indicated by cytochalasin is blocked when oocytes are treated simultaneously with cytochalasin and an antimicrotubule drug (colchicine: Colman *et al.*, 1981; nocodazole: M. W. Klymkowsky, unpublished observations), further evidence for a significant microtubule system within the oocyte (see above). In the subcortical cytoplasm of the oocyte, actin also has been reported to associate with yolk platelets (Colombo, 1983), where it may play a role in maintaining the asymmetry of yolk-platelet distribution (Danilchik and Gerhart, 1987) and in the contractions of the cytoplasm associated with cleavage (Section V).

A major structural feature of the oocyte cortex are the microvilli. Cortical MF bundles often appear to connect with the MF core of the microvilli (see Mooseker, 1985). During oocyte development the number and shape of these microvilli change dramatically, and these changes correspond to periods of intensive vitellogenin import and yolk-platelet formation (Danilchik and Gerhart, 1987). Changes in microvilli also occur in response to maturation and fertilization (Sections IV and V).

C. Intermediate Filaments

It is still not clear whether oocytes contain vimentin-type intermediate filaments. Franz *et al.* (1983) failed to detect vimentin within oocytes by either section-based immunofluorescence microscopy or two-dimensional gel electrophoresis. On the other hand, Godsave *et al.* (1984a) found vimentin immunoreactivity associated with the mitochondrial mass of early oocytes. We have examined oocytes using two monoclonal antibodies that react specifically with vimentin in *Xenopus* A_6 cells: 14h7, which was generated against A_6 cell residues, and RV202, which was generated against mammalian vimentin by Franz Ramaekers (University of Nijemgen). Both antibodies labeled the vimentin-containing follicle thecal cells that sometimes remain associated with oocytes; 14h7 strongly labeled the mitochondrial mass, whereas RV202 did not (Fig. 2A–D). Both antibodies react with an insoluble, 55-kDa polypeptide in A_6 cells; this polypeptide appears to be vimentin (Dent *et al.*, 1989). 14h7, but not RV202, reacts with an insoluble polypeptide of 57 kDa. Likewise, 14h7, but not RV202, reacts with polypeptides in the egg (Dent *et al.*, 1989). These results suggest that the egg, and presumably the oocyte as well, expresses a polypeptide distinct from, but immunologically related, to vimentin.

In any case, immunochemical staining with 14h7 allows the visualization of the reorganizing mitochondrial mass during stage II of oogenesis (Fig. 2E). First, the mitochondrial mass often begins to fragment into a number of smaller structures associated with the nuclear envelope. Next, strands of material are seen moving away from the mitochondrial mass, which is still the largest of these structures. These strands often form a kind of "basketwork" on the mitochondrial mass side of the oocyte (Fig. 2E). As the original mitochondrial mass completes its reorganization, the satellite structures associated with the nucleus disappear and strands of material are seen between the nucleus and the presumptive vegetal oocyte cortex. Similar results have been reported by both Godsave et al. (1984b), Wylie et al. (1985), and Palacek et al. (1985). The reorganization of the mitochondrial mass precedes the redistribution of both yolk (Danilchik and Gerhart, 1987) and Vg1 mRNA (Melton, 1987). Whether the reorganization of the mitochondrial mass is a prerequisite for, or the first readily observable response to a common process of cytoplasmic reorganization remains to be determined (see above). An unambiguous definition of the components recognized by antivimentin antibodies during these stages could well be useful in unraveling this process.

Cytokeratin-type IFs were first reported in *Xenopus* oocytes by Gall et al. (1983). Franz et al. (1983) demonstrated the presence of three cytokeratin proteins, a type II cytokeratin of 56,000 Da, and two type I cytokeratins of 46 and 42 kDa. The type II cytokeratin appears to be homologous to cytokeratins #8 (endo A) in mammals (Franz and Franke, 1986).

Godsave et al. (1984a) carried out a detailed study of the distribution of cytokeratin immunoreactivity through oogenesis. They found that cytokeratin immunoreactivity first appeared in stage I oocytes as sparse cortical threads. In later-stage previtellogenic oocytes, this cortical staining increased in intensity and some subcortical staining was found; cytokeratin staining was also found to surround and invade the mitochondrial mass. During vitellogenic stages, cytokeratin immunoreactivity is found in the cortex and as fine strands running radially through the endoplasm. They also noted a clear difference in the organization of these radial cytokeratin strands between animal and vegetal hemispheres. Godsave et al. (1984a) also found cytokeratin staining associated with the nuclear envelope during the early vitellogenic stages. During the later stages of oogenesis, the cytokeratin system becomes increasingly cortical and has a distinct animal–vegetal asymmetry that can be seen clearly using whole-mount immunocytochemistry (Klymkowsky et al., 1987). The cytokeratins of the vegetal hemisphere have a geodesic-type of organization, whereas the cytokeratins of the animal hemisphere appear largely disorganized.

The molecular mechanism underlying the reorganization of cytokeratin proteins during oogenesis is unknown. One possibility is that the pattern of cytokeratin gene expression itself changes during oogenesis. Alternatively, changes in posttranslational modification in either cytokeratin or cytokeratin-

associated proteins may be involved. The reorganization of cytokeratin filaments during oocyte maturation and fertilization, on the other hand, is clearly
mediated by posttranslational mechanisms (Section V). The redistribution of
cytokeratin during the later stages of oogenesis from the endoplasm to the
cortex may, in part, anticipate its use in the cortical–epithelial layer of the
embryo.

A recent report by Pondel and King (1987) suggests that the cytokeratin
system of the late-stage oocyte may play a role in the localization of maternal
mRNAs, specifically the Vg1 mRNA. Vg1 encodes a protein homologous to
transforming growth factor-β (TGF-β) (Weeks and Melton, 1987) and the
translated product of the Vg1 mRNA appears to play a role in the induction
of mesoderm (Kimmelman and Kirschner, 1987). In the late-stage oocyte,
Vg1 is localized to the vegetal cortex (Melton, 1987). When insoluble residues
are isolated from late-stage *Xenopus* oocytes, the Vg1 mRNA, unlike other
mRNAs, copurifies with the insoluble fraction (Pondel and King, 1987). The
major component of this insoluble fraction is the cytokeratins. During oocyte
maturation, cytokeratin organization breaks down (Klymkowsky *et al.*, 1987),
the cytokeratins themselves become soluble (Klymkowsky and Maynell, 1989),
and the Vg1 mRNA no longer associates with the insoluble fraction (M. L.
King, personal communication). In the egg, Vg1 mRNA is found to have moved
away from the cortex and occupies a sizable portion of the vegetal hemisphere
(Weeks and Melton, 1987). If Vg1 can be shown to physically interact with
the oocyte's cytokeratin system, then there would be a strong case that cytokeratin asymmetry has functional significance. That functionally significant
interactions between intermediate filaments and cytoplasmic components can
occur is demonstrated by the involvement of vimentin-type intermediate filaments in the organization of frog virus 3 cytoplasmic assembly sites (Murti
et al., 1988).

IV. THE TRANSFORMATION FROM OOCYTE TO EGG

In vivo, the stage VI oocyte is transformed into the egg in response to
pituitary hormones; maturation occurs *in vitro* in response to progesterone.
It is clear that new RNA transcription, protein synthesis, and changes in protein phosphorylation all play significant roles in oocyte maturation (Schuetz,
1985; Maller, 1985). Maturation is accompanied by a number of changes in
both the follicular layer and within the oocyte itself. The macrovilli of the
inner sheet of follicle cells disconnect from the oocyte, and the follicular sheet
ruptures to release the mature oocyte. Within the oocyte, maturation leads
to a general increase in cellular activity. The first major structural landmark

is the breakdown of the nuclear envelope. In somatic tissue, nuclear envelope breakdown is thought to be mediated by the hyperphosphorylation of the nuclear lamin proteins (Newport and Forbes, 1987). In the oocyte, there is a clear polarity to nuclear envelope breakdown, which begins at the basal surface of the nucleus. Breakdown is accompanied by the localization of RNA-containing granules (nuage) to the basal region and the migration of nucleoli to the apical surface (Brachet et al., 1970). Nuclear proteins released during the course of nuclear envelope breakdown are located primarily in the animal hemisphere, in regions of small- and medium-sized yolk platelets (Hausen et al., 1985). In addition, they seem to diffuse rapidly through the cortical region of the oocyte (Dreyer et al., 1983; Hausen et al., 1985). Some of these nuclear components become resequestered into the nuclei of specific cell types at various stages of embryogenesis, suggesting that they may play a role in the control of cellular differentiation (Dreyer et al., 1981).

Around the time of nuclear envelope breakdown, the block on MT polymerization is released, and a network of MTs appears at the basal region of the disintegrating nuclear envelope (Huchon and Ozon, 1985, as cited by Heidemann et al., 1985). These MTs reorganize to form the barrel-shaped meiotic spindle (Fig. 3A). The spindle poles of the oocyte spindle differ from those of mitotic cells in that they appear to lack centrioles and are more diffusely organized. The meiotic spindle moves toward the animal hemisphere cortex in a process that appears to depend on MFs; MTs do not appear to be involved, since chromosomes come to localize at the animal pole even in their absence (Ryabova et al., 1986). The movement of the meiotic spindle is accompanied by the migration of pigment out of the cortex which produces an unpigmented region within the animal hemisphere. Meiosis I ends with the extrusion of the first polar body, proceeds into meiosis II, and arrests at metaphase II. An important starting point for studying the regulation of the MT cytoskeleton during oocyte maturation may be the recently identified XMAP, a microtubule-associated protein from Xenopus eggs that appears with maturation, stimulates MT assembly, and is phosphorylated during mitosis (Gard and Kirschner, 1987b).

During oocyte maturation, the system of cortical cytokeratin filaments is disrupted (Klymkowsky et al., 1987; Klymkowsky and Maynell, 1987). The mechanism of this fragmentation is unclear. Cytokeratin filaments can change their organization (Franke et al., 1983) and are known to fragment (Franke et al., 1982; Lane and Klymkowsky, 1982; Lane et al., 1982; Tolle et al., 1987) during mitosis in a number of epithelial cell types. The binding of anticytokeratin antibodies can also cause the fragmentation of cytokeratin filaments (Klymkowsky, 1982; Lane and Klymkowsky, 1982; Klymkowsky et al., 1983; Tolle et al., 1985; Maynell and Klymkowsky, 1989), and IF organization is sensitive to the metabolic state of the cell (Tolle et al., 1987;

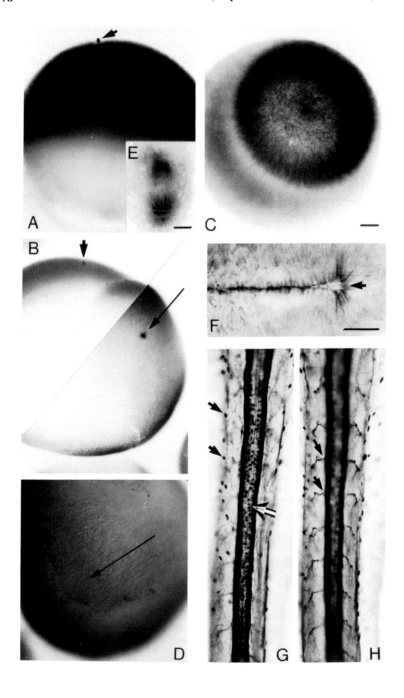

Klymkowsky, 1988). Cytokeratin fragmentation could also be due to changes in posttranslational modification, as appears to be the case with the nuclear lamins, and/or the binding of proteins released from the nucleus during nuclear envelope breakdown.

Microfilaments also undergo a significant change during maturation. The cortical–subcortical MF system, which is incompetent to undergo Ca^{2+}-mediated contraction in the oocyte, becomes competent to contract (Section V,A). It is tempting to speculate that protein phosphorylation, known to play a key role in oocyte maturation (Ezzell *et al.*, 1983; Gard and Kirschner, 1987b; see Maller, 1985; Ozon *et al.*, 1987), also regulates MT polymerization, MF-based contraction, and cytokeratin organization, resulting in the reorganization of the oocyte's cytoskeleton in preparation for fertilization and early development.

V. CYTOSKELETAL DYNAMICS AND THE DETERMINATION OF EMBRYONIC AXES

If the study of *Xenopus* early development has shown nothing else, it has demonstrated the developmental importance of the arrangement and rearrangement of cytoplasmic components within the egg. The cytoskeleton is clearly important in generating these rearrangements, although the detailed mechanisms by which it acts are for the most part obscure. In this section, we will examine what is known and hypothesized about the cytoskeleton in the first cell cycle.

←───

Fig. 3. Whole-mount labeling of the egg, early post fertilization and later-stage embryo with antitubulin and antiacetylated tubulin antibodies. Staining with anti-β-tubulin antibody reveals the meiotic spindle of the unfertilized egg (arrow in A). Within 15–20 min after fertilization (B), it is possible to visualize both the meiotic spindle (short arrow) and the nascent sperm aster (long arrow). The sperm aster expands dramatically (C). Later in the period leading up to first cleavage, a global system of oriented cortical MTs can be seen (D; orientation indicated by arrow). An antibody against acetylated α-tubulin reveals the presence of acetylated tubulin in mitotic spindles (E) derived from rapidly cleaving, stage 8 embryos. Acetylated tubulin is also found in the neural fold region of stage 20 embryos (arrow in F) and in ciliated epidermal cells (black arrows in G), nerve roots (arrows in H), and what appear to be neuronal cell bodies (black on white on black arrow in G) in the neural tube region of a stage 35 embryo. (G and H are different optical sections from the same embryo viewed from the dorsal side.) Bar in C, 100 μm for A–D; bar in E, 10 μm; bar in F, 100 μm.

A. Fertilization

The *Xenopus* egg is fertilized by a single sperm penetrating the animal hemisphere. The specificity for the animal hemisphere is as yet unexplained, but the animal hemisphere clearly differs from the vegetal hemisphere in a number of physical properties (see Section III,B), including a greater potential for contraction in response to Ca^{2+} (Ezzell *et al.*, 1985; Merriam *et al.*, 1983). In response to the penetration of the sperm, an activation wave travels from the animal to the vegetal pole (Hara and Tydeman, 1979). The activation wave is characterized by the lengthening of the microvilli, the contraction of the cortex, and the exocytosis of the cortical granules (Takeichi and Kubota, 1984). Apparently, it is initiated and propagated by a rise in intracellular Ca^{2+} (Gingell, 1970; Schroeder and Strickland, 1974; Busa and Nuccitelli, 1985; Kubota *et al.*, 1987).

Given the high concentration of MFs in the cortical region of the egg and the known involvement of MFs in contractile processes, one might suspect that the cortical contraction and the lengthening of microvilli are a direct result of Ca^{2+}-dependent, MF-mediated events. However, experiments designed to test this hypothesis have given contradictory results. Manes *et al.* (1978) reported that injection of cytochalasin B did not halt cortical contraction. On the other hand, Ezzell *et al.* (1985) reported that N-ethylmaleimide-treated heavy meromyosin (NEN-HMM), which inhibits myosin-mediated, MF-based contractile events, inhibited both the cortical contraction and lengthening of microvilli. Christensen *et al.* (1984) found that cortical contraction in bisected eggs is myosin dependent and inhibited by NEM-HMM. Given that cytochalasin B acts primarily by blocking the assembly of MFs and by causing the depolymerization of labile MFs, it seems that a stable population of MFs mediate cortical contraction in the fertilized egg.

Fertilization–activation also begins a process that results in the reorganization of the egg's cytokeratin system (Klymkowsky *et al.*, 1987; Klymkowsky and Maynell, 1987). The aggregated cytokeratin protein of the egg is first reorganized into a global system of oriented filaments. This initial directionality disappears by the time of first cleavage and may correspond to the MT-mediated, cortical–endoplasmic rotation (Fig. 3D; Section V,C). By second cleavage, the cortical cytokeratin filaments form a characteristic "fishnet" system that covers the surface of the vegetal hemisphere (Klymkowsky *et al.*, 1987). While apparently influenced by MTs during the period leading up to first cleavage, the reorganization of cytokeratins does not depend on MTs, as it occurs in the presence of nocodazole. Cytokeratin reorganization also occurs in the presence of cycloheximide (Klymkowsky and Maynell, 1987) and therefore appears to be due to changes in factors that regulate cytokeratin organization and not to the synthesis of new cytokeratin proteins.

B. Aster Formation and Pronuclear Migration

During pronuclear migration, which occurs during the first half of the first cell cycle, both pronuclei form MT-based asters and begin to migrate toward the center of the cell; the sperm pronucleus moves from the sperm entry point and the egg pronucleus descends from the animal pole (Ubbels *et al.*, 1983; Stewart-Savage and Grey, 1982) (Fig. 3B). The sperm aster is larger than the egg's, so large in fact that its expansion can be seen as a postfertilization contraction wave on the surface of the egg (Hara *et al.*, 1977; Ubbels *et al.*, 1983). By whole-mount immunocytochemistry, it is possible to visualize the initial formation (Fig. 3B) and growth of the sperm aster until it forms a dense cap of MTs (Fig. 3C) (A. Ellis and M. W. Klymkowsky, unpublished observation). We can visualize the meiotic spindle of the egg up to 20–25 min after fertilization (Fig. 3B). Migration of both pronuclei depends on the sperm aster (Manes and Barbieri, 1977; Subtelny and Bradt, 1963; Breidis and Elinson, 1982; Ubbels *et al.*, 1983); the function of the egg aster, if any, is unclear.

Migration of the sperm pronucleus is accompanied by the formation of a trail of pigment from the cortex that follows the path of the sperm pronucleus into the cytoplasm (Palacek *et al.*, 1978). This sperm trail forms after relaxation of the cortical contraction about 10 min into fertilization (Stewart-Savage and Grey, 1982). Its formation appears to depend on the sperm aster, since it also forms when eggs are injected with a sperm homogenate containing a MT center, but not when eggs are prick-activated.

C. Dorsal–Ventral Polarity

The dorsal–ventral polarity of the initially radially symmetric egg is determined by the rotation of the cortex relative to the underlying cytoplasm (Ancel and Vitemberger, 1948; Vincent *et al.*, 1986; Vincent and Gerhart, 1987). This rotation occurs in the middle (0.45–0.8 of the time to first cleavage) of the first cell cycle and results in the formation of a gray crescent on the prospective dorsal side of the embryo. The rotation depends on the MT cytoskeleton in that it is directed away from the site of the sperm entry and depends on the integrity of the MT system during the period of rotation.

Manes and Barbieri (1976, 1977) demonstrated that eggs injected with sperm homogenate form a gray crescent opposite the site of injection, but that buffer-injected eggs form a gray crescent in a position independent of the site of injection. Ubbels *et al.* (1983) noted that vinblastine and colchicine, drugs that inhibit MT assembly, inhibited the redistribution of yolk that normally accompanies fertilization. Yolk redistribution also does not occur in prick-activated eggs, which lack a sperm aster.

Evidence for a direct role of MTs in the production of dorsal–ventral polarity comes from several sources. Manes *et al.* (1978) showed that colchicine inhibits gray crescent formation even in prick-activated eggs, which have no sperm aster. Scharf and Gerhart (1983) found that the determination of the dorsal–ventral axis by the sperm is sensitive to cold, high pressure, and ultraviolet light during a critical period in the first cell cycle that corresponds to the period of cortical rotation. These treatments are known to affect MT integrity in other systems, and a role for MTs was further indicated by the observation that these treatments could be inhibited by heavy water, which is known to stabilize MTs. Furthermore, Elinson (1985) measured the level of polymerized tubulin in artificially activated eggs throughout the cell cycle and found that it dropped at fertilization, but rose to preactivation levels at the time of the cortical rotation.

The MT system involved in the cortical rotation can be visualized directly using whole-mount cytochemistry (Fig. 3D) (B. Rowning and J. C. Gerhart, personal communication). A system of parallel MTs, oriented with respect to the sperm entry point, forms and girdles the entire egg. It bears a striking similarity to the transient system of cortical MTs described in sea urchin eggs during the period leading up to first cleavage (Harris 1979; Harris *et al.*, 1980), suggesting that a spiral rotation of the cortex with respect to the endoplasm may play an important role in the establishment of embryonic axes in sea urchin. In normal development, this system of MTs appears to direct the initial orientation of cytokeratin filaments (Klymkowsky *et al.*, 1987). Whether it provides the force or simply the orientation for the cortical rotation remains unclear.

D. Preparation for First Cleavage

Hara *et al.* (1977) described a series of two surface contraction waves that occur 20 and 10 min before first cleavage in *Xenopus*. By observing carbon particles on the surface of the newt eggs, Sawai (1982) determined that the first of these precleavage waves is a relaxation wave and that the second corresponds to a wave of cortical stiffening. The occurrence of this second wave correlates with the acquisition by the cortex of the ability to form a cleavage furrow (Sawai and Yoneda, 1974; Sawai, 1972, 1982; see also Section V,E).

Early drug experiments were interpreted to mean that these waves are not mediated by the MT- and MF-based cytoskeleton, since they occur in the presence of colchicine, vinblastine, and cytochalasin B (Hara *et al.*, 1980; Christensen and Merriam, 1982). Christensen and Merriam (1982) suggested that these surface waves are related to an actin-independent contraction of the subcortical matrix triggered by Ca^{2+} (Merriam *et al.*, 1983; Merriam and Sauterer, 1983). However, Elinson (1983) found a cytochalasin B-sensitive stiffening of the cytoplasm about 20 min prior to cleavage. He suggested that this stiffening represents a MF-based gelation of the cytoplasm that might act to "fix" the structure of the reorganized cytoplasm.

E. First Cleavage

Much of our early knowledge regarding the role of MFs and MTs in cytokinesis was gained from manipulations of the first cleavage cycle of amphibian eggs. Their large size allows easy manipulation of the spindle orientation and the cleaving cortex. In *Xenopus*, the first cleavage plane often correlates with the left–right axis of the embryo (Klein, 1987). The relationship between the orientation of the spindle and its ability to induce a cleavage furrow in the cortex has been a fruitful area of study since Zotin's experiments of 1964. He showed that rotation of the spindle after first cleavage by 90° results in the formation of the second cleavage furrow parallel, rather than perpendicular, to the first. In addition, when multiple basal bodies (Heidemann and Kirschner, 1978) or centrioles (Maller *et al.*, 1976) are injected into the egg they form asters and induce multiple cleavage furrows. Thus, it is the aster, not the spindle or the chromosomes, that determines the plane of the cleavage furrow. The mechanism by which asters determine the position of the cleavage furrow remains obscure (see Asnes and Schroeder, 1979).

A number of elegant experiments have examined the ability of the cytoplasm to induce a furrow in the cortex and the ability of the cortex to respond. The initial studies on the cleavage furrow were done by Waddington (1952). He drained newt eggs of cytoplasm before first cleavage and intercalated a cellophane strip between the cortex and subcortical cytoplasm showing that the cortex, once induced, cleaves autonomously (see also Selman and Waddington, 1955). Working with the newt, Sawai *et al.* (1969) demonstrated that the cortex could be cut ahead of the furrow without arresting the furrow (see also Dan and Kojima, 1963), that displacement of the subcortical cytoplasm displaces the cleavage furrow, and that transplantation of subcortical cytoplasm, derived from a region in front of the cleavage furrow's path, would induce the formation of a cleavage furrow in regions that would not normally cleave. These results have been confirmed in *Xenopus* (Sawai, 1983). The ability of the cortex to form a cleavage furrow moves as a meridional band from the animal to the vegetal pole during cleavage (Sawai, 1972, 1974) and corresponds to the pre-cleavage wave of cortical contraction and cytoplasmic stiffness (see Section V,D). These waves of contraction precede all early cleavages, not just the first, and appear to be driven by a cytoplasmic clock (Sawai, 1979; Hara *et al.*, 1980).

Early electron microscopic studies on first cleavage showed the presence of filaments subjacent to the cleavage furrow (Bluemink, 1970; Selman and Perry, 1970; Kalt, 1971a,b; Singal and Sanders, 1974). To many, this observation suggested a "purse-string" model for cleavage furrow constriction. Support for this model came when Perry *et al.* (1971) were able to decorate the furrow filaments with the S_1 fragment of myosin, thus demonstrating that they are MFs. Unexpectedly, when Bluemink (1971a,b) treated eggs with cytochalasin B, he found that it caused the regression of the cleavage, but did not inhibit the initial formation of the cleavage furrow. The appearance of the cleavage furrow in cytochalasin B-treated embryos is due to the impermeability of the egg membrane to cytochalasin; once the new membrane of the cleavage furrow is formed by the fusion of membrane vesicles, cytochalasin enters and causes a regression of the initial furrow (DeLaat *et al.*, 1973, 1974; DeLaat and Bluemink, 1974). The intracellular injection of cytochalasin B completely blocks the formation of the cleavage furrow (Luchtel *et al.*, 1976). Ca^{2+} probably regulates MF-based contraction of the furrow, since there is a transient rise in intracellular Ca^{2+} just before cleavage (Baker and Warner, 1972).

VI. GASTRULATION, NEURULATION, AND BEYOND

Amphibia in general and *Xenopus* in particular are important systems for the study of morphogenetic movement in early development. In this section, we will briefly review the role of the cytoskeleton in these movements (see Keller, 1985, for a more complete review).

A. Microfilaments and Microtubules

Waddington and Perry (1966; Perry and Waddington, 1966) were the first to suggest that MFs and MTs play a role in the cell elongation and apical constriction necessary for bottle cell formation during gastrulation and neural fold formation. This idea was expanded on by Baker and Schroeder (1967), who argued strongly that the MFs in the apices of neural plate cells constrict by a purse-string mechanism and that the MTs are responsible for concommitant elongation of these cells. Support for this mechanism comes from the work of Burnside (1971, 1973), who showed that as the apices of newt cells constrict, their MFs become more dense, suggesting a sliding filament model of constriction. Burnside also pointed out that the total length of MTs in these cells is constant as the cells elongate. This observation supports the idea that MTs generate cellular elongation by sliding past each other.

Experimental manipulation of the cytoskeleton provided further evidence for the role of MTs and MFs in neurulation. Karfunkel (1971) used vinblastine to show that disruption of MTs correlates with the inhibition of changes in cell morphogenesis during gastrulation and neurulation. In particular, both vinblastine and heavy water stopped neural plate cells from elongating. Burnside (1973) reported that cytochalasin B halts neurulation. In an attempt to avoid the potential artifacts associated with the use of cytochalasin B, Messier and Seguin (1978) used pressure to disrupt MFs and MTs in *Xenopus* embryos. They were able to demonstrate that pressures of 4000 psi for 180 min disrupted apical MFs, but not MTs, and resulted in an expansion of the apical end of the cells; treatment for 330 min caused depolymerization of MTs and rounding up of normally elongate cells.

Similar experiments yielded different results when applied to the *Xenopus* gastrula. Cooke (1973) noted almost incidentally that the mitotic inhibitor Colcemid, while halting mitosis, allowed most of the morphogenetic movements of gastrulation to continue. By injecting cytochalasin B and colchicine into the blastocoel of blastula, Nakatsuji (1979) showed that cytochalasin B immediately inhibited gastrulation, but that colchicine inhibited gastrulation only as mitotic cells began to accumulate.

Microtubules are often posttranslationally modified by acetylation and detyrosylation (see Webster *et al.*, 1987 and ref. therein). Although these modifications appear to correlate with the stabilization of MTs, their physiological significance is unclear. To study the distribution and eventually the function of stable MTs, we have begun to use antibodies directed against acetylated α-tubulin (Piperno and Fuller, 1985) and detyrosylated α-tubulin (Gundersen *et al.*, 1984). Tubulin–acetylase activity is clearly present in the egg, as the meiotic spindle is heavily labeled by antiacetylated α-tubulin antibody (not shown); mitotic spindles and midbodies in the early embryo also contain acetylated tubulin (Fig. 3E). This is surprising, since the cells of the early blastula are dividing every 30 min and would not be expected to contain many long-lived MTs. In *Drosophila*, acetylated tubulin appears only after the period of rapid nuclear division has ended (Wolf *et al.*, 1988). In later-stage *Xenopus* embryos, we find no significant staining of mitotic spindles by antiacetylated α-tubulin antibody (not shown; Chu *et al.*, 1989).

Acetylated tubulin is concentrated in the neural folds during neurulation and within the ciliated cells of the epidermis (Fig. 3F–H). Neurons and their processes are heavily stained by the antiacetylated α-tubulin antibody (Fig. 3G and H), and the presence of acetylated tubulin could prove to be a useful marker for at least some classes of neurons (Chu *et al.*, 1989). Detyrosylated tubulin is found in the ciliated epiderminal cells and primarily in nonneuronal cells in later-stage embryos (not shown; Chu *et al.*, 1989).

B. Intermediate Filaments

The cytokeratin system of the early embryo, like that of the late-stage oocyte is composed of only three cytokeratin proteins. It has a distinct animal–vegetal asymmetry that persists past embryonic stage 6 (Klymkowsky *et al.*, 1987) (all embryonic stages according to Nieuwkoop and Faber, 1975). Well into gastrulation, the cytokeratin systems of animal hemisphere cells differ quantitatively and perhaps qualitatively from those of the much larger vegetal hemisphere blastomeres (Fig. 4). Cytokeratin asymmetry in the early embryo can be affected somewhat by inverting the fertilized egg, but the cytokeratin system retains its original asymmetry (Klymkowsky and Maynell, 1987). The function of cytokeratin asymmetry in the early embryo is unclear, as is the function of the early embryonic cytokeratin system itself.

The cytokeratin system of the later embryo is integrated into a supracellular system through desmosomal-type adherence junctions (Perry, 1975). The function of this cytokeratin system is unclear. It could provide mechanical stability to the epithelial layer of the *Xenopus* embryo during the course of gastrulation and afterward. However, cytokeratin filaments do not appear to be essential components in amphibian gastrulation, since they appear only during neurulation in the urodele *Triturus* (Perry, 1975). To examine the function of the embryonic cytokeratin system directly, we have carried out a series of experiments using the intraembryonic injection of anticytokeratin antibodies (Maynell and Klymkowsky, 1989). Fertilized eggs were injected with antibody during the first third of the first cell cycle. Embryos injected with the anti-β-tubulin antibody E7 (Chu and Klymkowsky, 1987) or the cytokeratin antibodies AE1 (Sun *et al.*, 1985) or 1h5 (Klymkowsky *et al.*, 1987) had no more than a transient disruption of their normal cytokeratin systems and developed normally. In contrast, both the monoclonal anticytokeratin antibodies AE3 (Sun *et al.*, 1985) and αIFA (Pruss *et al.*, 1981) caused specific defects in gastrulation when injected into fertilized eggs. Neither antibody had an obvious effect on

⟶

Fig. 4. Organization of cytokeratin filaments in the gastrulating embryo. A stage 10.5 embryo stained with the monoclonal anticytokeratin antibody AE3 is shown in three optical sections (A–C). A focuses on the cytokeratin system of the yolk plug, B focuses on the vegetal side of the blastopore, and C focuses on the animal side of the blastopore. The organization of cytokeratin filaments in the yolk-plug cells is visualized at higher power (D) or after squashing the embryo and viewing it using a 63× planapochromat lens (E). Bar in C, 100 μm for A–C; bar in D, 100 μm; bar in E, 10 μm.

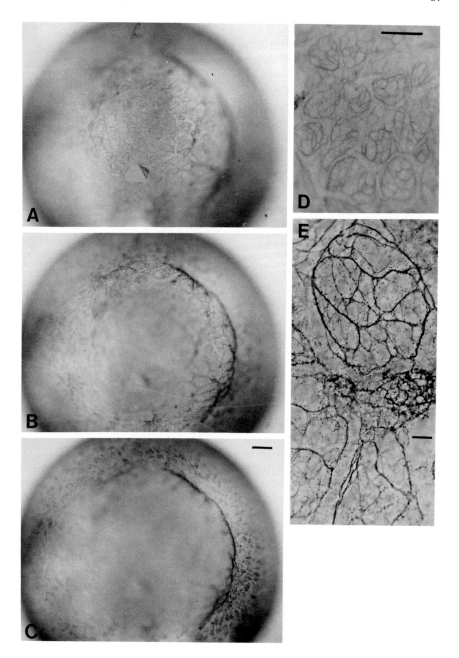

development prior to gastrulation and neither produced a dramatic disruption of cortical cytokeratin organization, leaving the mechanism of their action unclear.

During the course of our studies, we have found that in certain broods of embryos, the formation of the early embryonic cytokeratin system is delayed. Instead of forming by the end of the first or second cell cycle, a normal cytokeratin system did not appear until stage 5 to 7 (Klymkowsky and Maynell, 1987). This observation indicates that a transient disruption of cytokeratin organization during the early blastula stage of development produces no significant defect in development. Similar results have been obtained by Emerson & Pederson (1987) in the early mouse embryo.

VII. CYTOSKELETAL PROTEINS AS MARKERS OF DIFFERENTIATION

Cytoskeletal proteins have been used with good results as markers of differentiation in *Xenopus*. Encoded by multigene families, specific cytoskeletal subunit proteins are often expressed in a cell-type-specific manner. These proteins are generally major cellular proteins and therefore easy to detect by biochemical and immunochemical methods. Figure 5 illustrates some key events in early *Xenopus* embryogenesis and the timing of the appearance of various cytoskeletal proteins discussed below.

A. Skeletal Muscle Actin and Myosin

Gurdon *et al.* (1985) have used the skeletal muscle actin gene to study mesodermal differentiation. Using a cDNA probe, they demonstrated that a lo-

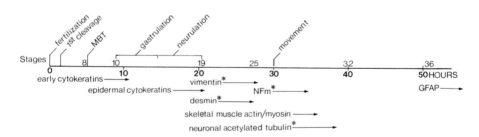

Fig. 5. *Xenopus* development and the expression of cytoskeletal proteins. Asterisks indicate work done in the laboratory of the authors.

calized determinant in the subequatorial region is necessary and sufficient for the generation of actin-producing cell lineages, i.e., the mesoderm of the somites. This determinant is apparently not dependent on cell–cell interactions, since dissociated blastomeres still produce skeletal muscle actin. To follow the expression of muscle-specific proteins, we have used the monoclonal muscle actin antibody B4 (supplied by J. Lessard, Cincinnati, Ohio), the antifast skeletal muscle myosin antibody F59, and the antislow skeletal muscle myosin antibodies S46 and S58 (supplied by J. B. Miller, Stanford University, Palo Alto, California). In whole-mount immunocytochemistry, both muscle actin and fast muscle myosin were first expressed at stages 17–18 and appeared in the somitic myotome prior to somitic segmentation and rotation (Chu *et al.*, 1989). Slow skeletal muscle isoforms appear significantly later than the fast type, accumulating in significant amount only after stage 46 (McMillan and Klymkowsky, unpublished observations).

B. Intermediate Filament Proteins

Intermediate filament subunit proteins are an obvious choice for studies of cellular differentiation, since different IF proteins are expressed in different tissues and are relatively easy to assay immunologically. The cytokeratin proteins provide a particularly good example. During development in *Xenopus*, the original three cytokeratins of the egg are joined by a second group around the time of the midblastula transition (Jonas *et al.*, 1985; Winkles *et al.*, 1985; Jamrich *et al.*, 1987). Molecular analysis indicates that the expression of at least one of these cytokeratins is independent of cell–cell interactions, suggesting that its expression may be mediated by determinants present in the egg, as in the case of skeletal muscle actin (Sargent *et al.*, 1986).

Immunocytochemical and biochemical studies indicate that there are two basic types of cytokeratin proteins in the early embryo. Those restricted to the embryonic epidermis and those found in both the epidermis and internally (Godsave *et al.*, 1986). In our studies, the monoclonal antibodies AE1 and AE3 appear to react specifically with epidermal cytokeratins in embryos of stage 35 and earlier. The monoclonal antibody 1h5 (Klymkowsky *et al.*, 1987) recognizes the 56-kDa cytokeratin of the egg and labels both epidermal and internal structures, including the luminal surface of the neural tube. It does not appear to label the ciliated cells of the epidermis (not shown). In addition, Godsave *et al.* (1986) have defined antibodies that reveal differences in cytokeratin expression between epidermis and neural fold regions in stage 17 embryos. Thus, cytokeratin-type intermediate filament proteins provide specific markers for a number of cell types within the early *Xenopus* embryo.

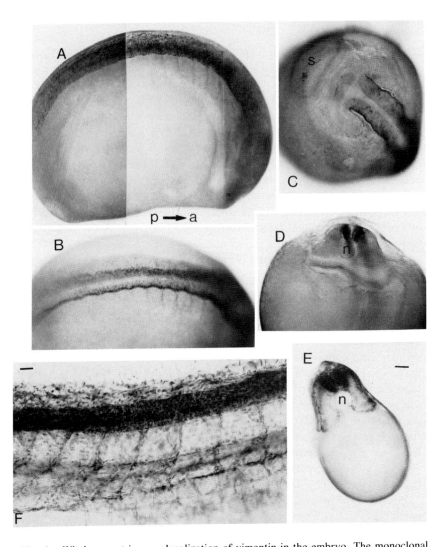

Fig. 6. Whole-mount immunolocalization of vimentin in the embryo. The monoclonal anti-mammalian vimentin antibody RV202 was used to visualize the expression of vimentin. In stage 22 embryos (A, side view; B, dorsal view; C, anterior view), vimentin immunoreactivity is found in the neural tube and head regions. Cross sections, prepared by cutting the embryo in half after immunolabeling, reveal that vimentin is located primarily on the menengial surface of the neural tube. s (in C), the location of the anterior cement gland. p → a (in A), Posterior-to-anterior axis for A, B, and F. n (in D and E), Notochord. In stage 28 embryos (E, cross section), vimentin is located in the menegial region, and vimentin-positive cells are found surrounding the somite surfaces. The distribution of vimentin in stage 35 embryos is quite complex. F illustrates this complexity in a close-up view of the dorsal region of a stage 35 embryo. The dark band at the top of the figure is the neural tube, the repeated divisions are the somites, and other vimentin-positive cells can be seen throughout the optical section. Bar in E, 100 μm for A–E; bar in F, 50 μm.

The IF protein vimentin is also a useful marker for early differentiation. Godsave *et al.* (1986) reported the first appearance of vimentin in the stage 25 embryo, where it is found at the margin of the neural tube. To further characterize the pattern of vimentin expression, we have used whole-mount immunocytochemistry to map vimentin expression during early *Xenopus* development (Dent *et al.*, 1989). We first find vimentin expression along the margins of the neural tube at stage 19 (Figs. 6A–D). The vimentin-positive cells of the neural tube appear to be radial glia (see also Godsave *et al.*, 1986). By stage 22–23, vimentin-positive cells are found overlaying the somites (Figs. 6D–F); vimentin-positive cells located ventral to the somites are observed beginning at stage 24.

Godsave *et al.* (1986) reported that the glial-specific IF protein GFAP first appears in the neural tube at stage 33, and that the large neurofilament protein appears at stage 48 (swimming tadpole). We have used antibodies directed against desmin and the low (NFl)- and middle (NFm)-molecular-weight neurofilament proteins to visualize the pattern of expression of these intermediate filament types (Chu *et al.*, 1989); we find that desmin is first expressed in the somitic myotome beginning at stage 20–21. The neural-specific protein NFm appears first at stage 25, whereas NFl does not appear until approximately stage 43.

The nuclear lamins are also expressed in a cell-type-specific manner in *Xenopus* (Krohne and Benavente, 1986). The oocyte and early embryo express a single lamin protein, lamin III. Lamin III eventually disappears at tadpole stage, but reappears in myocytes, neurons, and sertoli cells of the adult organism (Benavente *et al.*, 1985; Stick and Hausen, 1985). Lamin I is found in all somatic cells after the midblastula (stage IX) transition. Lamin II appears in all somatic cells after midgastrula stage. Lamin IV is found in sperm.

VIII. CONCLUSIONS

Whole-mount immunocytochemistry is a major advance in the tools available for the study of *Xenopus* development. It eliminates the time-consuming and often inadequate process of reconstructing serial sections in order to obtain accurate three-dimensional images of the oocyte, egg, or embryo. It is simple and inexpensive to carry out, yet it provides a dramatic new perspective from which to view development. Many samples can be prepared simultaneously and examined in short order. For example, we have been able to visualize the reorganization and asymmetry of cytokeratins within the oocyte and early embryo (Klymkowsky *et al.*, 1987); the reorganization of the nuclear envelope (Fig. 1D and E), annulate lamellae (Klymkowsky, unpublished observations), mitochondrial mass (Fig. 2A–E), and tubulin (Fig. 2F–H) during oogenesis; the reorganization of tubulin during the period of first cleavage (Fig. 3A–D); the expression of acetylated tubulin within the embryo (Fig. 3E–H; Chu *et al.*, 1989); the organization of cytokeratins during gastrulation (Fig. 4); and the appearance of vimentin during embryogenesis (Fig. 6; Dent *et. al.*, 1989). In addition, B. Rowning and J. C. Gerhart (personal communication) have used whole-mount immunocytochemistry to clarify the relationship between MTs and the cortical–endoplasmic rotation that determines the dorsal–ventral axis of the embryo. It appears that these same methods, together with β-galactosidase-based fusion proteins, can be used to study the factors regulating the temporal and spatial pattern of gene expression (M. W. Klymkowsky, unpublished observation); it may even be possible to use cleared oocytes and embryos in *in situ* hybridization analysis for mRNA localization.

All this is not to suggest that whole-mount staining is not without its drawbacks. It is still not possible to obtain the maximum theoretical resolution of the light microscope using whole-mount specimens of *Xenopus*, because of the difficulty in obtaining the high-resolution, long-working-distance objectives required to image through the 1.2-mm-thick *Xenopus* oocyte. A more serious drawback is the out-of-focus information that routinely contaminates standard microscopic images. While this is generally not a serious problem in studies of cultured cells, which are on the order of 20–50 μm thick, it is a significant problem in the 0.5- to 1.2-mm-thick *Xenopus* oocyte–embryo. There are three possible ways to circumvent this problem. The first is to simply squash the stained oocyte or embryo and view it using high-resolution objectives (Figs. 3E and 4E); this has the obvious disadvantage of destroying the three-dimensional quality of the specimen, but it is fast, inexpensive, and in many cases effective. The second is the use of computational methods to remove the out-of-focus information (Agard and Sedat, 1983). A third way may be the use of confocal microscopy (see White *et al.*, 1987).

The simplicity of whole-mount immunocytochemistry makes it much easier to examine the effects of various experimental manipulations on oogenesis and embryogenesis. Of the methods available for studying the role of the cytoskeleton in these processes, two seem to be particularly promising: the intracellular injection of monoclonal antibodies and the introduction of synthetic (dominant negative effect) mutations (Herskowitz, 1987). Both offer defined reagents with which to manipulate cytoskeletal organization and function. Antibodies directed against specific cytoskeletal or cytoskeletal-associated proteins can be generated in order to block particular functions without producing the global disruption of the cytoskeleton typically found with anti-cytoskeletal drugs. Similarly, synthetic mutations can produce specific defects in cytoskeletal function. For example, Levitt *et al.* (1987) found that the expression of a mutant β-actin induces specific changes in cellular morphology presumably by modifying the interactions between actin and other proteins within the cell. Albers and Fuchs (1987) found that the expression of a truncated cytokeratin protein causes the disruption of cytokeratin filament organization, similar to that produced by injected antibody. Both antibodies and synthetic mutations can be delivered into specific blastomeres and thereby produce phenotypic mosaics (see Moody, 1987 for fate map of 32-cell embryo). Antibodies can, in theory at least, be prepared against almost any part of the molecule of interest; synthetic mutants may not be so flexible. However, whereas the supply of antibody within the oocyte or embryo is limited by the amount initially introduced, a DNA-encoded synthetic mutant gene product could be maintained at high levels through ongoing transcription–translation. Whichever method proves appropriate in any particular case, the combination of these techniques together with experimental simplicity and global perspective offered by whole-mount immunocytochemistry opens a new realm in the experimental study of development mechanics.

ACKNOWLEDGMENTS

We thank our collaborators and colleagues at the University of Colorado at Boulder for their support; in particular, Laurie Maynell for her dedication and Anna Ellis for her dogged pursuit of tubulin within the early embryo. We also thank Jeff Bachant, Christine Bednarz, Bob Boswell, Chris Link, and Jon Van Blerkom for their comments on the manuscript. We thank Andrew Murray (University of California, San Francisco) for sharing his embryo-clearing method with us prior to publication. We thank Henry Sun, Larry Gerace, Franz Ramaekers, Gianni Piperno, Greg Gunderson, Chloe Bulinsky, Jeff Miller, and Jim Lessard for the gifts of monoclonal antibodies. J. A. D. is an NSF predoctoral fellow. M. W. K. is a Pew Biomedical Scholar. To a large extent, this work was made possible by the support of the Pew Trust, whose support we acknowledge and deeply appreciate. We have also been supported by grants from the NSF, the NIH, and the March of Dimes.

REFERENCES

Agard, D., and Sedat, J. W. (1983). Three-dimensional architecture of a polytene nucleus. *Nature (London)* **302,** 676–681.

Albers, K., and Fuchs, E. (1987). The expression of mutant epidermal keratin cDNAs transfected in simple epithelial and squamous cell carcinoma lines. *J. Cell Biol.* **105,** 245–250.

Al-Mukhtar, K. A., and Webb, A. C. (1971). An ultrastructural study of primordial germ cells, oogonia and early oocytes in *Xenopus laevis. J. Embryol. Exp. Morphol.* **26,** 195–217.

Ancel, P., and Vitemberger, P. (1948). Recherches sur le déterminisme de l'asymetric bilatérale dans l'oeuf des amphibiens. *Bull. Biol. Fr. Belg., Suppl.* **31,** 1–182.

Asnes, C. F., and Schroeder, T. E. (1979). Cell cleavage: Ultrastructural evidence against equatorial stimulation by aster microtubules. *Exp. Cell Res.* **122,** 327–338.

Baker, P. C., and Schoeder, T. E. (1967). Cytoplasmic filaments and morphogenetic movement in the amphibian neural tube. *Dev. Biol.* **15,** 432–450.

Baker, P. F., and Warner, A. E. (1972). Intracellular calcium and cell cleavage in the early embryos of *Xenopus laevis. J. Cell Biol.* **53,** 579–581.

Balinsky, B. I., and Devis, R. J. (1963). Origin and differentiation of cytoplasmic structures in the oocytes of *Xenopus laevis. J. Embryol. Exp. Morph.* **26,** 195–217.

Bartnik, E., Osborn, M., and Weber, K. (1985). Intermediate filaments in non-neuronal cells of invertebrates: Isolation and biochemical characterization of intermediate filaments from the esophageal epithelium of the mollusc *Helix pomatia. J. Cell. Biol.* **101,** 427–440.

Bartnik, E., Osborn, M., and Weber, K. (1986). Intermediate filaments in muscle and epithelial cells of nematodes. *J. Cell Biol.* **102,** 2033–2041.

Benavente, R., Krohne, G., and Franke, W. W. (1985). Cell type-specific expression of nuclear lamina proteins during development in *Xenopus laevis. Cell (Cambridge, Mass.)* **41,** 177–190.

Billet, F. S., and Adam, E. (1976). The structure of the mitochondrial cloud of *Xenopus laevis* oocytes. *J. Embryol. Exp. Morphol.* **33,** 697–710.

Bluemink, J. G. (1970). The first cleavage of the amphibian egg; an electron microscope study of the onset of cytokinesis in the egg of *Ambystoma mexicanum. J. Ultrastruct. Res.* **32,** 142–166.

Bluemink, J. G. (1971a). Effects of cytochalasin B on surface contractility and cell junction formation during egg cleavage in *Xenopus laevis. Cytobiology* **3,** 176–187.

Bluemink, J. G. (1971b). Cytokinesis and cytochalasin-induced furrow regression in first-cleavage zygote of *Xenopus laevis. Z. Zellforsch. Mikrosk. Anat.* **121,** 102–126.

Brachet, J., Hanocq, F., and Van Gansen, P. (1970). A cytochemical and ultrastructural analysis of *in vitro* maturation in amphibian oocytes. *Dev. Biol.* **21,** 157–195.

Bray, D., Heath, J., and Moss, D. (1985). The membrane-associated cortex of animal cells: Its structure and mechanical properties. *J. Cell Sci., Suppl.* **4,** 71–88.

Briedis, B., and Elinson, R. P. (1982). Suppression of male pronuclear movement in frog eggs by hydrostatic pressure and deuterium oxide yields androgenetic haploids. *J. Exp. Zool.* **222,** 45–57.

Brinkley, B. R. (1985). Microtubule organizing centers. *Annu. Rev. Cell Biol.* **1,** 145–172.

Burnside, B. (1971). Microtubules and microfilaments in newt neurulation. *Dev. Biol.* **26,** 416–441.

Burnside, B. (1973). Microtubules and microfilaments in amphibian neurulation. *Am. Zool.* **13,** 989–1006.

Burridge, K. (1986). Substrate adhesions in normal and transformed fibroblasts: Organization and regulation of cytoskeletal, membrane and extracellular matrix components at focal contacts. *Cancer Rev.* **4,** 18–78.

Busa, W. B., and Nuccitelli, R. (1985). An elevated free cytosolic Ca + + wave follows fertilization in the eggs of the frog, *Xenopus laevis. J. Cell Biol.* **100,** 1325–1329.

Campanella, C., and Andreuccetti, P. (1977). Structural observations on cortical endoplasmic reticulum and on residual cortical granules in the egg of *Xenopus laevis. Dev. Biol.* **56,** 1–10.

Capco, D. G., and Jeffery, W. R. (1982). Transient localization of messenger RNA in *Xenopus laevis* oocytes. *Dev. Biol.* **89,** 1–12.

Chu, D., and Klymkowsky, M. W. (1987). Experimental analysis of cytoskeletal function in early *Xenopus* embryos. *In* "The Cytoskeleton in Development and Differentiation." ICSU Press **8**, 331–333.

Christensen, K., and Merriam, R. W. (1982). Insensitivity to cytochalasin B of surface contractions keyed to cleavage in the *Xenopus* egg. *J. Embryol. Exp. Morphol.* **72**, 143–151.

Christensen, K., Sauterer, R., and Merriam, R. W. (1984). Role of soluble myosin in cortical contractions of *Xenopus* eggs. *Nature (London)* **310**, 150–151.

Clark, T. G., and Merriam, R. W. (1978). Actin in *Xenopus* oocytes. I. Polymerization and gelation in vitro. *J. Cell Biol.* **77**, 427–438.

Colman, A., Morser, J., Lane, C., Besley, J., Wylie, C., and Valle, G. (1981). Fate of secretory proteins trapped in oocytes of *Xenopus laevis* by disruption of the cytoskeleton or imbalanced subunit synthesis. *J. Cell Biol.* **91**, 770–780.

Colombo, R. (1983). Actin in *Xenopus* yolk platelets: A peculiar and debated presence. *J. Cell Sci.* **63**, 263–270.

Cooke, J. (1973). Properties of the primary organization field in the embryo of *Xenopus laevis*. Pattern formation and regulation following early inhibition by mitosis. *J. Embryol. Exp. Morphol.* **30**, 49–62.

Cowan, N. J., Lewis, S. A., Sarkar, S., and Gu, W. (1987). Functional versatility of mammalian β-tubulin isotypes. *In* "The Cytoskeleton in Development and Differentiation" (R. B. Maccioni and J. Arechaga, eds.), Vol. 8, pp. 157–166.

Dan, K., and Kojima, M. K. (1963). A study on the mechanism of cleavage in the amphibian egg. *J. Exp. Biol.* **40**, 7–14.

Danilchik, M. V., and Gerhart, J. C. (1987). Differentiation of the animal–vegetal axis in *Xenopus laevis* oocytes. I. Polarized intracellular translocation of platelets establishes the yolk gradient. *Dev. Biol.* **122**, 101–112.

Delaat, S. W., and Bluemink, J. G. (1974). New membrane formation during cytokinesis in normal and Cytochalasin B-treated eggs of *Xenopus laevis*. *J. Cell Biol.* **60**, 529–540.

Delaat, S. W., Luchtel, D., and Bluemink, J. G. (1973). The action of cytochalasin B during egg cleavage in *Xenopus laevis:* Dependence on cell membrane permeability. *Dev. Biol.* **31**, 163–177.

DeLaat, S. W., Buwalda, J. A., and Habets, A. M. M. C. (1974). Intracellular distribution, cell membrane permeability and membrane potential of the *Xenopus* egg during first cleavage. *Exp. Cell Res.* **89**, 1–14.

Dent, J. A., Polson, A. G., and Klymkowsky, M. W. (1989). A wholemount immunocytochemical analysis of the expression of the intermediate filament protein vimentin in *Xenopus*. (submitted).

Dictus, W. J. A. G., Van Zoelen, E. J. J., Tetteroo, P. A. T., Tertoolen, L. G. J., De Laat, S. W., and Bluemink, J. G. (1984). Lateral mobility of plasma membrane lipids in *Xenopus* eggs: Regional differences related to animal/vegetal polarity become extreme upon fertilization. *Dev. Biol.* **101**, 201–211.

Dreyer, C., Singer, H., and Hausen, P. (1981). Tissue specific nuclear antigens in the germinal vesicle of *Xenopus laevis* oocytes. *Wilhelm Roux's Arch. Dev. Biol.* **190**, 197–207.

Dumont, J. N. (1972). Oogenesis in *Xenopus laevis* (Daudin). I. Stages of oocyte development in laboratory maintained animals. *J. Morphol.* **136**, 153–180.

Dumont, J. N., and Wallace, R. A. (1972). The effects of vinblastine on isolated *Xenopus* oocytes. *J. Cell Biol.* **53**, 605–610.

Dustin, P. (1984). *Microtubules.* 2nd Ed.

Elinson, R. P. (1983). Cytoplasmic phases in the first cell cycle of the activated frog egg. *Dev. Biol.* **100**, 440–451.

Elinson, R. P. (1985). Changes in levels of polymeric tubulin associated with activation and dorso–ventral polarity of the frog egg. *Dev. Biol.* **109**, 224–233.

Emerson, J. A., and Pederson, R. A. (1987). The expression of cytokeratins in the preimplantation mouse embryo. *J. Cell Biol.* **105**, 285a.

Ezzell, R. M., Brothers, A. J., and Cande, W. Z. (1983). Phosphorylation-dependent contraction of actomyosin gels from amphibian eggs. *Nature (London)* **306**, 620–622.

Ezzell, R. M., Cande, W. Z., and Brothers, A. J. (1985). Ca^{++}-ionophore-induced microvilli and cortical contractions in *Xenopus* eggs. Evidence for involvement of actomyosin. *Wilhelm Roux's Arch. Dev. Biol.* **194**, 140–147.

Franke, W. W. (1987). Nuclear lamins and cytoplasmic intermediate filament proteins: A growing multigene family. *Cell (Cambridge, Mass.)* **48**, 3–4.

Franke, W. W., Rathke, P. C., Seib, E., Trendelenburg, M. F., Osborn, M., and Weber, K. (1976). Distribution and mode of arrangement of microfilamentous structures and actin in the cortex of the amphibian oocyte. *Cytobiologie* **14**, 111–130.

Franke, W. W., Schmid, E., Grund, C., and Geiger, B. (1982). Intermediate filamentous structures: Transient disintegration and inclusion of subunit proteins in granular aggregates. *Cell (Cambridge, Mass.)* **30**, 103–113.

Franke, W. W., Schmid, E., Wellsteed, J., Grund, C., Gig, O., and Geiger, B. (1983). Change of cytokeratin filament organization during the cell cycle: selective masking of an immunologic determinant in interphase *PtK$_2$* cells. *J. Cell Biol.* **97**, 1255–1260.

Franz, J. K., and Franke, W. W. (1986). Cloning of cDNA and amino acid sequence of a cytokeratin expressed in oocytes of *Xenopus laevis. Proc. Natl. Acad. Sci. U.S.A.* **83**, 6475–6479.

Franz, J. K., Gall, L., Williams, M. A., Picheral, B., and Franke, W. W. (1983). Intermediate-size filaments in a germ cell: Expression of cytokeratins in oocytes and eggs of the frog *Xenopus. Proc. Natl. Acad. Sci. U.S.A.* **80**, 6254–6258.

Gall, L., Picheral, B., and Gounon, P. (1983). Cytochemical evidence for the presence of intermediate filaments and microfilaments in the egg of *Xenopus laevis. Biol. Cell* **47**, 331–342.

Gard, D. L., and Kirschner, M. W. (1987a). Microtubule assembly in cytoplasmic extracts of *Xenopus* oocytes and eggs. *J. Cell Biol.* **105**, 2191–2201.

Gard, D. L., and Kirschner, M. W. (1987b). A microtubule-associated protein from *Xenopus* eggs that specifically promotes assembly at the plus-end. *J. Cell Biol.* **105**, 2203–2215.

Gawlitta, W., Osborn, M., and Weber, K. (1981). Coiling of intermediate filaments induced by the microinjection of a vimentin-specific antibody does not interfere with locomotion and mitosis. *Eur. J. Cell Biol.* **26**, 83–90.

Gerhart, J. C. (1980). Mechanisms regulating pattern formation in the amphibian egg and early embryo. *Biol. Regul. Devel.* **2**, 133–316.

Gerhart, J. C., and Keller, R. (1986). Region-specific cell activities in amphibian gastrulation. *Annu. Rev. Cell Biol.* **2**, 201–229.

Gingell, D. (1970). Contractile response at the surface of an amphibian egg. *J. Embryol. Exp. Morphol.* **23**, 583–609.

Godsave, S. F., Wylie, C. C., Lane, E. B., and Anderton, B. H. (1984a). Intermediate filaments in the *Xenopus* oocyte: The appearance and distribution of cytokeratin-containing filaments. *J. Embryol. Exp. Morphol.* **83**, 157–167.

Godsave, S. F., Anderton, B. H., Heasman, J., and Wylie, C. C. (1984b). Oocytes and early embryos of *Xenopus laevis* contain intermediate filaments which react with anti-mammalian vimentin antibodies. *J. Embrol. Exp. Morphol.* **83**, 169–184.

Godsave, S. F., Anderton, B. H., and Wylie, C. C. (1986). The appearance and distribution of intermediate filament proteins during differentiation of the central nervous system, skin and notochord of *Xenopus laevis. J. Embryol. Exp. Morphol.* **97**, 201–223.

Guidice, G. J., and Fuchs, E. (1987). The transfection of epidermal keratin genes into fibroblasts and simple epithelial cells: Evidence for inducing a type I keratin by a type II gene. *Cell (Cambridge, Mass.)* **48**, 453–463.

Gundersen, G. G., Kalnoski, M. H., and Bulinsky, J. C. (1984). Distinct populations of microtubules: Tyrosinated and nontyrosinated alpha tubulin are distributed differently in vivo. *Cell (Cambridge, Mass.)* **38**, 789–799.

Gurdon, J. B. (1968). Changes in somatic cell nuclei inserted into growing and maturing amphibian oocytes. *J. Embryol. Exp. Morphol.* **20**, 401–441.

Gurdon, J. B., Mohun, T. J., Brennan, S., and Cascio, S. (1985). Actin genes in *Xenopus* and their developmental control. *J. Embryol. Exp. Morphol.* **89**, Suppl., 125–136.

Hara, K., and Tydeman, P. (1979). Cinematographic observation of an "activation wave" (AW) on the locally inseminated egg of *Xenopus laevis. Wilhelm Roux's Arch. Dev. Biol.* **186**, 91–94.

Hara, K., Tydeman, P., and Hengst, R. T. M. (1977). Cinematographic observation of "postfertilization waves" (PFW) on the zygote of *Xenopus laevis. Wilhelm Roux's Arch. Dev. Biol.* **181**, 189–192.

Hara, K., Tydeman, P., and Kirschner, M. (1980). A cytoplasmic clock with the same period as the division cycle in *Xenopus* eggs. *Proc. Natl. Acad. Sci. U.S.A.* **77**, 462–466.

Harris, P. (1979). A spiral cortical fiber system in fertilized sea urchin eggs. *Dev. Biol.* **68**, 525–532.

Harris, P., Osborn, M., and Weber, K. (1980). A spiral array of microtubules in the fertilized sea urchin egg cortex examined by indirect immunofluorescence and electron microscopy. *Exp. Cell Res.* **126**, 227–236.

Hausen, P., Wang, Y. H., Dreyer, C., and Stick, R. (1985). Distribution of nuclear proteins during maturation of the *Xenopus* oocyte. *J. Embryol. Exp. Morphol.* **89**, Suppl., 17–34.

Heasman, J., Quarmby, J., and Wylie, C. C. (1984). The mitochondrial cloud of *Xenopus* oocytes: The source of germinal granule material. *Dev. Biol.* **105**, 458–469.

Heidemann, S. R., and Kirschner, M. W. (1975). Aster formation in eggs of *Xenopus laevis:* Induction by isolated basal bodies. *J. Cell Biol.* **67**, 105–117.

Heidemann, S. R., and Kirschner, M. W. (1978). Induced formation of asters and cleavage furrows in oocytes of *Xenopus laevis* during in vitro maturation. *J. Exp. Zool.* **204**, 431–444.

Heidemann, S. R., Hamborg, M. A., Balasz, J. E., and Lindley, S. (1985). Microtubules in immature oocytes of *Xenopus laevis. J. Cell Sci.* **77**, 129–141.

Herskowitz, I. (1). Functional inactivation of genes by dominant negative mutations. *Nature (London)* **329**, 219–222.

Jamrich, M., Sargent, T. D., and Dawid, I. B. (1987). Cell-type-specific expression of epidermal cytokeratin genes during gastrulation of *Xenopus laevis. Genes Dev.* **1**, 124–132.

Jonas, E., Sargent, T. D., and Dawid, I. B. (1985). Epidermal keratin gene expressed in embryos of *Xenopus laevis. Proc. Natl. Acad. Sci. U.S.A.* **82**, 5413–5417.

Kalt, M. R. (1971a). The relationship between cleavage and blastocoel formation in *Xenopus laevis.* I. Light microscopic observations. *J. Embryol. Exp. Morphol.* **26**, 37–49.

Kalt, M. R. (1971b). The relationship between cleavage and blastocoel formation in *Xenopus laevis.* II. Electron microscopic observations. *Dev. Biol.* **25**, 30–56.

Karsenti, E., Newport, J., Hubble, R., and Kirschner, M. (1984). Interconversion of metaphase and interphase microtubule arrays, as studied by the injection of centrosomes and nuclei into *Xenopus* eggs. *J. Cell Biol.* **98**, 1730–1745.

Keller, R. (1985). The cellular basis of amphibian gastrulation. *In* "Developmental Biology: A Comprehensive Synthesis" (L. W. Browder, ed.), Vol. 2, pp. 241–327. Plenum, New York.

Kimmelman, D., and Kirschner, M. (1987). Synergistic induction of mesoderm by FGF and TFGβ and the identification of an mRNA coding for FGF in the early *Xenopus* embryo. *Cell* **51**, 869–877.

Kirschner, M., and Mitchison, T. (1986). Beyond self-assembly: From microtubules to morphogenesis. *Cell (Cambridge, Mass.)* **45**, 329–342.

Klein, S. L. (1987). The first cleavage furrow demarcates the dorsal-ventral axis in *Xenopus* embryos. *Dev. Biol.* **120**, 299–304.

Klymkowsky, M. W. (1981). Intermediate filaments in 3T3 cells collapse after the intracellular injection of a monoclonal anti-intermediate filament antibody. *Nature (London)* **291**, 249–251.

Klymkowsky, M. W. (1982). Vimentin and keratin intermediate filament systems in cultured *PtK₂* epithelial cells are interrelated. *EMBO J.* **1**, 161–165.

Klymkowsky, M. W. (1988). Metabolic inhibitors and intermediate filament organization in human fibroblasts. *Exp. Cell Res.* **174**, 282–290.

Klymkowsky, M. W., and Maynell, L. A. (1987). Reorganization and function of the cortical cytokeratin system in *Xenopus* oocytes, eggs, and early embryos. *J. Cell Biol.* **105**, 265a.

Klymkowsky, M. W., and Maynell, L. A. (1989). In preparation.

Klymkowsky, M. W., Miller, R. H., and Lane, E. B. (1983). Morphology, behavior, and interaction of cultured epithelial cells after the antibody-induced disruption of keratin filament organization. *J. Cell Biol.* **96**, 494–509.

Klymkowsky, M. W., Maynell, and Polson, A. G. (1987). Polar asymmetry in the organization of the cortical cytokeratins system of *Xenopus laevis* oocytes and embryos. *Development.* **100**, 543–557.

Kreis, T. E., Geiger, B., Schmid, E., Jorcano, J. L., and Franke, W. W. (1983). *De novo* synthesis and specific assembly of keratin filaments in nonepithelial cells after microinjection of mRNA for epidermal keratin. *Cell (Cambridge, Mass.)* **32**, 1125–1137.

Krohne, G., and Benavente, R. (1986). The nuclear lamins: A multigene family of proteins in evolution and development. *Exp. Cell Res.* **162**, 1–10.

Kubota, H. Y., Yoshimoto, Y., Yoneda, M., and Hiramoto, Y. (1987). Free calcium wave upon activation in *Xenopus* eggs. *Dev. Biol.* **119**, 129–136.

Lane, E. B., and Klymkowsky, M. W. (1982). Epithelial tonofilaments: Investigating their form and function using monoclonal antibodies. *Cold Spring Harbor Symp. Quant. Biol.* **46**, 387–402.

Lane, E. B., Goodman, S. L., and Trejdosiewicz, L. K. (1982). Disruption of the keratin filament network during epithelial cell division. *EMBO J.* **1**, 1365–1372.

Leonard, D. G. B., Gorham, J. D., Cole, P., Greene, L. A., and Ziff, E. B. (1988). A nerve growth factor-regulated messenger RNA encodes a new intermediate filament protein. *J. Cell Biol.* **106**, 181–193.

Lessman, C. A. (1987). Germinal vesicle migration and dissolution in *Rana pipiens* oocytes: Effects of steroids and microtubule poisons. *Cell Differ.* **20**, 238–251.

Lessman, C. A., Marshall, W. S., and Habibi, H. R. (1986). Movement and dissolution of the nucleus (germinal vesicle) during *Rana* oocyte meiosis: Effects of demecolcine (colcemid) and centrifugation. *Gamete Res.* **14**, 11–23.

Levitt, J., Ng, S. Y., Aebi, U., Varma, M., Latter, G., Burbeck, S., Kedes, L., and Gunning, P. (1987). Expression of transfected mutant β-actin genes: Alterations of cell morphology and evidence for autoregulation in actin pools. *Mol. Cell. Biol.* **7**, 2457–2466.

Lin, J. J. C., and Feramisco, J. R. (1981). Disruption of the *in vivo* distribution of intermediate filaments in fibroblasts through the microinjection of a specific monoclonal antibody. *Cell (Cambridge, Mass.)* **24**, 185–193.

Lopata, M. A., and Cleveland, D. W. (1987). *In vivo* microtubules are copolymers of available β-tubulin isotypes: Localization of each of six vertebrate β-tubulin isotypes using polyclonal antibodies elicited by synthetic peptide antigens. *J. Cell Biol.* **105**, 1707–1720.

Luchtel, D., Bluemink, J. G., and DeLaat, S. W. (1976). The effect of injected cytochalasin B on filament organization in the cleaving egg of *Xenopus laevis*. *J. Ultrastruct. Res.* **54**, 406–419.

Maller, J. L. (1985). Oocyte maturation in amphibians. *In* "Developmental Biology: A Comprehensive Synthesis." (L. W. Browder, ed.), Vol. 1, pp. 289–311. Plenum, New York.

Maller, J. L., Poccia, D., Nishioka, D., Kidd, P., Gerhart, J., and Hartman, H. (1976). Spindle formation and cleavage in *Xenopus* eggs injected with centriole containing fraction of sperm. *Exp. Cell Res.* **99**, 285–294.

Manes, M. E., and Barbieri, F. D. (1976). Symmetrization in the amphibian egg by disrupted sperm cells. *Dev. Biol.* **53**, 138–141.

Manes, M. E., and Barbieri, F. D. (1977). On the possibility of sperm aster involvement in dorsal–ventral polarization and pronuclear migration in the amphibian egg. *J. Embryol. Exp. Morphol.* **40**, 187–197.

Manes, M. E., Elinson, R. P., and Barbieri, F. D. (1978). Formation of the amphibian grey crescent: Effects of colchicine and cytochalasin B. *Wilhelm Roux's Arch. Dev. Biol.* **185**, 99–104.

Maynell, L. A., and Klymkowsky, M. W. (1989). In preparation.

Melton, D. A. (1987). Translocation of a localized maternal mRNA to the vegetal pole of *Xenopus* oocytes. *Nature* **328**, 80–83.

Merriam, R. W., and Clark, T. G. (1978). Actin in *Xenopus* oocytes. II. Intracellular distribution and polymerizability. *J. Cell Biol.* **77**, 439–447.

Merriam, R. W., and Sauterer, R. A. (1983). Localization of a pigment-containing structure near the surface of *Xenopus* eggs which contracts in response to calcium. *J. Embryol. Exp. Morphol.* **76**, 51–65.

Merriam, R. W., Sauterer, R. A., and Christensen, K. (1983). A subcortical, pigment-containing structure in *Xenopus* eggs with contractile properties. *Dev. Biol.* **95**, 439–446.

Messier, P. E., and Seguin, C. (1978). The effects of high pressure on microfilaments and microtubules in *Xenopus laevis*. *J. Embryol. Exp. Morphol.* **44**, 281–295.

Moody, S. A. (1987). Fates of the blastomeres of the 32-cell-stage *Xenopus* embryo. *Dev. Biol.* **122**, 300–319.

Mooseker, M. S. (1985). Organization, chemistry and assembly of the cytoskeletal apparatus of the intestinal brush border. *Annu. Rev. Cell Biol.* **1**, 209–241.

Murti, G., Goorha, R., and Klymkowsky, M. W. (1988). Intermediate filaments play a role in the assembly of frog virus 3 cytoplasmic assembly sites. *Virology* **162**, 264–269.

Nakatsuji, N. (1979). Effects of injected inhibitors of microfilament and microtubule function on the gastrulation movement in *Xenopus laevis*. *Dev. Biol.* **68**, 140–150.

Newport, J. W., and Forbes, D. J. (1987). The nucleus: Structure, function and dynamics. *Ann. Rev. Biochem.* **56**, 535–565.

Nieuwkoop, P. D., and Faber, J. (1975). "Normal Table of *Xenopus* Development (Daudin)," 2nd ed. North-Holland Publ., Amsterdam.

Nieuwkoop, P. D., and Sutasurya, L. A. (1979). "Primordial Germ Cells in the Chordates: Embryogenesis and Phylogenesis." Cambridge Univ. Press, London and New York.

Nieuwkoop, P. D., Johnen, A. G., and Abler, B. (1985). "The Epigenetic Nature of Early Chordate Development: Inductive Interaction and Competence." Cambridge Univ. Press, London and New York.

Olmstead, J. B. (1986). Microtubule-associated proteins. *Annu. Rev. Cell Biol.* **2**, 421–457.

Ozun, R., Mulner, O., Boyer, J., and Belle, R. (1987). Role of protein phosphorylation in

Xenopus oocyte meiotic maturation. *In* "Molecular Regulation of Nuclear Events in Mitosis and Meiosis." (R. A. Schlegel, M. S. Halleck, and P. N. Rao, eds.), pp. 111–130. Academic Press, Orlando.

Palacek, J., Ubbels, G. A., and Rzehak, K. (1978). Changes in the external and internal pigment pattern upon fertilization in the egg of *Xenopus laevis*. *J. Embryol. Exp. Morphol.* **45**, 203–214.

Palacek, J., Habrova, V., Nedvidek, J., and Romanovsky, A. (1985). Dynamics of tubulin structures in *Xenopus laevis* oogenesis. *J. Embryol. Exp. Morphol.* **87**, 75–86.

Perry, M. M. (1975). Microfilaments in the external surface layer of the early amphibian embryo. *J. Embryol. Exp. Morphol.* **33**, 127–146.

Perry, M. M., and Waddington, C. H. (1966). Ultrastructure of the blastopore cells in the newt. *J. Embryol. Exp. Morphol.* **15**, 317–330.

Perry, M. M., John, H. A., and Thomas, N. S. T. (1971). Actin-like filaments in the cleavage furrow of newt egg. *Exp. Cell Res.* **65**, 249–253.

Pestell, R. Q. (1975). Microtubule protein synthesis during oogenesis and early embryogenesis in *Xenopus laevis*. *Biochem. J.* **145**, 527–534.

Peterson, S. P., and Berns, M. W. (1980). The centriolar complex. *Int. Rev. Cytol.* **64**, 81–106.

Piperno, G., and Fuller, M. T. (1985). Monoclonal antibodies specific for an acetylated form of α-tubulin recognizes the antigen in cilia and flagella in a variety of organisms. *J. Cell Biol.* **101**, 2085–2094.

Pollard, T. D., and Cooper, J. A. (1986). Actin and actin-binding proteins: A critical evaluation of mechanisms and functions. *Annu. Rev. Biochem.* **55**, 987–1035.

Pondel, M., and King, M. L. (1987). A localized maternal mRNA is associated with the cytoskeleton. *J. Cell Biol.* **105**, 335a.

Preston, S. F., Deanin, G. G., Hanson, R. K., and Gordon, M. W. (1981). Tubulin : tyrosine ligase in oocytes and embryos of *Xenopus laevis*. *Dev. Biol.* **81**, 36–42.

Pruss, R. M., Mirsky, R. M., Raff, M. C., Thorpe, R., Dowling, A. J., and Anderton, B. H. (1981). All classes of intermediate filaments share a common antigenic determinant defined by a monoclonal antibody. *Cell* **27**, 419–428.

Rebagliati, M. R., Weeks, D. L., Harvey, R. P., and Melton, D. A. (1985). Identification and cloning of localized maternal RNAs from *Xenopus* eggs. *Cell (Cambridge, Mass.)* **42**, 769–777.

Robinson, K. R. (1979). Electrical currents through full-grown and maturing *Xenopus* oocytes. *Proc. Natl. Acad. Sci. U.S.A.* **76**, 837–841.

Ryabova, L. V., Betina, M. I., and Vassetzky, S. G. (1986). Influence of cytochalasin B on oocyte maturation in *Xenopus laevis*. *Cell Differ.* **19**, 89–96.

Sargent, T. D., Jamrich, M., and Dawid, I. B. (1986). Cell interactions and the control of gene activity during early development of *Xenopus laevis*. *Dev. Biol.* **114**, 238–246.

Sawai, T. (1972). Roles of cortical and subcortical components in the cleavage furrow formation in amphibia. *J. Cell Sci.* **11**, 543–556.

Sawai, T. (1974). Furrow formation on a piece of cortex transplanted to the cleavage plane of the newt egg. *J. Cell Sci.* **15**, 259–267.

Sawai, T. (1979). Cyclic changes in the cortical layer of non-nucleated fragments of the newt's egg. *J. Embrol. Exp. Morphol.* **53**, 183–193.

Sawai, T. (1982). Wavelike propagation of stretching and shrinkage in the surface of the newt's egg before the first cleavage. *J. Exp. Zool.* **222**, 59–68.

Sawai, T. (1983). Cytoplasmic and cortical factors participating in cleavage furrow formation in eggs of three amphibian genera; *Ambystoma, Xenopus* and *Cynops*. *J. Embryol. Exp. Morphol.* **77**, 243–254.

Sawai, T., and Yoneda, M. (1974). Wave of stiffness propagating along the surface of the newt egg during cleavage. *J. Cell Biol.* **60**, 1–7.

Sawai, T., Kubota, T., and Kojima, M. K. (1969). Cortical and subcortical changes preceding furrow formation in the cleavage of newt eggs. *Dev. Growth Differ.* **11**, 246–254.

Scharf, S. R., and Gerhart, J. C. (1983). Axis determination in eggs of *Xenopus laevis:* A critical period before first cleavage, identified by the common effects of cold, pressure and ultraviolet irradiation. *Dev. Biol.* **99**, 75–87.

Schatz, P. J., Solomon, F., and Botstein, D. (1986). Genetically essential and nonessential α-tubulin genes specify functionally interchangeable proteins. *Mol. Cell. Biol.* **6**, 3722–3733.

Scheer, U. (1986). Injection of antibodies into the nucleus of amphibian oocytes: An experimental means of interfering with gene expression in the living cell. *J. Embryol. Exp. Morphol.* **97**, Suppl., 223–242.

Scheer, U., and Dabauvalle, M.-C. (1985). Functional organization of the amphibian oocyte nucleus. *In* "Developmental Biology: A Comprehensive Synthesis" (L. W. Browder, ed.), Vol. 1, pp. 385–430. Plenum, New York.

Scheer, U., Hinssen, H., Franke, W. W., and Jockusch, B. M. (1984). Microinjection of actin-binding proteins and actin antibodies demonstrates involvement of nuclear actin in transcription of lampbrush chromosomes. *Cell (Cambridge, Mass.)* **39**, 111–122.

Schliwa, M., and Van Blerkom, J. (1981). Structural interactions of cytoskeletal components. *J. Cell Biol.* **90**, 222–235.

Schroeder, T. E., and Strickland, D. L. (1974). Ionophore A23187, calcium and contractility in frog eggs. *Exp. Cell Res.* **83**, 139–142.

Schuetz, A. W. (1985). Local control mechanisms during oogenesis and folliculogenesis. *In* "Developmental Biology: A Comprehensive Synthesis" (L. W. Browder, ed.), Vol. 1, pp. 1–83.

Selman, G. G., and Perry, M. M. (1970). Ultrastructural changes in the surface layers of the newt's egg in relation to the mechanism of its cleavage. *J. Cell Sci.* **6**, 207–227.

Selman, G. G., and Waddington, C. H. (1955). The mechanism of cell division in the cleavage of the newt's egg. *J. Exp. Biol.* **32**, 700–733.

Singal, P. K., and Sanders, E. J. (1974). An ultrastructure study of the first cleavage of *Xenopus* embryos. *J. Ultrastruct. Res.* **47**, 433–451.

Singer, S. J., and Kupfer, A. (1986). The directed migration of eukaryotic cells. *Annu. Rev. Cell Biol.* **2**, 337–365.

Slack, J. M. W. (1983). "From Egg to Embryo." Cambridge Univ. Press, London and New York.

Slack, J., ed. (1985). "Early Amphibian Development." J. Embryol. Exp. Morphol. Vol. 89, Suppl.

Smith, R. C., Neff, A. W., and Malacinski, G. M. (1986). Accumulation, organization and deployment of oogenetically derived *Xenopus* yolk/nonyolk proteins. *J. Embryol. Exp. Morphol.* **97**, Suppl., 45–64.

Snow, C. M., Senior, A., and Gerace, A. (1987). Monoclonal antibodies identify a group of nuclear pore complex glycoproteins. *J. Cell Biol.* **104**, 1143–1156.

Steinert, P. M., and Perry, D. A. D. (1985). Intermediate filaments: Conformity and diversity of expression and structure. *Annu. Rev. Cell Biol.* **1**, 41–65.

Stewart-Savage, J., and Grey, R. D. (1982). The temporal and spatial relationships between cortical contraction, sperm trail formation, and pronuclear migration in fertilized *Xenopus* eggs. *Wilhelm Roux's Arch. Dev. Biol.* **191**, 241–245.

Stick, R., and Hausen, P. (1985). Changes in the nuclear lamina composition during early development of *Xenopus laevis*. *Cell (Cambridge, Mass.)* **41**, 191–200.

Stossel, T. P., Chaponnier, C., Ezzell, R. M., Hartwig, J. H., Janmey, P. A., Kwiatkowski, D. J., Lind, S. E., Smith, D. B., Southwick, F. S., Yin, H. L., and Zaner, K. S. (1985). Nonmuscle actin binding proteins. *Annu. Rev. Cell Biol.* **1,** 353–402.

Subtelny, S., and Bradt, C. (1963). Cytological observations on the early developmental stages of activated *Rana pipiens* eggs receiving a transplanted blastula nucleus. *J. Morphol.* **112,** 45–60.

Sun, T.-T., Tseng, S. C. G., Huang, A. J.-W., Cooper, D., Schermer, A., Lynch, M. H., Weiss, R., and Eichner, R. (1985). Monoclonal antibodies studies of mammalian epithelial keratins: A review. *Ann. N.Y. Acad. Sci.* **455,** 307–329.

Takeichi, T., and Kubota, H. Y. (1984). Structural basis of the activation wave in the egg of *Xenopus laevis. J. Embryol. Exp. Morphol.* **81,** 1–16.

Thyberg, J., and Moskalewski, S. (1985). Microtubules and the organization of the golgi complex. *Exp. Cell Res.* **159,** 1–16.

Tolle, H. G., Weber, K., and Osborn, M. (1985). Microinjection of monoclonal antibodies specific for one intermediate filament protein in cells containing multiple keratins allow insight into the composition of particular 10nm filaments. *Eur. J. Cell Biol.* **38,** 234–244.

Tolle, H. G., Weber, K., and Osborn, M. (1987). Keratin filament disruption in interphase and mitotic cells–how is it induced? *Eur. J. Cell Biol.* **43,** 35–47.

Traub, P. (1985). "Intermediate Filaments." Springer-Verlag, Berlin and New York.

Trinkaus, J. P. (1984). "Cells into Organs: The Forces that Shape the Embryos," 2nd ed. Prentice-Hall, Englewood Cliffs, New Jersey.

Ubbels, G. A., Hara, K., Koster, C. H., and Kirschner, M. W. (1983). Evidence for a functional role of the cytoskeleton in determination of the dorsoventral axis in *Xenopus laevis* eggs. *J. Embryol. Exp. Morphol.* **77,** 15–37.

Vacquier, V. D. (1981). Dynamic changes of the egg cortex. *Dev. Biol.* **84,** 1–26.

Vincent, J.-P., and Gerhart, J. C. (1987). Subcortical rotation in the *Xenopus* egg: An early step in embryonic axis specification. *Dev. Biol.* **123,** 526–539.

Vincent, J.-P., Oster, G. F., and Gerhart, J. C. (1986). Kinematics of grey crescent formation in *Xenopus* eggs: The displacement of subcortical cytoplasm relative to egg surface. *Dev. Biol.* **113,** 484–500.

Waddington, C. H. (1952). Preliminary observation on the mechanism of cleavage in the amphibian egg. *J. Exp. Biol.* **29,** 484–489.

Waddington, C. H., and Perry, M. M. (1966). A note on the mechanisms of cell deformation in the neural folds of amphibia. *Exp. Cell Res.* **41,** 691–693.

Wallace, R. A., and Misulovin, Z. (1978). Long-term growth and differentiation of *Xenopus* oocytes in a defined medium. *Proc. Natl. Acad. Sci. U.S.A.* **75,** 5534–5538.

Webster, D. R., Gunderen, G. G., Bulinski, J. C., and Bonsy, G. G. (1987). Assembly and turnover of detyrosinated tubulin *in vivo. J. Cell Biol.* **105,** 265–276.

Weeks, D. L., and Melton, D. A. (1987). A maternal mRNA localizing to the vegetal hemisphere in *Xenopus* eggs codes for a growth factor related to TGFβ. *Cell* **51,** 861–867.

White, J. G., Amos, W. B., and Fordham, M. (1987). An evaluation of confocal versus conventional imaging of biological structures by fluorescence light microscopy. *J. Cell Biol.* **105,** 41–48.

Winkles, J. A., Sargent, T. D., Parry, D. A. D., Jonas, E., and Dawid, I. B. (1985). Developmentally regulated cytokeratin gene in *Xenopus laevis. Mol. Cell Biol.* **5,** 2575–2581.

Wolf, N., Regan, C., and Fuller, M. T. (1987). Temporal and spatial pattern of differences in microtubule behavior during *Drosophila* embryogenesis revealed by distribution of tubulin isoforms. *Development* **102,** 311–324.

Wylie, C. C., Brown, D., Godsave, S. F., Quarmby, J., and Heasman, J. (1985). The cytoskeleton of *Xenopus* oocytes and its role in development. *J. Embryol. Exp. Morphol.* **89,** Suppl., 1–15.

Wylie, C. C., Heasman, J., Parke, J. M., Anderton, B., and Tang, P. (1986). Cytoskeletal changes during oogenesis and early development in of *Xenopus laevis. J. Cell Sci.,* Suppl. **5,** 329–341.

Zotin, A. I. (1964). The mechanism of cleavage in amphibian and sturgeon eggs. *J. Embryol. Exp. Morphol.* **12,** 247–262.

5

Egg Cortical Architecture

FRANK J. LONGO

Department of Anatomy
College of Medicine
The University of Iowa
Iowa City, Iowa 52242

As indicated by Schroeder (1981), the widely used term "cortex" is fundamentally misleading, as it cannot be specifically defined. It is rarely acknowledged to be an operational term, and the precise meaning varies according to usage. In general, animal eggs develop a cortical layer of variable thickness (1–5 μm) delimited externally by the plasma membrane (Figs. 1a–d and 2a–c). This layer is, for the most part, free of yolk, but in many species, contains granules that are formed during oogenesis; these organelles come to lie only within the egg cortex and, hence, are called cortical granules. In addition to cortical granules, this layer contains cisternae of endoplasmic reticulum, ribosomes, and cytoskeletal elements. Demarcation of the cortex from the deeper cytoplasm or subcortex has not been satisfactorily identified.

Despite difficulties in defining the cortex as a structural entity having precise limits, the term has been used in a conceptual sense by numerous cellular and developmental biologists, who continue to demonstrate the importance of this layer for many developmental processes. The cortex is first to react to the contact of sperm, and during processes of maturation, fertilization, and

THE CELL BIOLOGY OF
FERTILIZATION

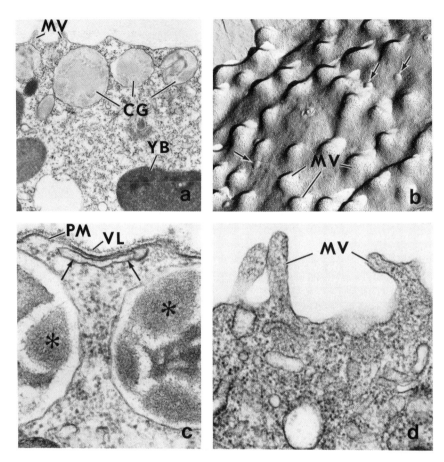

Fig. 1. Cortices of unfertilized (a, b, and c) and fertilized (d) sea urchin (*Arbacia punc-tulata*) eggs. (a) Cortex of an *Arbacia* egg depicting cortical granules (CG) and microvilli (MV). YB, Yolk body. (b) Freeze-fracture replica of the P-face of the plasma member of an *Arbacia* egg containing numerous intramembraneous particles and dome-shaped elevations (arrows), which indicate the presence of underlying cortical granules. MV, Microvilli. (c) Portion of an *Arbacia* egg cortex showing the vitelline layer (VL), cortical endoplasmic reticulum (arrows), which is closely associated with the plasma membrane (PM) and cortical granules (∗). (d) Cortex of a fertilized *Arbacia* egg possessing elongated microvilli (MV).

early development, it undergoes dynamic changes in structure and function. Following insemination in many species, it has been shown to undergo changes in viscoelastic–mechanical properties (reviewed by Vacquier, 1981). It may also demonstrate streaming movements in which "information" or "mor-phogenetic determinants" become localized to specific cellular regions and where specific localizing activities, such as those associated with the yellow

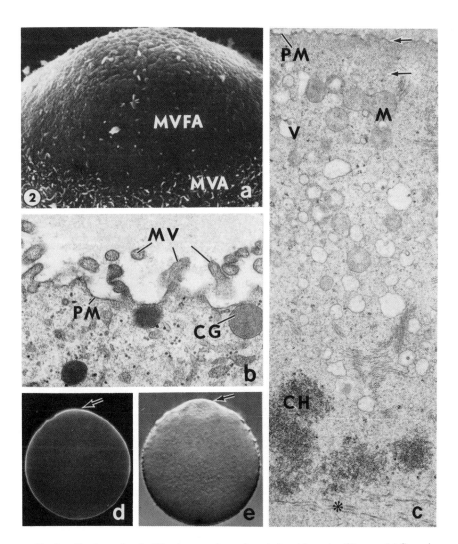

Fig. 2. Cortices of unfertilized mouse (a, c, d, and e) and hamster (b) eggs. (a) Scanning electron micrograph of a mature, oviductal mouse egg showing the microvillus-free area (MVFA) that is associated with the underlying meiotic apparatus and a portion of the microvillus area (MVA) that is characteristic of the remainder of the egg surface. (b) Surface of a hamster egg along with the microvillus area. Cortical granules (CG) are associated with the plasma membrane (PM). MV, Microvilli. (c) Section through the microvillus-free area of a mouse egg depicting a portion of the underlying meiotic apparatus (∗). Although the plasma membrane (PM) in the region of the meiotic spindle undulates, it is not projected into microvilli. Vesicles (V) and mitochondria (M) are restricted from the cortical region, which is rich in actin and indicated by the arrows. CH, Meiotic chromosomes. (d) Mature mouse egg incubated in rabbit antiactin serum followed by FITC-labeled goat anti-rabbit antibody demonstrating cortically located actin. The fluorescent layer is enhanced in the region associated with the microvillus-free area (arrow). (e) Same preparation shown in d photographed with differential contrast optics. The arrow points to the meiotic spindle, which underlies the microvillus-free area.

crescent of ascidian eggs and the gray crescent of amphibian eggs, are manifest (reviewed by Davidson, 1986).

In this review, structural and functional changes of the egg cortex, as a result of gamete fusion and cortical granule dehiscence, are presented and discussed. Processes related to aspects presented here have been reviewed and provide additional perspectives of dynamic changes the egg cortex undergoes at fertilization (Schroeder, 1979, 1981; Gulyas, 1980; Shapiro *et al.*, 1980; Elinson, 1980; Vacquier, 1981; Schuel, 1985; Longo, 1985; Maro *et al.*, 1986a).

I. ARCHITECTURE OF THE EGG CORTEX

The cortices of sea urchin eggs, which provide the basis of much that is known regarding egg cortical structure, are lined with a layer of cortical granules about 1 μm in diameter (Fig. 1a). In *Strongylocentrotus*, there are about 18,000 of these organelles/egg (Vacquier, 1981). They are manufactured by the Golgi complex and become closely associated with the plasma membrane during oocyte development (Anderson, 1968; see also Szollosi, 1967). The cortices of some ophuroid, anthozoa, and polychaete eggs contain a layer of cortical granules, five to six deep (Pasteels, 1966; Holland, 1979; Fallon and Austin, 1967; Dewel and Clark, 1974). Eggs of the ascidian *Ciona* and the salamander *Triturus* do not have cortical granules (Hope *et al.*, 1963; Rosati *et al.*, 1977; see also Elinson, 1986). In *Ciona*, a population of granules located in the subcortex reportedly functions as cortical granules (Rosati *et al.*, 1977).

Ultrastructurally, cortical granules of sea urchin eggs display variations in organization depending on the species (Schuel, 1985). The patterns that are observed reflect the organization of its components that localize to specific portions of the extracellular coats that develop following cortical granule dehiscence (Anderson, 1968; Schuel, 1985). The content of *Arbacia* cortical granules is distinguished by a crenated, central mass, surrounded by some lenticular material (Fig. 1a). In *Strongylocentrotus*, there is a spiral of electron-dense material, which is associated with an amorphous mass. The organization of cortical granule internal components of *Mytilus* (Humphreys, 1967) and other organisms has also been described. The cortical granules in amphibian and mammalian eggs do not show unusually complex patterns and are filled with electron-dense granular material (Fig. 2c; Kemp and Istock, 1967; Gulyas, 1980).

Cortical granule contents from sea urchins, amphibians, and mammals have been examined directly by biochemical and cytochemical techniques and indirectly by analysis of the medium following their discharge (for a review, see Schuel, 1985). Calcium, serine protease, and sulfated mucopolysaccharides appear to be universal components of these structures. Peroxidase, $\beta 1,3$-gluconase, hyaline protein, β-glucuronidase, and other proteins are also present in the cortical granules of some organisms.

Although the structure of the sea urchin egg cortex has been analyzed by a number of techniques, the nature of the association of the cortical granules and the plasma membrane remains an enigma (Millonig, 1969; Detering et al., 1977; Longo, 1981, 1985). Electron microscopy of conventionally prepared eggs shows the cytoplasmic region associated with the cortical granules and plasma membrane as relatively unspecialized, i.e., lacking any apparent modification which might serve to attach the two structures (Anderson, 1968; Millonig, 1969). However, with quick-frozen and freeze-substituted preparations, filaments join cortical granules to the plasma membrane and extend from the granules into the cytoplasm (Chandler, 1984). These filaments may contribute to the attachment of the cortical granules to the plasma membrane.

The connection of the cortical granules to the oolemma is sufficiently strong to survive forces encountered during the isolation of plasma membrane–cortical granule complexes (Detering et al., 1977) and during preparation of cortical lawns, where the cytoplasm is sheared away (Vacquier, 1975). The normal attachment of cortical granules to the overlying plasma membrane can be disrupted by choatropic agents, urethane and tertiary amines, reinforcing the idea that a special attachment exists between the two structures (Longo and Anderson, 1970; Vacquier, 1975; Hylander and Summers, 1981; Decker and Kinsey, 1983).

Modifications of the sea urchin egg plasma membrane have been observed in areas occupied by cortical granules with freeze-fracture replicas, by scanning electron microscopy, and with filipin staining for the demonstration of 3β-hydroxysterol components (Longo, 1981; Carron and Longo, 1983; Zimmerberg et al., 1985). The plasma membrane modifications seen with freeze-fracture replication are dome-shaped areas lacking intramembranous particles (Fig. 1b). These modifications appear to form as a result of the association of the cortical granule with the plasmalemma, i.e., they are lacking in the plasma membrane of fertilized and immature eggs in which the cortical granules are absent or not localized to the cortex (Longo, 1981). Furthermore, the dome-shaped areas disappear when the mature egg is treated with amines (Longo, 1981). These specializations are believed to represent specific contacts between the plasma membrane and the cortical granule, which produce or stimulate conditions required for bilayer fusion.

Unique patterning of intramembranous particles possibly induced by structures subjacent to the plasma membrane have been described in numerous cells having secretory activities (Satir et al., 1973; Friend and Fawcett, 1974; Satir, 1976; Beisson et al., 1976; Weiss et al., 1977a,b; Kinsey and Koehler, 1978). In addition, clearings of intramembranous particles have been observed in portions of the plasma membrane associated with secretory vesicles and are generally considered to represent areas depleted of membrane proteins at the fusion zone (Chi *et al.*, 1975; Lawson *et al.*, 1977; Orci *et al.*, 1977; Orci and Perrelet, 1977; Amherdt *et al.*, 1978; Swift and Murkherjee, 1978; Theodosis *et al.*, 1978).

Unlike the situation that exists in most other cells, in *Xenopus* oocytes, the E-face of the plasma membrane is endowed with a higher concentration of intramembranous particles than the P-face (Bluemink and Tertoolen, 1978). Although the overall density of intramembranous particles is the same in the animal hemisphere versus the vegetal hemisphere, there is a preponderance of small intramembranous particles in the animal hemisphere that Bluemink and Tertoolen (1978) suggested might be involved with cortical morphogenesis following insemination. Freeze-fracture replicas of the plasma membrane of mammalian eggs have also been examined and a lower intramembranous particle density has been noted at those sites of presumptive cortical granule exocytosis (Koehler *et al.*, 1982; Suzuki and Yanagimachi, 1983; see also Koehler *et al.*, 1985).

The plasma membrane of the sea urchin *(Arbacia)* egg is reflected into short microvilli, which lack a core of actin microfilaments (Fig. 1b; Carron and Longo, 1982). The underlying cortical granules tend to be situated in areas which lack microvilli (Schroeder, 1979; Longo, 1981). In amphibian and mammalian eggs, the microvilli are relatively longer and contain a microfilamentous core (Phillips *et al.*, 1985; Longo and Chen, 1985). Attached to the sea urchin oolemma is a glycocalyx or vitelline layer (Fig. 1c; Anderson, 1968; Kidd, 1978). It is this structure to which sperm bind via bindin (Vacquier and Moy, 1977) and which, at the time of cortical granule exocytosis, becomes detached from the egg surface to form the fertilization membrane (see Kay and Shapiro, 1985).

In many organisms, particularly sea urchins, the structure of the cortex is virtually identical along all regions of the egg surface. However, in eggs that are fertilized at metaphase I or II of meiosis (e.g., molluscs, annelids, and vertebrates), the meiotic spindle is positioned within the cortex and the cytoplasm associated with this structure usually differs in its composition to other regions of the egg cortex (Fig. 2a–e). For example, in mammalian eggs the region which overlies the meiotic spindle is distinguished by the absence of microvilli and cortical granules, by a diminished affinity to the plant lectin concanavalin A, and by the presence of a dense layer of actin filaments (Johnson *et al.*, 1975; Eager *et al.*, 1976; Nicosia *et al.*, 1977, 1978; Phillips and Shalgi, 1980; Albertini, 1984; Ebensperger and Barros, 1984; Maro *et al.*, 1984, 1986a,b; Longo and Chen, 1985; Karasiewiez and Soltynska, 1985; Van Blerkom and Bell, 1986; Capco and McGaughey, 1986). This region has been referred to as the microvillus-free area (Fig. 1a and c; Longo and Chen, 1985).

In mammalian oocytes as the meiotic spindle develops, it moves to the animal pole. In hamster oocytes, a cortical granule-free area appears in the cortex with the arrival of the metaphase spindle as a result of the peripheral migration and exocytosis of cortical granules (Okada *et al.*, 1986). Concomitant with the cortical localization of the meiotic spindle is the formation of a thickened layer of actin filaments and the disappearance of microvilli (Longo, 1985; Van Blerkom and Bell, 1986), suggesting that localization of the meiotic spindle and associated cortical modifications are linked (Fig. 1d and e). Experimental support for this contention has been presented (Longo and Chen, 1985; Van Blerkom and Bell, 1986; Maro *et al.*, 1986a,b).

Although a meiotic spindle is not formed in mouse oocytes treated with colchicine, the chromosomes move to the egg cortex and a microvillus-free area forms in the region of the cortex associated with the chromosomes (Longo and Chen, 1985). Moreover, when the meiotic chromosomes are prevented from moving to the egg cortex, a microvillus-free area does not develop (Longo and Chen, 1984, 1985; Van Blerkom and Bell, 1986; see also Maro *et al.*, 1986a,b). Bivalent chromosomes transferred to germinal vesicle-intact or maturing oocytes are also capable of inducing a localized thickening of actin, the loss of microvilli, and a reduction of surface glycoproteins (Van Blerkom and Bell, 1986). These observations show that interaction of the meiotic chromosomes with the egg cortex brings about the formation of the microvillus-free area and that the capacity for this transformation occurs prior to germinal vesicle breakdown. Moreover, the cortical and plasma membrane polarity that is established at metaphase I and II in mouse oocytes is induced rather than preexisting; the entire surface of the oocyte is capable of differentiation in response to the presence of chromosomes (Van Blerkom and Bell, 1986).

Subsequent to its migration to the egg cortex, the meiotic spindle becomes anchored to the plasma membrane (Chambers, 1917; Conklin, 1917; Shimizu, 1981a). In *Tubifex,* the meiotic apparatus appears to be tethered to the egg surface by structural connections between a filamentous cortical layer, possibly actin, and microtubules of the peripheral aster (Shimizu, 1981a). A similar morphology is also seen in *Ilyanassa* oocytes (Burgess, 1977). Interestingly, meiotic spindle attachment to the egg surface in *Chaetopterus* is colchicine sensitive and unaffected by cytochalasin B (Hamaguchi *et al.*, 1983). In contrast to observations with the eggs of some invertebrates (Longo, 1972; Peaucellier *et al.*, 1974), cytochalasin B prevents the cortical localization of the meiotic spindle in maturing mouse oocytes, suggesting that a cytochalasin B-sensitive component of the cytoskeletal system is involved in this movement (Wassarman *et al.*, 1976; Longo and Chen, 1985). The roles that microtubules, microfilaments, and possibly other cytoskeletal structures may serve in cortical localization and attachment of the meiotic spindle to the egg surface warrants further investigation. With respect to studies in mammals, actin has been demonstrated in nuclei and in mitotic spindles and has been implicated in force production of chromosome movements during mitosis (Zimmerman and Forer, 1981). In light of these investigations and studies demonstrating the disruptive effects of cytochalasin B on actin (Yahara *et al.*, 1982; Schliwa, 1981), an actin-based system may be responsible for the cortical localization of the meiotic spindle in mouse oocytes (Longo and Chen, 1984, 1985).

II. CORTICAL GRANULE REACTION

Exocytosis of cortical granules has been studied in eggs of invertebrates, vertebrates, and mammals and appears to involve similar processes (Fig. 3; Anderson, 1968; Elinson, 1980; Gulyas, 1980). In sea urchins, sperm–egg binding is followed by the dehiscence of cortical granules that underlie the plasmalemma, and exocytosis spreads from the point of gamete contact in a wave to the opposite pole of the egg (Afzelius, 1956; Endo, 1961; Wolpert and Mercer, 1961; Anderson, 1968; Millonig, 1969; see also Holland, 1980; Longo *et al.*, 1982). In some species of pelecypods and annelids, cortical granules are present, but do not undergo exocytosis or a change at fertilization (Pasteels and de Harven, 1962; Rebhun, 1962; Humphreys, 1967). The fate of cortical granule contents and the development of extracellular layers surrounding activated eggs have been reviewed (Kay and Shapiro, 1985).

The mechanisms by which the egg plasma and the cortical granule membranes fuse and the nature of the intermediates in this process are unclear. Using transmission electron microscopy, Anderson (1968) and Millonig (1969) indicated that opening of cortical granules may occur via multiple fusions between the cortical granule membrane and oolemma, and thereby, a series of vesicles, composed of membrane derived from both the cortical granules and oolemma, are released to the perivitelline space. Using freeze-fracture replicas, Chandler and Heuser (1979) were unable to find intermediate stages of cortical granule membrane–plasma membrane fusion, suggesting that the fusion process is completed very rapidly. They indicated that a single pore is formed, which increases in size to allow dehiscence of the cortical granule contents. This suggests that all of the membrane delimiting a cortical granule is incorporated into the egg plasma membrane when the two structures fuse.

The current paradigm regarding the mechanisms of cortical granule discharge is that calcium functions as an essential intracellular messenger (for reviews, see Shen, 1983; Whitaker and Steinhardt, 1985; Jaffe, 1985). The release of calcium from different cellular compartments has been demonstrated for sea urchin eggs (Zucker *et al.*, 1978; Zucker and Steinhardt, 1978), and the store involved with cortical granule discharge appears to be associated with the

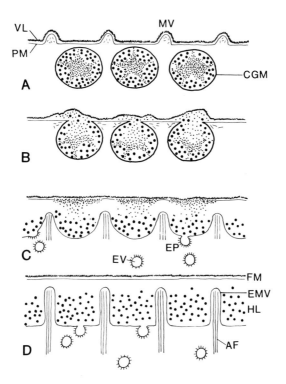

Fig. 3. Diagramatic representation of cortical changes in a fertilized sea urchin egg. (A) Cortex of an unfertilized egg depicting cortical granules, plasma membrane (PM), vitelline layer (VL), and short microvilli (MV). The cortical endoplasmic reticulum, normally present in association with the plasma membrane, is not depicted. CGM, Cortical granule membrane. (B) Cortical granule discharge and vitelline layer elevation. (C and D) A portion of the cortical granule contents has joined with the vitelline layer to form the fertilization membrane (FM). The remaining cortical granule material remains in the perivitelline space to become the hyaline layer (HL). Immediately following cortical granule discharge, portions of the plasma membrane become involved in endocytosis as evidenced by endocytotic pits and vesicles (EP and EV). The surface of the fertilized egg is projected into elongate microvilli (EMV) containing a core of actin filaments (AF).

endoplasmic reticulum located within the egg cortex (Fig. 1c; see Luttmer and Longo, 1985, for discussion). Almost all calcium-binding ability of the unfertilized sea urchin egg is found in a large particle fraction (microsomes) isolated by differential centrifugation (Steinhardt and Epel, 1974). Consistent with this observation are demonstrations that preparations of vesicles derived from *Xenopus* eggs are able to sequester calcium in an ATP-dependent manner (Cartaud *et al.*, 1984) and that calcium is associated with the plasma membrane and cortical granules, as well as cortical endoplasmic reticulum of sea urchin eggs (Cardasis *et al.*, 1978; Sardet, 1984). In addition, electron microscopic studies have demonstrated specialized regions of the egg endoplasmic reticulum that are associated with the cortical granules in *Xenopus*, sea urchin, and mouse ova (Campanella and Andreuccetti, 1977; Gardiner and Grey, 1983; Sardet, 1984; Luttmer and Longo, 1985). These observations, as well as (1) the striking morphological similarity of the plasma membrane–endoplasmic reticulum association observed in *Xenopus*, mouse, and sea urchin eggs to the transverse tubule and sarcoplasmic reticulum of muscle cells (Endo, 1977; Gardiner and Grey, 1983; Sardet, 1984; Luttmer and Longo, 1985) and (2) the temporal correlation of cortical endoplasmic reticulum development and the capacity of *Xenopus* eggs to propagate a wave of cortical granule exocytosis (Charbonneau and Grey, 1984; Campanella *et al.*, 1984), suggest that the close association of the plasma membrane and cortical endoplasmic reticulum transduces the interaction of gametes into an intracellular calcium release, which then triggers the cortical granule reaction and the activation of development.

At the completion of the cortical granule reaction in the eggs of sea urchins, amphibians, and fish, virtually all of the cortical granules have been discharged. In mice, a substantial number of cortical granules (about 25% of the population) are exocytosed before sperm–egg fusion; the remainder are dehisced at fertilization (Nicosia *et al.*, 1977). In *Sabellaria*, the cortical granule reaction is initiated when the eggs are spawned into sea water (Pasteels, 1965). In *Urechis*, a subset of cortical granules is released at insemination; the remainder are discharged later with the elevation of the vitelline layer (Paul, 1975).

For organisms whose eggs do not possess cortical granules nor undergo a cortical granule reaction, it is clear that cortical granule exocytosis is not a required feature for fertilization, nor for egg activation and development. Furthermore, inhibition of the cortical granule reaction in sea urchins does not impair events of fertilization and cleavage (Longo and Anderson, 1970; Vacquier, 1975; Schmidt and Epel, 1983).

III. MEMBRANE CHANGES AT FERTILIZATION

As a consequence of the cortical granule reaction, there is the externalization of cortical granule contents which have profound structural and physiological effects on the egg (for reviews, see Shapiro and Eddy, 1980; Shapiro *et al.*, 1981; Kay and Shapiro, 1985). Insertion of cortical granule membrane into the egg plasma membrane is followed by dramatic structural changes of the cortex and oolemma (Fig. 3). The resultant membrane of the fertilized egg has been referred to as a mosaic, indicating that it is derived from several sources; i.e., the egg plasma membrane, the cortical granule membrane, and the sperm plasmalemma (Colwin and Colwin, 1967; Anderson, 1968; Schroeder, 1979). There is essentially a doubling of the surface area of the activated echinoid egg as a result of the cortical granule reaction, i.e., the sum total of membrane delimiting all of the cortical granules within the egg is equivalent to the surface area of the egg plasma membrane and both sources of membranes are believed to be completely incorporated with the cortical granule reaction (Schroeder, 1979; Vacquier, 1981). That all of the membrane delimiting the cortical granules and the plasma membrane of the unfertilized egg become a part of the plasmalemma of the activated ovum has not been established (see Anderson, 1968; Millonig, 1969).

Since many aspects of fertilization are membrane-mediated events leading to egg activation, it is not unlikely that a change in the plasma membrane is an obligatory step in cellular activation (Pardee et al., 1974; Campisi and Scandella, 1978). Potential changes in the organization of membrane lipids following insemination have been studied, and the results are controversial. Using electron spin resonance spectroscopy, Campisi and Scandella (1978, 1980a) demonstrated an increase in bulk membrane fluidity of sea urchin eggs after fertilization. However, because the spin label (fatty acid) was equilibrated among all subcellular membrane fractions, it could not be determined whether (1) ovum activation is accompanied by a change in total cellular membranes to a more fluid state or (2) more specialized membranes (such as the plasmalemma) entered a more fluid state, and the probe was showing the average change of altered and unaltered membranes. Changes of membrane lipids accompanying activation are probably not a result of the cortical granule reaction, as eggs partially activated by ammonia showed a similar effect. In experiments with cortical fractions, it has been shown that the fluidity of the fertilized egg cortex is less than that of the unfertilized egg cortex (Campisi and Scandella, 1980b). Adding calcium to cortical fractions from unfertilized eggs resulted in a fluidity decrease *in vitro*. It has been suggested that this change may represent an alteration in membrane structure rather than a direct interaction of calcium with phospholipid groups (Campisi and Scandella, 1980b).

Another approach to the question of possible organizational changes in egg plasma membrane lipids has been explored with the fluorescent probe, merocyanine 540 (Freidus et al., 1984). These studies indicate that cortical granule fusion results in changes in plasma membrane lipid organization, i.e., membrane lipids become more loosely organized.

Analyses of membrane lipid changes in sea urchin and mouse eggs using fluorescence photobleaching recovery suggest that fertilization is not accompanied by a change in bulk membrane viscosity; rather it is associated with alterations in the ensemble of lipid domains (Wolf et al., 1981a,b; Wolf and Ziomek, 1983). The different lipid analogs employed by Wolf et al. (1981a,b) indicated the existence of lipid domains, differing in composition or physical states from the average for the plasma membrane. These results suggest the existence of gel and fluid lipid domains within the egg plasma membrane, the proportion and composition of which change upon fertilization. At fertilization, there may be a reordering of lipid domains, which release inactive proteins from gel regions of the plasma membrane into fluid regions, where they would become active. Changes in lipid composition and gel–fluid transformations at fertilization could activate protein functions not requiring the synthesis or insertion of new materials into the membrane.

Studies have been performed in sea urchin *(Arbacia)* eggs treated with filipin to detect alterations in membrane sterols at activation (Carron and Longo, 1983). The plasma membranes of treated, unfertilized eggs possess numerous filipin–sterol complexes, while fewer complexes are associated with membranes delimiting cortical granules, indicating that the plasma membrane is relatively rich in β-hydroxysterols (de Kruijff and Demel, 1974; Carron and Longo, 1983). This dichotomy does not appear to be related to a filipin impermeability, and differences in filipin staining of the plasma and cortical granule membranes may represent differences in sterol content. Biochemical analysis (Decker and Kinsey, 1983), however, indicate that the cholesterol content of cortical granules is significantly higher than that of the egg plasma membrane, suggesting that, following the cortical granule reaction, there would be a substantial increase in plasmalemma cholesterol. Analyses of fertilized egg plasma membranes failed to confirm this expectation (Decker and Kinsey, 1983).

Following its fusion with the plasmalemma, membrane formerly delimiting cortical granules undergoes a rapid increase in the number of filipin–sterol complexes (Carron and Longo, 1983). Other than regions involved in endocytosis, the plasma membrane of the zygote possesses a homogeneous distribution of filipin–sterol complexes and appears structurally similar to that of the unfertilized ovum.

How the cortical granule membrane might acquire an increase in filipin–sterol complexes has not been determined. Lateral displacement of sterols from membranous regions derived from the original egg plasma membrane may be involved (Friend, 1982); however, there is no evidence documenting such a process in activated eggs. Sterols have been shown to diffuse rapidly in bilayers (Träuble and Sackermann, 1972), which is consistent with an extremely rapid lateral displacement of sterols into membrane patches derived from cortical granules (Carron and Longo, 1983).

Fluorescence photobleaching recovery experiments have been performed with mouse eggs using probes to membrane proteins, suggesting that interactions with cytoskeletal components may regulate membrane protein diffusion (Wolf and Ziomek, 1983). As with membrane lipids, the proteins probed demonstrated a heterogeneous distribution. Moreover, although "new" membranes (i.e., cortical granule and sperm plasma membranes) are added to the egg plasmalemma at fertilization, there is no generalized effect on the diffusion of membrane protein in the mouse egg.

Binding studies using plant lectins also have been utilized in an effort to demonstrate possible membrane changes between fertilized and unfertilized eggs. Investigations with mouse and hamster eggs have shown that concanavalin A-binding sites change quantitatively following fertilization (Yanagimachi and Nicolson, 1976; DeFelici and Siracusa, 1981). In ascidian eggs, both the agglutinability and number of concanavalin A receptors increase following activation (O'Dell et al., 1973). These changes in lectin binding following fertilization may reflect modifications in the nature and/or structure of the binding sites themselves. Alterations in lectin binding may also be influenced by membrane fluidity and functional states of the cytoskeleton (Karsenti et al., 1977; Marshall and Heiniger, 1979). In the sea urchin Strongylocentrotus, two classes of concanavalin A-binding sites have been identified: a high-affinity site associated with the vitelline layer and a low-affinity site associated with the plasma membrane. The number of low-affinity sites doubles at fertilization, possibly as a result of the insertion of cortical granule membrane (Veron and Shapiro, 1977). Although the increase in low-affinity binding sites may be due to the appearance of cryptic sites, there is no doubling when eggs are activated with ammonia, supporting the notion that the increase in number of sites is caused by the addition of cortical granule membrane to the egg plasmalemma.

Examination of freeze-fracture replicas of unfertilized sea urchin eggs demonstrates a significant difference in the number of intramembranous particles within the plasmalemma and the cortical granule membrane. In Arbacia, the number of intramembranous particles within the P-face of the cortical granule membrane is about 30% of that in the P-face of the egg plasma membrane (Longo, 1981). Studies have been carried out to determine what happens to this dichotomy following cortical granule exocytosis; i.e., whether localized areas, corresponding to patches of cortical granule membrane, are present within the plasma membrane of the fertilized egg or whether particles within the plasma membrane of the activated egg are homogeneously distributed. A homogeneous distribution of particles would suggest an intermixing of components within the mosaic membrane. The mosaic pattern of the fertilized egg plasmalemma, in terms of intramembranous particles, is temporary; recognizable differences between the original egg plasma membrane and cortical granule membrane are lost soon after cortical granule exocytosis (Pollack, 1978; Chandler and Heuser, 1979; Longo, 1981). Patches, containing a reduced number of intramembranous particles and corresponding to the former cortical granule membrane, are not found in the plasma membrane of the activated egg. This indicates a rapid alteration in the composition of cortical granule membrane following its fusion with the plasma membrane. By 4-min postin-semination, the density of intramembranous particles in the P-face of the plas-ma membrane of the fertilized egg is slightly reduced from that of the membrane of the unfertilized egg, suggesting a possible "flow" of intramembranous par-ticles from the oolemma into membrane derived from the cortical granules. This suggestion is in keeping with the fluid character of membranes and is consistent with schemes reported for other cells (Frye and Edidin, 1970; Singer and Nicolson, 1972).

Changes in the distribution of intramembranous particles occur in the plasma membrane of *Spisula* eggs, which do not have a cortical granule reaction. Following activation, there is an approximate 2-fold increase in density of intramembranous particles within the plasma membrane of *Spisula* eggs (Lon-go, 1976a,b). The functional significance of this membrane change and whether it is related to the development of a block to polyspermy have not been de-termined.

IV. INTEGRATION OF THE SPERM AND EGG PLASMA MEMBRANE

That all of the sperm plasma membrane is incorporated into the egg plasma membrane at fertilization is assumed in many instances, although experimental evidence has not verified this unequivocally. Electron microscopic studies of sperm incorporation in some invertebrates and mammals have demonstrated membranous elements at the site of gamete fusion that appear to be derived from the fused sperm and/or the egg plasmalemma (Franklin, 1965; Colwin and Colwin, 1967; Pikó, 1969; Zamboni, 1971; Bedford and Cooper, 1978). Insertion of sperm plasma membrane components into the egg plasma mem-brane has been demonstrated by O'Rand (1977), who used isoantiserum against whole rabbit sperm. Fertilized rabbit eggs lyse when incubated with the an-tibody and compliment. Unfertilized eggs are unaffected, suggesting that sperm antigen are exposed on the surface of the fertilized egg.

Investigations examining the integration of the sperm and egg plasma mem-branes at fertilization, where one of the gametes has been labeled, have been carried out in both invertebrates and mammals (Yanagimachi *et al.*, 1973; Gabel *et al.*, 1979; Longo, 1982). Prior to sperm–egg fusion in hamsters, the sperm plasma membrane of the postacrosomal region does not bind colloidal iron hydroxide. Once gamete fusion has been initiated, however, the former sperm plasma membrane is able to bind this marker (Yanagimachi *et al.*, 1973). The rapid increase in colloidal iron hydroxide binding on the incorporating sperm head is believed to be a result of intermixing of sperm–egg membrane components comparable to the intermingling of antigenic determinants after fusion of somatic cells (Frye and Edidin, 1970). These observations, however, do not exclude the possibility that colloidal iron hydroxide-binding receptors are enzymatically added to sperm plasma membrane oligosaccharides after fusion or that colloidal iron hydroxide-binding membrane components are inserted into the sperm plasma membrane following fertilization.

Similar experiments have been carried out with the surf clam, *Spisula,* in which concanavalin A binding to the egg, but not to the sperm plasma membrane, has been demonstrated by the horseradish peroxidase–diaminobenzidine reaction (HRP–DAB; Longo, 1982). Because of this dichotomy in lectin binding, changes in the affinity of the sperm plasmalemma following its fusion and integration with components of the egg plasma membrane can be followed. By 1-min postinsemination, the plasma membranes of fertilized *Spisula* eggs react with concanavalin A–HRP–DAB and are associated uniformly with enzymatic precipitate, except at sites of sperm incorporation. These portions of unstained plasma membrane are derived from the sperm and are localized to the apex of the fertilization cone. From 2- to 4-min postinsemination, HRP–DAB reaction product gradually becomes associated with all of the membrane delimiting the fertilization cone. By 4-min postinsemination, no difference in staining of plasma membranes derived from the egg or the sperm is detected. These observations are consistent with the movement of concanavalin A-binding sites from the egg plasmalemma into the sperm plasma membrane. Similar results have also been obtained using cationized ferritin-labeled gametes of the sea urchin, *Arbacia,* which also showed that significant rearrangements occur in the egg and sperm plasma membranes following gamete fusion, giving rise to asymmetries in membrane topography (Longo, 1986a). Components of both membranes are redistributed within the bilayer adjacent to and delimiting the fertilization cone.

Not all components of the sperm and egg plasma membrane appear to intermix rapidly following gamete fusion. Sea urchin and mouse eggs fertilized with fluorescent- or ^{125}I-labeled sperm reportedly retain a topographically mosaic surface, as if the lateral mobility of sperm plasma membrane components were restricted and retained as a discreet patch (Gabel *et al.*, 1979; Shapiro *et al.*, 1982). It is remarkable that some incorporated sperm-surface proteins persist within a localized area, despite rearrangements of the egg plasma membrane involving exocytosis of cortical granules and endocytosis. Similar experiments have also indicated that labeled sperm components (surface and mitochondrial) are internalized after fertilization (Gundersen *et al.*, 1982); some proteins persist intact, while others undergo a specific, limited degradation (Gundersen and Shapiro, 1984; Gundersen *et al.*, 1986). These results are consistent with electron microscopic studies, indicating the incorporation of portions of the sperm plasma membrane (Bedford, 1972; Colwin and Colwin, 1967; Bedford and Cooper, 1978; Yanagimachi and Noda, 1970).

V. FERTILIZATION CONE FORMATION

At the site of gamete fusion, a protuberance forms, which has been referred to as the fertilization or incorporation cone (Fig. 4a and b; Longo and Anderson, 1968; Longo, 1973; Schatten and Schatten, 1980; Tilney and Jaffe, 1980). Formation of this structure in many invertebrates involves a movement of egg cytoplasm into the region surrounding the sperm nucleus, mitochondria, and axonemal complex, resulting in a protrusion at the site of sperm entry (Longo, 1973). The fertilization cone increases in size as more of the egg cytoplasm surrounds incorporated sperm components, which in turn move deeper into the ovum. Based on scanning electron microscopic observations, it has been claimed that microvilli in the region of gamete fusion cluster and engulf the spermatozoon (Cline *et al.,* 1983; Schatten and Schatten, 1980). Although microvillar elongation is first recognized in the vicinity of gamete fusion, their involvement in sperm incorporation was not apparent by transmission electron microscopy (Longo, 1980).

During early stages of sperm incorporation, the fertilization cone is relatively small; it increases greatly in size after the sperm nucleus passes through it and comes to rest within the egg cortex (Longo, 1980; Schatten and Mazia, 1976). The maximum size of fertilization cones varies depending on the organism in question; in mature *Arbacia* eggs, they measure approximately 6 µm in length by 4 µm in diameter, 5- to 7-min postinsemination. They then regress and are reabsorbed by 10-min postinsemination. Interestingly, in *Arbacia,* the fertilization cones that form on immature eggs are much larger than those that develop on mature ova, e.g., sizes of 25 µm in length by 10 µm in diameter are not unusual (Fig. 4a and b; Seifriz, 1926; Tilney and Jaffe, 1980; Dale and Santella, 1985).

Surface area measurements of sea urchin oocyte fertilization cones indicate that considerably more membrane delimits this cortical projection of cytoplasm that can be accommodated by the sperm plasma membrane (Longo, 1986a). That is, if all of the sperm plasma membrane becomes a part of the delimiting membrane of the fertilization cone, it would comprise less than 10% of the surface area. This and the absence of evidence demonstrating a contribution from the spermatozoon other than its plasmalemma indicate that most of the membrane (90% plus) delimiting the fertilization cone is derived from the oocyte. Qualitative assessment of oocytes at different times following insemination with different degrees of polyspermy indicate that the presence of microvilli is inversely related to the number and the size of fertilization cones. This suggests that microvilli are retracted into the oocyte surface; the membrane thus produced accommodates fertilization cone expansion (Dale and Santella, 1985; Longo, 1986a). These conclusions are consistent with observations in somatic cells, indicating that surface material stored in microex-tensions of suspended cultured cells is used during spreading (see Trinkaus, 1980; Erickson and Trinkaus, 1976; Rovensky and Vasiliev, 1984).

Recent studies have also demonstrated that the fertilization cones of in-seminated sea urchin oocytes (Longo, 1986a,b) have a distinctive crenulated appearance which differs from that of the remainder of the oocyte (Fig. 4b). Membrane-delimiting fertilization cones also has a much lower affinity for agents that stain negatively charged and carbohydrate moieties (Fig. 4b). This difference in surface properties of membrane delimiting the site of sperm–egg fusion is not due solely to incorporated sperm plasma membrane and does not occur when inseminated oocytes are incubated in cytochalasin B. These observations indicate that following insemination significant rearrangements of surface molecules takes place within the egg plasmalemma that give rise to asymmetries in membrane topography. They are also consistent with reports (Elgsaeter and Branton, 1974; Sheetz *et al.*, 1980; Wu *et al.*, 1982; Tank *et al.*, 1982; Jacobson *et al.*, 1984) that elements of the cytoskeletal system affect lateral movement of membrane components and implicate actin in the re-cruitment of membrane to the fertilization cone. That oocyte microvilli may be involved with fertilization cone expansion implies this asymmetry may occur as a result of (1) the migration of specific components from the egg plasma membrane into the domain of the fertilization cone or (2) the modi-fication of a more general pool of plasma membrane components as they be-come a part of the fertilization cone.

Migration of surface components, both sperm and egg derived, within the plasma membrane of the zygote is coincident with dramatic changes in the functional state of the egg, which in turn could be a result of functional al-terations of the membrane. The movement of surface molecules may be in-volved in new membrane functions similar, for example, to the activation of adenylate cyclase, resulting from the lateral mobility of membrane molecules (Martin, 1983).

In sea urchin eggs, fertilization cones are filled with numerous bundles of actin filaments that show a polarity when reacted with heavy meromyosin or S1 (Tilney and Jaffe, 1980), i.e., the arrowhead complexes that form are di-rected to the center of the egg. The microfilaments found in the fertilization cone are polymerized *in situ* from cortical, monomeric actin as few actin fil-aments are present in the egg cortex (Spudich and Spudich, 1979; Cline *et al.*, 1983). Whether and how actin in the fertilization cone might function to effect sperm incorporation is not entirely clear. Actin filaments of the sperm acrosomal process are also polarized with the heavy meromyosin–actin ar-rowheads pointing to the sperm nucleus. Consequently, egg myosin could not bridge sliding actin filaments of both the fertilization cone and the acrosomal process to bring about sperm nucleus incorporation (Tilney and Kallenbach,

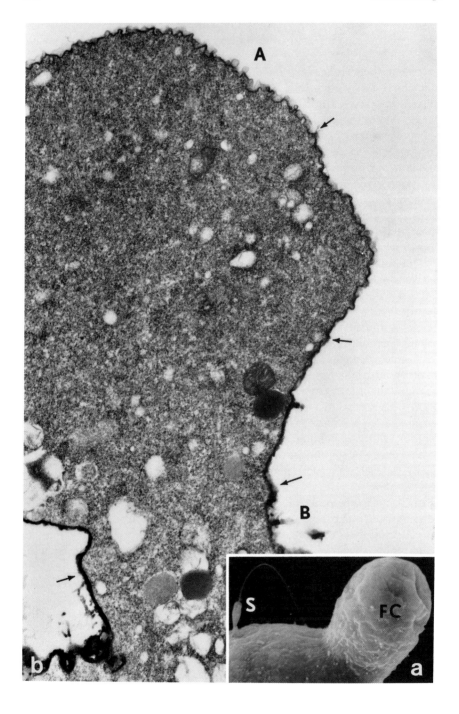

1979); both sets of actin filaments are polarized in the wrong direction, when compared to the orientation of myosin and actin of a sarcomere. It is possible that actin filaments present in the fertilization cone might be primarily involved in the elevation and enlargement of this cortical projection of cytoplasm.

Cytochalasin B, a drug that disrupts actin microfilaments, has been shown to inhibit surface activity of fertilized sea urchin eggs, such as microvillar elongation and fertilization cone formation (Longo, 1980; see also Schatten *et al.*, 1986). Cytochalasin B-treated eggs undergo a cortical granule reaction, elevate a fertilization membrane, and are metabolically activated (Gould-Somero *et al.*, 1977; Longo, 1978). These observations are consistent with the suggestion that, at the site of gamete fusion, there is a localized polymerization of actin that participates in the formation of the fertilization cone. In addition, experiments with cytochalasin B also indicate that treated sea urchin eggs can be activated by sperm, but sperm fail to enter the egg (Longo, 1978; Gould-Somero *et al.*, 1977). How the sperm is capable of activating the egg in this instance without entering it has not been determined. It is possible that the acrosomal process fuses with the egg plasma membrane, but since actin and its polymerization are impaired, the bridge linking the fused sperm and egg is weak, and the sperm is removed from the egg surface by exocytosing cortical granules. Another possibility, consistent with recent observations that ionic activation of *Lytechinus* eggs precedes sperm–egg plasma membrane fusion (Longo *et al.*, 1986), is that cytochalasin B inhibits fusion of the egg and sperm and that gamete contact–binding in this instance is sufficient for egg activation.

In teleosts, the fertilizing spermatozoon enters a region of the egg cytoplasm that appears to be highly specialized. At the base of the micropyle, the plasma membrane of the unfertilized egg is differentiated into a structure apparently designed for sperm binding. It is a short projection in *Fundulus* (Brummett and Dumont, 1979), a cluster of microvilli in *Cyrinus* (Kudo, 1980), *Rhodeus* (Ohta and Iwamatsu, 1983), and *Brachydanio* (Hart and Donovan, 1983). In the eggs of some teleosts, the site of sperm incorporation may lack cortical granules (Brummett and Dumont, 1979; Kobayashi and Yamamoto, 1981; Hart and Donovan, 1983), or granules may be present, but they are smaller than those in other regions of the cortex (Gilkey *et al.*, 1978). Changes in the teleost

←———————————————————————————————————

Fig. 4. Fertilization cones of inseminated sea urchin (*Arbacia punctulata*) oocytes. (a) Scanning electron micrograph of an inseminated oocyte depicting a fertilization cone (FC) and surrounding area possessing relatively few microvilli. S, Sperm. Specimen fixed 12-min postinsemination. (b) Fertilization cone of an oocyte (8-min postinsemination) incubated in concanavalin A. The amount of peroxidase–diaminobenzidine reaction product bound to the plasma membrane (arrows) increases from the apex (A) to the base (B) of the fertilization cone. Note the crenulated surface of the fertilization cone.

egg cortex associated with fertilization cone formation, as determined by scanning electron microscopy, have been described (Brummett *et al.*, 1985; Kudo, 1980; Kudo and Sato, 1985; Ohta and Iwamatsu, 1983; Iwamatsu and Ohta, 1978).

At the site of sperm entry in anurans, a microvillus-free bleb of cytoplasm forms, presumably functionally equivalent to a fertilization cone (Picheral, 1977; Elinson and Manes, 1978; Picheral and Charbonneau, 1982). Eventually it disappears and is replaced by a small clump of elongate microvilli. The microvillus-free bleb is believed to be pinched off in *Pleurodeles,* leaving the microvilli on the surface of the fertilized egg (Picheral, 1977; Picheral and Charbonneau, 1982). If this is the case, then it is possible that plasma membrane components, as well as other sperm-derived structures, are eliminated from the egg. The site of sperm entry reportedly remains detectable as a clump of microvilli for at least 2 hr.

Following the fusion of the egg and sperm plasma membranes in mammals, tongues of cytoplasm surround the anterior portion of the sperm head (Pikó, 1969), forming a vesicle that is present for a time within the zygote (Yanagimachi and Noda, 1970). At the site of gamete fusion, a protrusion of cytoplasm forms which is homologous to the fertilization cones seen in invertebrate eggs and is often referred to as an incorporation cone (Shalgi *et al.*, 1978; Zamboni, 1971; for reviews, see Gaddum-Rosse, 1985; Maro *et al.*, 1986a). As in sea urchins, the incorporation cone of fertilized mammalian eggs reaches its maximum only after the sperm head has entered the cortex (Gaddum-Rosse, 1985). In mouse eggs, this protrusion is filled with cytoplasmic organelles found in other regions of the zygote, and along its plasmalemma is a prominent layer of actin (Maro *et al.*, 1984, 1986b). Surface movements also occur in fertilized rat and mouse oocytes (Gaddum-Rosse *et al.*, 1984; Battaglia and Gaddum-Rosse, 1984; Waksmundzka *et al.*, 1984). This transient activity involves localized elevations of the cortical cytoplasm, which disappear following incorporation of the sperm tail. Formation of cortical elevations is sensitive to cytochalasin B and occurs when eggs are activated with the calcium ionophore A23187. These observations suggest that the cortical elevations are a manifestation of cytoskeletal changes of the oocyte involving actin and are characteristic of the activation processes and not dependent on the presence of sperm components.

VI. MICROVILLAR ELONGATION

It has been shown that the total surface area of cortical granule membrane in a *Strongylocentrotus* egg is greater than that of the plasmalemma (Schroeder, 1979). Hence, if all the cortical granule membrane is incorporated into the egg plasmalemma, there would be at least a 2-fold increase in surface area of the egg at fertilization. However, by 16-min postinsemination the surface area of the activated egg is only slightly larger than that of the unactivated ovum, indicating a rapid accommodation in surface membrane. The microvillar elongation that occurs following insemination may be one means of accommodating a surface increase in the activated sea urchin egg (Eddy and Shapiro, 1976; Schroeder, 1979). However, surface area measurements indicate that elongated microvilli cannot compensate for all the cortical granule membrane that might be incorporated, and membrane internalization has been proposed as a mechanism to quantitatively modify the surface area of activated eggs (Schroeder, 1979).

Rapid elongation of microvilli is believed to occur primarily in areas occupied by the original plasma membrane (Chandler and Heuser, 1981) and may take place only at sites on the egg surface, where cortical granules have exocytosed (Fisher et al., 1982). However, more recent investigations (Fisher et al., 1985) indicate that it may occur in sea urchin eggs in which cortical granule exocytosis has been inhibited, but after some delay. By 2-min postinsemination, cytoplasmic upheavals develop at the bases of elongating microvilli and form mounds which possess two to four microvilli projecting from their apices. By 5-min postinsemination, the mounds shrink and are interconnected by ruffles of cytoplasm (Chandler and Heuser, 1981). Similar morphological changes along the bases of microvilli have also been described for Spisula eggs (Longo, 1976a). In sea urchins, these changes are a result of the reorganization of the cortical cytoskeletal system, which gives rise to a cortex that is projected into elongate microvilli and contains cytoskeletal elements, endoplasmic reticulum, and ground substance (Fig. 1d).

Although present in high concentration in unfertilized ova, few or no actin filaments are found associated with the cortices of unfertilized echinoderm eggs and relatively little (9–20%) of the total egg actin pool is present in the polymerized form (Spudich and Spudich, 1979; Otto et al., 1980; Coffe et al., 1982; Yonemura and Kinoshita, 1986). Biochemical studies (Spudich and Spudich, 1979; Otto et al., 1980) and fluorescent microscopic observations with NBD–phallicidin (Yonemura and Kinoshita, 1986) demonstrate that monomeric, cortical actin is induced to polymerize into filaments at fertilization. Actin has also been demonstrated in the cortices of amphibian and mammalian eggs (Clark and Merriam, 1978; Maro et al., 1984; Reima and Lehtonen, 1985; Longo and Chen, 1985).

Investigations, with both intact eggs and isolated cortices exposed to different ionic conditions, demonstrate that microvillar elongation is stimulated by the calcium flux characteristic of egg activation (Carron and Longo, 1980, 1982; Begg et al., 1982). As a consequence of this process, microvilli increase three to four times their original length and obtain polarized bundles of actin filaments (Burgess and Schroeder, 1977). Microvillar elongation does not occur when eggs are incubated in media such as ammonia, which induces an increase in intracellular pH. However, actin filament bundle formation is triggered by an increase in intracellular pH. Formation of actin filament bundles is not necessary for microvillar elongation, but is required for rigid support of microvilli. It has been suggested that events of activation prior to the intracellular pH increase induce the formation of cortical microfilament networks and microvillar elongation (Carron and Longo, 1982). The microfilaments provide the structural and/or contractile framework for support of the egg surface, which is undergoing extensive rearrangement. Microfilament organization within the microvilli, i.e., bundle formation, may then be a consequence of cytoplasmic alkalinization. Hence, actin filament bundle formation in the cortex of the fertilized sea urchin egg appears to be a two-step process (Tilney and Jaffe, 1980): (1) the polymerization of actin to form filaments randomly oriented, but in most cases, with one end in contact with the plasma membrane, followed by (2) the association of filaments by macromolecular bridges to form bundles.

Microvillar elongation also occurs in fertilized medaka eggs (Iwamatsu and Ohta, 1976). Microvilli elongate starting at the opening of dehisced cortical granules; this is followed by a propagation of microvilli along the surface of the dehisced cortical granule. This change is accompanied by the formation of an electron-dense layer, possibly actin, that underlies the former cortical granule membrane.

The mechanisms of cortical reorganization are not known, but are likely to involve actin-binding proteins as described in other systems (Craig and Pollard, 1982). The distribution of α-actinin during fertilization has been investigated by microinjection of rhodamine-labeled α-actinin into living sea urchin eggs (Mabuchi et al., 1985). This probe is uniformly distributed in the cytoplasm of unfertilized eggs. Upon fertilization, however, it concentrates in the zygote cortex including the fertilization cone. Migration of fluorescently labeled α-actinin into microvilli apparently does not occur. Aggregation of actin filaments and their association with bundling protein, e.g., fascin, may give rise to microfilament bundles in egg microvilli (Spudich and Amos, 1979; Otto et al., 1980; Tilney and Jaffe, 1980; DeRosier and Edds, 1980; Mabuchi and Nonomura, 1981). A profilin-like protein may prevent actin from polymerizing in the unfertilized egg (Mabuchi, 1981; Hosoya et al., 1982). Although fascin is found in the unfertilized sea urchin egg and has been localized in microvilli of fertilized ova, its interaction with actin has not been shown to be calcium or pH sensitive (Bryan and Kane, 1982). Hence, other actin-binding proteins may be instrumental in microvillar elongation; cytoplasmic alkalinization may give rise to microfilament bundle formation by promoting interactions between actin and actin-binding proteins. In this context, Otto and Schroeder (1984) have shown that 1-methyladenine stimulates starfish oocytes to undergo major organizational changes involving actin, fascin, and a 220-kDa protein.

In addition to changes in microvillar conformation, the eggs of a number of different animals undergo changes in cortical rigidity and contraction that appear to involve an actomyosin system (see Vacquier, 1981). Cyclical changes in surface tension and contraction have been correlated with cytoskeletal alterations (Coffe *et al.*, 1982) and also occur in anucleate egg fragments with the same cycle as in normal embryos (Yoneda *et al.*, 1978; Shimizu, 1981b; Yamamoto and Yoneda, 1983). These observations indicate that egg activation initiates processes that are autonomous of the nucleus, that regulate, in a cyclical manner, cortical–cytoskeletal components and cytoplasmic contraction.

In addition to contractile processes, the cortical cytoskeleton of fertilized and unfertilized eggs may also be important for other functions. For example, comparison of sea urchin eggs and zygotes indicate a correlation between the activation of protein synthesis and the association of polysomes with the cortical cytoskeleton (Moon *et al.*, 1983).

VII. ENDOCYTOSIS

Immediately following the cortical granule reaction and concomitant with the elongation of microvilli is the development of endocytotic pits and vesicles (Anderson, 1968; Chandler and Heuser, 1979, 1981; Donovan and Hart, 1982; Fisher and Rebhun, 1983; Carron and Longo, 1984; Sardet, 1984). Endocytosis in sea urchin eggs commences as a burst 3- to 5-min postinsemination in which portions of the plasma membrane are taken into the cytoplasm. Whether portions of the original plasmalemma or the cortical granule membrane are preferentially endocytosed has not been determined. In light of observations demonstrating significant changes in the composition of the egg plasma membrane at fertilization, it seems unlikely that discrete patches of membrane persist intact to be selectively endocytosed.

When the cortical granule reaction is inhibited by high pressure, the endocytotic burst that immediately follows is inhibited (Fisher *et al.*, 1985). Interestingly, in such cases, endocytosis occurs much later than observed in untreated zygotes. These results suggest that cortical granule exocytosis is not the only cause of surface transformations involving endocytosis.

That endocytosis follows the cortical granule reaction suggests a mechanism for both surface-area reduction and cell-surface remodeling, which may be relevant to physiological changes characteristic of fertilized eggs. That endocytosis follows the cortical granule reaction is consistent with observations in secretory cells where, after exocytosis, excess membrane may be removed from the cell surface in the form of endocytotic vesicles (Pelletier, 1973; Orci *et al.*, 1973; Kalina and Robinovitch, 1975; Oliver and Hand, 1978). The extent of membrane internalized by endocytosis at fertilization appears to be extensive and persists up to the time of cleavage. Whether pinocytosis remains constant over this period has not been established; however, it has been estimated that about 26,300 μm^2 of surface membrane/*Strongylocentrotus* egg is resorbed by endocytosis during the first 4 min of fertilization (Fisher and Rebhun, 1983). This represents approximately 46% of the membrane presumably added to the egg surface by cortical granule exocytosis. The relationship between cortical granule exocytosis and endocytosis, in terms of the quantity of membrane in flux, is unclear since (1) the rate of membrane interiorization is unknown, (2) the amount of cortical granule membrane added to the zygote surface has not been established, and (3) mechanisms other than endocytosis, which may contribute to the reduction of surface area, have not been eliminated.

Following the appearance of tracer in endocytosis vesicles of fertilized *Arbacia* eggs, label has been observed in lysosomes (Carron and Longo, 1984). This transition indicates that the tracer travels from one cellular compartment to another. That label is localized to lysosomes of zygotes examined up to 60-min postinsemination also suggests that surface membrane may be degraded or modified. Membrane components may then reenter cytoplasmic precursor pools by traversing the lysosomal membrane to be utilized at later stages of embryogenesis (De Duve and Wattiaux, 1966; Holtzman, 1976).

REFERENCES

Afzelius, B. A. (1956). The ultrastructure of the cortical granules and their products in the sea urchin egg as studied with the electron microscope. *Exp. Cell Res.* **10**, 257–285.

Albertini, D. F. (1984). Novel morphological approaches for the study of oocyte maturation. *Biol. Reprod.* **30**, 13–28.

Amherdt, M., Baggiolini, M., Perrelet, A., and Orci, L. (1978). Freeze fracture of membrane fusions in phagocytosing polymorphonuclear leukocytes. *Lab. Invest.* **39**, 398–404.

Anderson, E. (1968). Oocyte differentiation in the sea urchin, *Arbacia punctulata* with particular reference to the origin of cortical granules and their participation in the cortical reaction. *J. Cell Biol.* **37**, 514–539.

Battaglia, D. E., and Gaddum-Rosse, P. (1984). Rat eggs normally exhibit a variety of surface phenomena during fertilization. *Gamete Res.* **10**, 107–118.

Bedford, J. M. (1972). An electron microscopic study of sperm penetration into the rabbit egg after natural mating. *Am. J. Anat.* **133**, 213–254.

Bedford, J. M., and Cooper, G. W. (1978). Membrane fusion events in the fertilization of vertebrate eggs. *Cell Surf. Rev.* **5**, 65–125.

Begg, D. A., Rebhun, L. I., and Hyatt, H. (1982). Structural organization of actin in the sea urchin egg cortex: Microvillar elongation in the absence of actin filament bundle formation. *J. Cell Biol.* **93**, 24–32.

Beisson, J., Lefort-Tran, M., Pouphile, M., Rossignal, M., and Satir, B. (1976). Genetic analysis of membrane differentiation in *Paramecium*. *J. Cell Biol.* **69**, 126–143.

Bluemink, J. G., and Tertoolen, L. G. J. (1978). The plasma membrane IMP pattern as related to animal/vegetal polarity in the amphibian egg. *Dev. Biol.* **62**, 334–343.

Brummett, A., and Dumont, J. (1979). Initial stages of sperm penetration into the egg of *Fundulus heteroclitus*. *J. Exp.Zool.* **210**, 417–434.

Brummett, A., Dumont, J., and Richter, C. (1985). Later stages of sperm penetration and second polar body formation in the egg of *Fundulus heteroclitus*. *J. Exp. Zool.* **234**, 423–439.

Bryan, J., and Kane, R. E. (1982). Actin gelation in sea urchin egg extracts. *Methods Cell Biol.* **25**, 175–199.

Burgess, D. R. (1977). Ultrastructure of meiosis and polar body formation in the egg of the mud snail, *Ilyanassa obsoleta. In* "Cell Shape and Surface Architecture" (J. P. Revel, U. Henning, and F. Fox, eds.), pp. 569–579. Alan R. Liss, New York.

Burgess, D. R., and Schroeder, T. E. (1977). Polarized bundles of actin filaments within microvilli of fertilized sea urchin eggs. *J. Cell Biol.* **74**, 1032–1037.

Campanella, C., and Andreuccetti, P. (1977). Ultrastructural observations on cortical endoplasmic reticulum and on residual cortical granules in the egg of *Xenopus laevis*. *Dev. Biol.* **56**, 1–10.

Campanella, C., Andreuccetti, P., Taddei, C., and Talevi, R. (1984). The modification of cortical endoplasmic reticulum during *in vitro* maturation of *Xenopus laevis* oocytes and its involvement in cortical granule exocytosis. *J. Exp. Zool.* **229**, 283–293.

Campisi, J., and Scandella, C. J. (1978). Fertilization-induced changes in membrane fluidity of sea urchin eggs. *Science* **199**, 1336–1337.

Campisi, J., and Scandella, C. J. (1980a). Bulk membrane fluidity increases after fertilization or partial activation of sea urchin eggs. *J. Biol. Chem.* **255**, 5411–5419.

Campisi, J., and Scandella, C. J. (1980b). Calcium-induced decrease in membrane fluidity of sea urchin egg cortex after fertilization. *Nature (London)* **286**, 185–186.

Capco, D. G., and McGaughey, R. W. (1986). Cytoskeletal reorganization during early mammalian development: Analysis using embedment-free sections. *Dev. Biol.* **115**, 446–458.

Cardasis, C., Schuel, H., and Herman, L. (1978). Ultrastructural localization of calcium in unfertilized sea-urchin eggs. *J. Cell Sci.* **31**, 101–115.

Carron, C. P., and Longo, F. J. (1980). Relationship of intracellular pH and pronuclear development in the sea urchin, *Arbacia punctulata. Dev. Biol.* **79**, 478–487.

Carron, C. P., and Longo, F. J. (1982). Relation of cytoplasmic alkalinization to microvillar elongation and microfilament formation in the sea urchin egg. *Dev. Biol.* **89**, 128–137.

Carron, C. P., and Longo, F. J. (1983). Filipin/sterol complexes in fertilized and unfertilized sea urchin egg membranes. *Dev. Biol.* **99**, 482–488.

Carron, C. P., and Longo, F. J. (1984). Pinocytosis in fertilized sea urchin (*Arbacia punctulata*) eggs. *J. Exp. Zool.* **231**, 413–422.

Cartaud, A., Boyer, J., and Ozon, R. (1984). Calcium sequestering activities of reticulum vesicles from *Xenopus laevis* oocytes. *Exp. Cell Res.* **155**, 565–574.

Chambers, R. (1917). Microdissection studies. II. The cell aster: A reversal gelation phenomenon. *J. Exp. Zool.* **23**, 483–505.

Chandler, D. E., and Heuser, J. (1979). Membrane fusion during secretion: Cortical granule exocytosis in sea urchin eggs as studied by quick-freezing and freeze-fracture. *J. Cell Biol.* **83**, 91–108.

Chandler, D. E., and Heuser, J. (1981). Postfertilization growth of microvilli in the sea urchin egg: New views from eggs that have been quick-frozen, freeze-fractured and deeply etched. *Dev. Biol.* **92**, 393–400.

Chandler, D. R. (1984). Exocytosis *in vitro:* Ultrastructure of the isolated sea urchin egg cortex as seen in platinum replicas. *J. Ultrastruct. Res.* **89**, 198–211.

Charbonneau, M., and Grey, R. D. (1984). The onset of activation responsiveness during maturation coincides with the formation of the cortical endoplasmic reticulum in oocytes of *Xenopus laevis. Dev. Biol.* **102**, 90–97.

Chi, E. Y., Lagunoff, D., and Koehler, J. K. (1975). Electron microscopy of freeze-fractured rat peritoneal mast cells. *J. Ultrastruct. Res.* **57**, 46–54.

Clark, T. G., and Merriam, R. W. (1978). Actin in *Xenopus* oocytes. I. Polymerization and gelation *in vitro. J. Cell Biol.* **77**, 427–438.

Cline, C. A., Schatten, H., Balczon, R., and Schatten, G. (1983). Actin-mediated surface motility during sea urchin fertilization. *Cell Motil.* **13**, 513–524.

Coffe, G., Foucault, G., Soyer, M. O., DeBilly, F., and Pudles, J. (1982). State of actin during the cycle of cohesiveness of the cytoplasm in parthenogenetically activated sea urchin egg. *Exp. Cell Res.* **142**, 365–372.

Colwin, L. H., and Colwin, A. L. (1967). Membrane fusion in relation to sperm–egg association. *In* "Fertilization" (C. B. Metz and A. Monroy, eds.), Vol. 1, pp. 295–367. Academic Press, New York.

Conklin, E. G. (1917). Effects of centrifugal force on the structure and development of the eggs of *Cripedula. J. Exp. Zool.* **22**, 311–419.

Craig, S. W., and Pollard, T. D. (1982). Actin-binding proteins. *Trends Biochem. Sci.* **7**, 55–58.

Dale, B., and Santella, L. (1985). Sperm–oocyte interaction in the sea urchin. *J. Cell Sci.* **74**, 153–167.

Davidson, E. H. (1986). "Gene Activity in Early Development," 3rd ed. Academic Press, New York.

Decker, S. J., and Kinsey, W. H. (1983). Characterization of cortical secretory vesicles from the sea urchin egg. *Dev. Biol.* **96**, 37–45.

De Duve, C., and Wattiaux, R. (1966). Function of lysosomes. *Annu. Rev. Physiol.* **28**, 435–493.

DeFelici, M., and Siracusa, G. (1981). Fertilization-induced changes in concanavalin A binding to mouse eggs. *Exp. Cell Res.* **132**, 41–45.

de Kruijff, B., and Demel, R. A. (1974). Polyene antibiotic–sterol interactions in membranes of *Acholeplasma laidlawii* cells and lecithin lysosomes. III. Molecular structure of the polyene antibiotic-cholesterol complexes. *Biochim. Biophys. Acta* **339**, 57–70.

DeRosier, D. J., and Edds, K. T. (1980). Evidence for fascin cross links between the actin filaments and coelomocyte filopidia. *Exp. Cell Res.* **126**, 490–494.

Detering, N. K., Decker, G. L., Schmell, E. D., and Lennarz, W. J. (1977). Isolation and characterization of plasma membrane-associated cortical granules from sea urchin eggs. *J. Cell Biol.* **75**, 899–914.

Dewel, W. C., and Clark, W. H. (1974). A fine structural investigation of surface specializations and the cortical reaction in eggs of the cnidarian *Bunodosoma cavernata. J. Cell Biol.* **69**, 78–91.

Donovan, M., and Hart, N. H. (1982). Uptake of ferritin by the mosaic egg surface of *Brachydanio. J. Exp. Zool.* **223**, 229–304.

Eager, D. D., Johnson, M. H., and Thurley, K. W. (1976). Ultrastructural studies on the surface membrane of the mouse egg. *J. Cell Sci.,* **22**, 345–368.

Ebensperger, C., and Barros, C. (1984). Changes at the hamster oocyte surface from the germinal vesicle stage of ovulation. *Gamete Res.* **9**, 387–397.

Eddy, E. M., and Shapiro, B. M. (1976). Changes in the topography of the sea urchin egg after fertilization. *J. Cell Biol.* **71**, 35–48.

Elgsaeter, A., and Branton, D. (1974). Intramembrane particle aggregation in erythrocyte ghosts. I. Effects of protein removal. *J. Cell Biol.* **63,** 1018–1030.

Elinson, R. (1980). The amphibian egg cortex in fertilization and early development. *Symp. Soc. Dev. Biol.* **38,** 217–234. Academic Press, New York.

Elinson, R. (1986). Fertilization in amphibians: The ancestry of the block to polyspermy. *Int. Rev. Cytol.* **101,** 59–100.

Elinson, R. P., and Manes, M. E. (1978). Morphology of the site of sperm entry on the frog egg. *Dev. Biol.* **63,** 67–75.

Endo, M. (1977). Calcium release from the sarcoplasmic reticulum. *Physiol. Res.* **57,** 71–108.

Endo, Y. (1961). Changes in the cortical layer of sea urchin eggs at fertilization as studied with the electron microscope. I. *Clypeaster japonicus. Exp. Cell Res.* **25,** 383–397.

Erickson, C. A., and Trinkaus, J. P. (1976). Microvilli and blebs as sources of reserve surface membrane during cell spreading. *Exp. Cell Res.* **99,** 375–384.

Fallon, J. F., and Austin, C. R. (1967). Fine structure of gametes of *Nereis limbata* (Annelida) before and after interaction. *J. Exp. Zool.* **166,** 225–242.

Fisher, G. W., and Rebhun, L. I. (1983). Sea urchin egg cortical granule exocytosis is followed by a burst of membrane retrieval via uptake into coated vesicles. *Dev. Biol.* **99,** 456–472.

Fisher, G. W., Summers, R. G., and Rebhun, L. I. (1982). Cortical transformation in fertilized sea urchin eggs in the absence of cortical exocytosis. *J. Cell Biol.* **95,** 164a.

Fisher, G. W., Summers, R. G., and Rebhun, L. I. (1985). Analysis of sea urchin egg cortical transformation in the absence of cortical granule exocytosis. *Dev. Biol.* **109,** 489–503.

Franklin, L. E. (1965). Morphology of gamete membrane fusion and of sperm entry into oocytes of the sea urchin. *J. Cell Biol.* **25,** 81–100.

Freidus, D. J., Schlegel, R. A., and Williamson, P. (1984). Alteration of lipid organization following fertilization of sea urchin eggs. *Biochim. Biophys. Acta* **803,** 191–196.

Friend, D. S. (1982). Plasma-membrane diversity in a highly polarized cell. *J. Cell Biol.* **93,** 243–249.

Friend, D. S., and Fawcett, D. W. (1974). Membrane differentiations in freeze-fractured mammalian sperm. *J. Cell Biol.* **63,** 641–664.

Frye, L. D., and Edidin, M. (1970). The rapid intermixing of cell surface antigens after formation of mouse-human heterokaryons. *J. Cell Sci.* **7,** 319–335.

Gabel, C. A., Eddy, E. M., and Shapiro, B. M. (1979). After fertilization, sperm surface components remain as a patch in sea urchin and mouse embryos. *Cell (Cambridge, Mass.)* **18,** 207–215.

Gaddum-Rosse, P. (1985). Mammalian gamete interactions: What can be gained from observations on living eggs? *Am. J. Anat.* **174,** 347–356.

Gaddum-Rosse, P., Blandau, R. J., Langley, L. B., and Battaglia, D. E. (1984). *In vitro* fertilization in the rat: Observations on living eggs. *Fertil. Steril.* **42,** 285–292.

Gardiner, D. M., and Grey, R. D. (1983). Membrane junctions in *Xenopus* eggs: Their distribution suggests a role in calcium regulation. *J. Cell Biol.* **96,** 1159–1163.

Gilkey, J. C., Jaffe, L. F., Ridgeway, E. G., and Reynolds, G. T. (1978). A free calcium wave traverses the activating egg of the medaka, *Oryzias latipes. J. Cell Biol.* **76,** 448–466.

Gould-Somero, M., Holland, L., and Paul, M. (1977). Cytochalasin A inhibits sperm penetration into eggs of *Urechis caupo* (Echira). *Dev. Biol.* **58,** 11–22.

Gulyas, B. J. (1980). Cortical granules of mammalian eggs. *Int. Rev. Cytol.* **63,** 357–392.

Gundersen, G. G., and Shapiro, B. M. (1984). Sperm surface proteins persist after fertilization. *J. Cell Biol.* **99,** 1343–1353.

Gundersen, G. G., Gabel, C. A., and Shapiro, B. M. (1982). An intermediate state of fertilization involved in internalization of sperm components. *Dev. Biol.* **93**, 59–72.

Gundersen, G. G., Medill, L., and Shapiro, B. M. (1986). Sperm surface proteins are incorporated into the egg membrane and cytoplasm after fertilization. *Dev. Biol.* **113**, 207–217.

Hamaguchi, Y., Lutz, D. A., and Inoué, S. (1983). Cortical differentiation, asymmetric positioning and attachment of the meiotic spindle in *Chaetopterus pergamentaceous* oocytes. *J. Cell Biol.* **97**, 254a.

Hart, N., and Donovan, M. (1983). Fine structure of the chorion and site of sperm entry in the egg of *Brachydanio. J. Exp. Zool.* **227**, 277–296.

Holland, N. D. (1979). Electron microscopic study of the cortical reaction of an ophiuroid echinoderm. *Tissue Cell* **11**, 445–455.

Holland, N. D. (1980). Electron microscopic study of the cortical reaction in eggs of the starfish *(Patiria miniata). Cell Tissue Res.* **205**, 67–76.

Holtzman, E. (1976). "Lysosomes: A Survey." Springer-Verlag, Berlin and New York.

Hope, J., Humphries, A. A., and Bourne, G. H. (1963). Ultrastructural studies on developing oocytes of the salamander *(Triturus viridescens).* I. The relationships between follicle cells and developing oocytes. *J. Ultrastruct. Res.* **9**, 302–324.

Hosoya, H., Mabuchi, I., and Sakai, H. (1982). Actin modulating proteins in the sea urchin egg. I. Analysis of G-actin-binding proteins by DNase I-affinity chromatography and purification of a 17,000 molecular weight component. *J. Biochem. (Tokyo)* **92**, 1853–1862.

Humphreys, W. J. (1967). The fine structure of cortical granules in eggs and gastrulae of *Mytilus edulis. J. Ultrastruct. Res.* **17**, 314–326.

Hylander, B. L., and Summers, R. G. (1981). The effect of local anesthetics and ammonia on cortical granule-plasma membrane attachment in the sea urchin egg. *Dev. Biol.* **86**, 1–11.

Iwamatsu, T., and Ohta, T. (1976). Breakdown of the cortical alveoli of medaka eggs at the time of fertilization with a particular reference to the possible role of spherical bodies in the alveoli. *Wilhelm Roux's Arch. Dev. Biol.* **180**, 297–309.

Iwamatsu, T., and Ohta, T. (1978). Electron microscopic observations on sperm penetration and pronuclear formation in the fish egg. *J. Exp. Zool.* **205**, 157–180.

Jacobson, K., O'Dell, T. D., and August, T. (1984). Lateral diffusion of an 80,000-dalton glycoprotein in the plasma membrane of murine fibroblasts: Relationships to cell structure and function. *J. Cell Biol.* **99**, 1624–1633.

Jaffe, L. F. (1985). The role of calcium explosions, waves, and pulses in activating eggs. *In* "Biology of Fertilization" (C. B. Metz and A. Monroy, eds.), Vol. 3, pp. 128–165. Academic Press, New York.

Johnson, M. H., Eager, D., and Muggleton-Harris, A. (1975). Mosaicism in organization of concanavalin A receptors on surface membrane of mouse egg. *Nature (London)* **25**, 321–322.

Kalina, M., and Robinovitch, R. (1975). Exocytosis couples to endocytosis of ferritin in parotid acinar cells from isoprenalin stimulated rats. *Cell Tissue Res.* **163**, 373–382.

Karasiewicz, J., and Soltynska, M. S. (1985). Ultrastructural evidence of the presence of actin filaments in mouse eggs at fertilization. *Wilhelm Roux's Arch. Dev. Biol.* **194**, 369–372.

Karsenti, E., Bornens, M., and Avrameas, S. (1977). Control of density and microredistribution of concanavalin-A receptors in rat thymocytes at 4°C. *Eur. J. Biochem.* **75**, 251–256.

Kay, E. S., and Shapiro, B. M. (1985). The formation of the fertilization membrane of the sea urchin egg. *In* "Biology of Fertilization" (C. B. Metz and A. Monroy, eds.), Vol. 3, pp. 45–80. Academic Press, New York.

Kemp, N. E., and Istock, N. L. (1967). Cortical changes in growing oocytes and in fertilized or pricked eggs of *Rana pipiens*. *J. Cell Biol.* **34**, 111–122.

Kidd, P. (1978). The jelly and vitelline coats of the sea urchin egg: New ultrastructural features. *J. Ultrastruct. Res.* **64**, 204–215.

Kinsey, W. H., and Koehler, J. K. (1978). Cell surface changes associated with *in vitro* capacitation of hamster sperm. *J. Ultrastruct. Res.* **64**, 1–13.

Kobayashi, W., and Yamamoto, T. (1981). Fine structure of the micropylar apparatus of the chum salmon egg, with a discussion of the mechanism for blocking polyspermy. *J. Exp. Zool.* **217**, 265–275.

Koehler, J. K., DeCurtis, I., Stenchever, M. A., and Smith, D. (1982). Interaction of human sperm with zona free hamster eggs: A freeze-fracture study. *Gamete Res.* **6**, 371–386.

Koehler, J. K., Clark, J. M., and Smith, D. (1985). Freeze-fracture observations on mammalian oocytes. *Am. J. Anat.* **174**, 317–330.

Kudo, S. (1980). Sperm penetration and the formation of a fertilization cone in the common carp egg. *Dev., Growth Differ.* **22**, 403–414.

Kudo, S., and Sato, A. (1985). Fertilization cone of carp eggs as revealed by scanning electron microscopy. *Dev., Growth Differ.* **27**, 121–128.

Lawson, D., Raff, M. R., Gomperts, B., Fewtrell, C., and Gilula, N. G. (1977). Molecular events during membrane fusion: A study of exocytosis in rat peritoneal mast cells. *J. Cell Biol.* **72**, 242–259.

Longo, F. J. (1972). The effects of cytochalasin B on the events of fertilization in the surf clam. *Spisula solidissima*. I. Polar body formation. *J. Exp. Zool.* **182**, 321–344.

Longo, F. J. (1973). Fertilization: A comparative ultrastructural review. *Biol. Reprod.* **9**, 149–215.

Longo, F. J. (1976a). Cortical changes in *Spisula* eggs upon insemination. *J. Ultrastruct. Res.* **56**, 226–232.

Longo, F. J. (1976b). Ultrastructural aspects of fertilization in spiralian eggs. *Am. Zool.* **16**, 375–394.

Longo, F. J. (1978). Effects of cytochalasin B on sperm–egg interactions. *Dev. Biol.* **67**, 259–265.

Longo, F. J. (1980). Organization of microfilaments in sea urchin (*Arbacia punctulata*) eggs at fertilization: Effects of cytochalasin B. *Dev. Biol.* **74**, 422–433.

Longo, F. J. (1981). Morphological features of the surface of the sea urchin (*Arbacia punctulata*) egg. Oolemma–cortical granule association. *Dev. Biol.* **84**, 173–182.

Longo, F. J. (1982). Integration of sperm and egg plasma membrane components at fertilization. *Dev. Biol.* **89**, 409–416.

Longo, F. J. (1985). Fine structure of the mammalian egg cortex. *Am. J. Anat.* **174**, 303–315.

Longo, F. J. (1986a). Surface changes at fertilization: Integration of sea urchin (*Arbacia punctulata*) sperm and oocyte plasma membranes. *Dev. Biol.* **116**, 143–159.

Longo, F. J. (1986b). Fertilization cones of inseminated sea urchin (*Arbacia punctulata*) oocytes: Development of an asymmetry in plasma membrane topography. *Gamete Res.* **15**, 137–151.

Longo, F. J., and Anderson, E. (1968). The fine structure of pronuclear development and fusion in the sea urchin, *Arbacia punctulata*. *J. Cell Biol.* **39**, 339–368.

Longo, F. J., and Anderson, E. (1970). A cytological study of the relation of the cortical reaction to subsequent events of fertilization in urethane-treated eggs of the sea urchin, *Arbacia punctulata*. *J. Cell Biol.* **47**, 646–665.

Longo, F. J., and Chen, D. Y. (1984). Development of surface polarity in mouse eggs. *Scanning Electron Microsc.* pp. 703–716.

Longo, F. J., and Chen, D.-Y. (1985). Development of cortical polarity in mouse eggs: Involvement of the meiotic apparatus. *Dev. Biol.* **107**, 382–394.

Longo, F. J., So, F., and Schuetz, A. W. (1982). Meiotic maturation and the cortical granule reaction in starfish eggs. *Biol. Bull. (Woods Hole, Mass.)* **163**, 465–476.

Longo, F. J., Lynn, J. W., McCulloh, D. H., and Chambers, E. L. (1986). Correlative ultrastructural and electrophysiological studies of sperm–egg interactions of the sea urchin, *Lytechinus variagatus. Dev. Biol.* **118**, 155–166.

Luttmer, S., and Longo, F. J. (1985). Ultrastructural and morphometric observations of cortical endoplasmic reticulum in *Arbacia, Spisula* and mouse eggs. *Dev., Growth Differ.* **27**, 349–359.

Mabuchi, I. (1981). Purification from starfish eggs of a protein that depolymerizes actin. *J. Biochem. (Tokyo)* **89**, 1341–1344.

Mabuchi, I., and Nonomura, Y. (1981). Formation of actin paracrystals from sea urchin egg extract under actin polymerizing conditions. *Biomed. Res.* **2**, 143–153.

Mabuchi, I., Hamaguchi, Y., Kobayashi, T., Hosoya, H., Tsukita, S., and Tsukita, S. (1985). Alpha-actinin from sea urchin eggs: Biochemical properties, interaction with actin, and distribution in the cell during fertilization and cleavage. *J. Cell Biol.* **100**, 375–383.

Maro, B., Johnson, M. H., Pickering, S. J., and Flach, G. (1984). Changes in actin distribution during fertilization of the mouse egg. *J. Embryol. Exp. Morphol.* **81**, 211–237.

Maro, B., Howlett, S. H., and Johnson, M. H. (1986a). Cell and molecular interpretation of mouse early development: The first cell cycle. *In* "Gametogenesis and the Early Embryo" (J. G. Gall, ed.), pp. 389–407. Alan R. Liss, New York.

Maro, B., Johnson, M. H., Webb, M., and Flach, G. (1986b). Mechanism of polar body formation in the mouse oocyte: An interaction between the chromosomes, the cytoskeleton and the plasma membrane. *J. Embryol. Exp. Morphol.* **92**, 11–32.

Marshall, J. D., and Heiniger, H. J. (1979). High affinity concanavalin A binding to sterol-depleted cells. *J. Cell. Physiol.* **100**, 539–550.

Martin, B. R. (1983). Hormone receptors and the adenylate cyclase system: Historical overview. *Curr. Top. Membr. Transp.* **18**, 3–19.

Millonig, G. (1969). Fine structure analysis of the cortical reaction in the sea urchin egg: After normal fertilization and after electric induction. *J. Submicrosc. Cytol.* **1**, 69–84.

Moon, R. I., Nicosia, R. F., Olsen, C., Hille, M. G., and Jeffery, W. R. (1983). The cytoskeletal framework of sea urchin eggs and embryos: Developmental changes in the association of messenger RNA. *Dev. Biol.* **95**, 447–458.

Nicosia, S. V., Wolf, D. P., and Inoue, M. (1977). Cortical granule distribution and cell surface characterization in mouse eggs. *Dev. Biol.* **57**, 56–74.

Nicosia, S. V., Wolf, D. P., and Mastroianni, L. (1978). Surface topography of mouse eggs before and after insemination. *Gamete Res.* **1**, 145–155.

O'Dell, D. S., Ortolani, G., and Monroy, A. (1973). Increased binding of radioactive con A during maturation of ascidian eggs. *Exp. Cell Res.* **83**, 408–411.

Ohta, R., and Iwamatsu, T. (1983). Electron microscopic observations on sperm entry into eggs of the rose bitterling, *Rhodeus ocellatus. J. Exp. Zool.* **227**, 109–119.

Okada, A., Yanagimachi, R., and Yanagimachi, H. (1986). Development of a cortical granule-free area of cortex and the perivitelline space in the hamster oocyte during maturation and following ovulation. *J. Submicrosc. Cytol.* **18**, 233–247.

Oliver, C., and Hand, A. R. (1978). Uptake and fate of luminally administered horseradish peroxidase in resting and isoproterenol-stimulated rat parotid acinar cells. *J. Cell Biol.* **76**, 207–220.

O'Rand, M. G. (1977). The presence of sperm-specific surface isoantigens on the egg following fertilization. *J. Exp. Zool.* **212**, 267–273.

Orci, L., and Perrelet, A. (1977). Morphology of membrane systems in pancreatic islets. *In* "The Diabetic Pancreas" (B. W. Volk and K. F. Wellman, eds.), pp. 171–210. Plenum, New York.

Orci, L., Malaisse-Lage, F., Ravazzola, M., Amherdt, M., and Reynold, A. E. (1973). Exocytosis–endocytosis coupling in pancreatic beta cell. *Science* **181**, 561–562.

Orci, L., Perrelet, A., and Friend, D. S. (1977). Freeze-fracture of membrane fusions during exocytosis in pancreatic β-Cells. *J. Cell Biol.* **75**, 23–30.

Otto, J. J., and Schroeder, T. S. (1984). Assembly–disassembly of actin bundles in starfish oocytes: An analysis of actin-associated proteins in the isolated cortex. *Dev. Biol.* **101**, 263–273.

Otto, J. J., Kane, R. E., and Bryan, J. (1980). Redistribution of actin and fascin in sea urchin eggs after fertilization. *Cell Motil.* **1**, 31–40.

Pardee, A. B., De Asua, J., and Rozengurt, E. (1974). Functional membrane changes and cell growth significance and mechanism. *In* "Control of Proliferation in Animal Cells" (B. Clarkson and R. Baserga, eds.), pp. 547–561. Cold Spring Harbor Lab., Cold Spring Harbor, New York.

Pasteels, J. J. (1965). Etude au microscope électronique de la réaction corticale. *J. Embryol. p. Morphol.* **13**, 327–339.

Pasteels, J. J. (1966). La réaction cortical de fécondation de l'oeuf de *Nereis diversicolor*, édutiée au microscope électronique. *Acta Embryol. Morphol. Exp.* **9**, 155–163.

Pasteels, J. J., and de Harven, E. (1962). Etude au microscope életronique du cortex de l'oeuf de *Barnea candida* (Mollusque Bivalve), et son évolution au moment de la fécondation, de la maturation, et de la segmentation. *Arch. Biol.* **73**, 465–490.

Paul, M. (1975). Release of acid and changes in light-scattering properties following fertilization of *Urechis caupo* eggs. *Dev. Biol.* **43**, 299–312.

Peaucellier, G., Guerrier, P., and Bergerard, J. (1974). Effects of cytochalasin B on meiosis and development of fertilized and activated eggs of *Sabellaria alveolata* (Polychaete Annelid). *J. Embryol. Exp. Morphol.* **31**, 61–74.

Pelletier, G. (1973). Secretion and uptake of peroxidase by rat adenophypophyseal cells. *J. Ultrastruct. Res.* **43**, 445–459.

Phillips, D. M., and Shalgi, R. (1980). Surface architecture of the mouse and hamster zona pellucida and oocyte. *J. Ultrastruct. Res.* **72**, 1–12.

Phillips, D. M., Shalgi, R., and Dekel, N. (1985). Mammalian fertilization as seen with the scanning electron microscope. *Am. J. Anat.* **174**, 357–372.

Picheral, B. (1977). La fécondation chez le triton Pleurodèle. II. La pénétration des spermatozoïdes et la réaction locale de l'oeuf. *J. Ultrastruct. Res.* **60**, 181–202.

Picheral, B., and Charbonneau, M. (1982). Anuran fertilization: A morphological reinvestigation of some early events. *J. Ultrastruct. Res.* **81**, 306–321.

Pikó, L. (1969). Gamete structure and sperm entry in mammals. *In* "Fertilization" (C. B. Metz and A. Monroy, eds.), Vol. 2, pp. 325–403. Academic Press, New York.

Pollack, E. G. (1978). Fine structural analysis of animal cell surfaces: Membranes and cell surface topography. *Am. Zool.* **18**, 25–69.

Rebhun, L. I. (1962). Electron microscope studies on the vitelline membrane of the surf clam, *Spisula solidissima*. *J. Ultrastruct. Res.* **6**, 107–122.

Reima, I., and Lehtonen, E. (1985). Localization of nonerythroid spectrin and actin in mouse oocytes and preimplantation embryos. *Differentiation* **30**, 68–75.

Rosati, F., Monroy, A., and de Prisco, P. (1977). Fine structural study of fertilization in the ascidian, *Ciona intestinalis*. *J. Ultrastruct. Res.* **58**, 261–270.

Rovensky, Y. A., and Vasiliev, J. M. (1984). Surface topography of suspended tissue cells. *Int. Rev. Cytol.* **90**, 372–307.

Sardet, C. (1984). The ultrastructure of the sea urchin egg cortex isolated before and after fertilization. *Dev. Biol.* **105**, 196–210.

Satir, B. (1976). Genetic control of membrane mosaicism. *J. Supramol. Struct.* **5**, 381–389.

Satir, B., Schooley, C., and Satir, P. (1973). Membrane fusion in a model system. *J. Cell Biol.* **56**, 153–176.

Schatten, G., and Mazia, D. (1976). The surface events at fertilization: the movements of the spermatozoan through the sea urchin egg surface and the roles of the surface layers. *J. Supramol. Struct.* **5**, 343–369.

Schatten, G., Schatten, H., Spector, I., Cline, C., Paweletz, N., Simerly, C., and Petzelt, C. (1986). Latrunculin inhibits the microfilament-mediated processes during fertilization, cleavage and early development in sea urchins and mice. *Exp. Cell Res.* **116**, 191–208.

Schatten, H., and Schatten, G. (1980). Surface activity at the egg plasma membrane during sperm incorporation and its cytochalasin B sensitivity. *Dev. Biol.* **78**, 435–449.

Schliwa, M. (1981). Proteins associated with cytoplasmic actin. *Cell (Cambridge, Mass.)* **25**, 587–590.

Schmidt, T., and Epel, D. (1983). High hydrostatic pressure and the dissection of fertilization responses. I. The relationship between cortical granule exocytosis and protein efflux during fertilization of the sea urchin egg. *Exp. Cell Res.* **146**, 235–248.

Schroeder, T. E. (1979). Surface area change at fertilization: Resorption of the mosaic membrane. *Dev. Biol.* **70**, 306–326.

Schroeder, T. E. (1981). Interrelations between the cell surface and the cytoskeleton in cleaving sea urchin eggs. *In* "Cytoskeletal Elements and Plasma Membrane Organization" (G. Poste and G. L. Nicolson, eds.), pp. 170–216. Elsevier/North-Holland Biomedical Press, New York.

Schuel, H. (1985). Functions of egg cortical granules. *In* "Biology of Fertilization" (C. B. Metz and A. Monroy, eds.), Vol. 3, pp. 1–43. Academic Press, New York.

Seifriz, W. (1926). Protoplasmic papillae of *Echinarachnus* oocytes. *Protoplasma* **1**, 1–14.

Shalgi, R., Phillips, D. M., and Kraicer, P. F. (1978). Observations on the incorporation cone in the rat. *Gamete Res.* **1**, 27–37.

Shapiro, B. M., and Eddy, E. M. (1980). When sperm meets egg: Biochemical mechanisms of gamete interaction. *Int. Rev. Cytol.* **66**, 257–302.

Shapiro, B. M., Schackmann, R. W., Gabel, C. A., Foerder, C. A., Farance, M. L., Eddy, E. M., and Klebanoff, S. J. (1980). Molecular alterations in gamete surfaces during fertilization and early development. *Symp. Soc. Dev. Biol.* **38**, 257–302.

Shapiro, B. M., Schackmann, R. W., and Gabel, C. A. (1981). Molecular approaches to the study of fertilization. *Annu. Rev. Biochem.* **50**, 815–843.

Shapiro, B. M., Gundersen, G. G., Gabel, C. A., and Eddy, E. M. (1982). Fate of sperm surface components in the embryo. *In* "Cellular Recognition," (W. A. Frazier, L. Glazer, and D. I. Gottleib, eds.) pp. 833–844. Alan R. Liss, New York.

Sheetz, M. P., Schindler, M., and Koppel, D. E. (1980). Lateral mobility of integral membrane proteins is increased in spherocytic erythrocytes. *Nature (London)* **285**, 510–512.

Shen, S. S. (1983). Membrane properties and intracellular ion activities of marine invertebrate eggs and their changes during activation. *In* "Mechanism and Control of Animal Fertilization" (J. F. Hartmann, ed.), pp. 213–267. Academic Press, New York.

Shimizu, T. (1981a). Cortical differentiation of the animal pole during maturation division in fertilized eggs of *Tubifex* (Annelida, Oligochaeta). I. Meiotic apparatus formation. *Dev. Biol.* **85**, 65–76.

Shimizu, T. (1981b). Cortical differentiation of the animal pole during maturation division in fertilized eggs of Tubifex (Annelida, Oligochaeta). II. Polar body formation. *Dev. Biol.* **85**, 77–88.

Singer, S. J., and Nicolson, G. L. (1972). The fluid model of the structure of cell membranes. *Science* **178**, 720–731.

Spudich, A., and Spudich, J. A. (1979). Actin in triton-treated cortical preparations of unfertilized and fertilized sea urchin eggs. *J. Cell Biol.* **82**, 212–226.

Spudich, J. A., and Amos, L. A. (1979). Structure of actin filament bundles from microvilli of sea urchin eggs. *J. Mol. Biol.* **129**, 319–331.

Steinhardt, R. A., and Epel, D. (1974). Activation of sea urchin eggs by calcium ionophore. *Proc. Natl. Acad. Sci. U.S.A.* **71**, 1915–1919.

Suzuki, F., and Yanagimachi, R. (1983). Freeze-fracture observations of ovulated hamster oocytes with their cumulus cells. *Cell Tissue Res.* **231**, 365–374.

Swift, J. G., and Murkherjee, T. M. (1978). Membrane changes associated with mucus production in intestinal goblet cells. *J. Cell Sci.* **33**, 301–316.

Szollosi, D. (1967). Development of cortical granules and the cortical granule reaction in rat and hamster eggs. *Anat. Rec.* **159**, 431–446.

Tank, D. W., Wu, E. S., and Webb, W. W. (1982). Enhanced molecular diffusibility in muscle membrane blebs: Release of lateral constraints. *J. Cell Biol.* **92**, 207–212.

Theodosis, D. T., Dreifuss, J. J., Jacques, J., and Orci, L. (1978). A freeze fracture study of membrane events during neurophyophysis secretion. *J. Cell Biol.* **78**, 542–553.

Tilney, L. G., and Jaffe, L. A. (1980). Actin, microvilli and the fertilization cone of sea urchin eggs. *J. Cell Biol.* **87**, 771–782.

Tilney, L. G., and Kallenbach, N. (1979). Polymerization of actin. VI. The polarity of the actin filaments in the acrosomal process and how it might be determined. *J. Cell Biol.* **81**, 608–623.

Träuble, H., and Sackermann, E. (1972). Studies of the crystalline–liquid crystalline phase transition of lipid model membranes. III. Structure of a steroid–lecithin system below and above the lipid phase transition. *J. Am. Chem. Soc.* **94**, 4499–4510.

Trinkaus, J. P. (1980). Formation of protrusions of the cell surface during cell movement. *Prog. Clin. Biol. Res.* **44**, 887–906.

Vacquier, V. D. (1975). The isolation of intact cortical granules from sea urchin eggs: Calcium ions trigger granule discharge. *Dev. Biol.* **43**, 62–74.

Vacquier, V. D. (1981). Dynamic changes of the egg cortex. *Dev. Biol.* **84**, 1–26.

Vacquier, V. D., and Moy, G. W. (1977). Isolation of bindin: The protein responsible for adhesion of sperm to sea urchin eggs. *Proc. Natl. Acad. Sci. U.S.A.* **74**, 2456–2460.

Van Blerkom, J., and Bell, H. (1986). Regulation of development in the fully grown mouse oocyte: Chromosome-mediated temporal and spatial differentiation of the cytoplasm and plasma membrane. *J. Embryol. Exp. Morphol.* **93**, 213–238.

Veron, M., and Shapiro, B. M. (1977). Binding of concanavalin A to the surface of sea urchin eggs and its alteration upon fertilization. *J. Biol. Chem.* **252**, 1286–1292.

Waksmundzka, M., Krysiak, E., Karasiewicz, J., Czolowska, R., and Tarkowski, A. K. (1984). Autonomous cortical activity in mouse eggs controlled by a cytoplasmic clock. *J. Embryol. Exp. Morphol.* **79**, 77–96.

Wassarman, P. M., Josetowicz, W. J., and Letourneau, G. Z. (1976). Meiotic maturation of mouse oocytes *in vitro*: Inhibition of maturation at specific stages of nuclear progression. *J. Cell Sci.* **22**, 531–545.

Weiss, R. L., Goodenough, D. A., and Goodenough, U. W. (1977a). Membrane differentiations at sites specialized for cell fusion. *J. Cell Biol.* **72**, 144–160.

Weiss, R. L., Goodenough, D. A., and Goodenough, U. W. (1977b). Membrane particle arrays associated with the basal body and with contractile vacuole secretion in *Chlamydomonas*. *J. Cell Biol.* **72**, 144–160.

Whitaker, M. J., and Steinhardt, R. A. (1985). Ionic signaling in the sea urchin egg at fertilization. *In* "Biology of Fertilization" (C. B. Metz and A. Monroy, eds.), Vol. 3, pp. 168–211. Academic Press, New York.

Wolf, D. E., and Ziomek, C. A. (1983). Regionalization and lateral diffusion of membrane proteins in unfertilized and fertilized mouse eggs. *J. Cell Biol.* **96,** 1786–1790.

Wolf, D. E., Kinsey, W., Lennarz, W., and Edidin, M. (1981a). Changes in the organization of the sea urchin egg plasma membrane upon fertilization: Indications from the lateral diffusion rates of lipid-soluble fluorescent dyes. *Dev. Biol.* **81,** 133–138.

Wolf, D. E., Edidin, M., and Handyside, A. H. (1981b). Changes in the organization of the mouse egg plasma membrane upon fertilization and first cleavage: Indications from the lateral diffusion rates of fluorescent lipid analogs. *Dev. Biol.* **85,** 195–198.

Wolpert, L., and Mercer, E. H. (1961). An electron microscope study of fertilization of the sea urchin egg *Psammechinus milliaris*. *Exp. Cell Res.* **22,** 45–55.

Wu, E. S., Tank, D. W., and Webb, W. W. (1982). Unconstrained lateral diffusion of concanavalin A receptors of bulbous lymphocytes. *Proc. Natl. Acad. Sci. U.S.A.* **79,** 4962–4966.

Yahara, I., Harada, F., Sekita, S., Yoshihira, K., and Natori, S. (1982). Correlation between effects of 24 different cytochalasins on cellular structures and cellular events and those on actin *in vitro*. *J. Cell Biol.* **92,** 69–78.

Yamamoto, K., and Yoneda, M. (1983). Cytoplasmic cycle in meiotic division of starfish oocytes. *Dev. Biol.* **96,** 166–172.

Yanagimachi, R., and Nicolson, G. L. (1976). Lectin-binding properties of hamster egg zona pellucida and plasma membrane during maturation and preimplantation development. *Exp. Cell Res.* **100,** 249–257.

Yanagimachi, R., and Noda, Y. D. (1970). Electron microscope studies of sperm incorporation into the golden hamster egg. *Am. J. Anat.* **128,** 429–462.

Yanagimachi, R., Nicolson, G. L., Noda, Y. D., and Fujimoto, M. (1973). Electron microscopic observations of the distribution of acidic anionic residues on hamster spermatozoa and eggs before and during fertilization. *J. Ultrastruct. Res.* **43,** 344–353.

Yoneda, M., Ikeda, M., and Washitani, S. (1978). Periodic change in the tension at the surface of activated non-nucleate fragments of sea-urchin eggs. *Dev., Growth Differ.* **20,** 329–330.

Yonemura, S., and Kinoshita, S. (1986). Actin filament organization in the sand dollar egg cortex. *Dev. Biol.* **115,** 171–183.

Zamboni, L. (1971). "Fine Morphology of Mammalian Fertilization." Harper & Row, New York.

Zimmerberg, J., Sardet, C., and Epel, D. (1985). Exocytosis of sea urchin egg cortical vesicles *in vitro* is retarded by hyperosmotic sucrose: Kinetics of fusion monitored by quantitative light-scattering microscopy. *J. Cell Biol.* **101,** 2398–2410.

Zimmerman, A. M., and Forer, A., eds. (1981). "Mitosis/Cytokinesis." Academic Press, New York.

Zucker, R. S., and Steinhardt, R. A. (1978). Prevention of the cortical reaction in fertilized sea urchin eggs by injection of calcium-chelating ligands. *Biochim. Biophys. Acta* **541,** 459–466.

Zucker, R. S., Steinhardt, R. A., and Winkler, M. M. (1978). Intracellular calcium release and the mechanisms of parthenogenetic activation of the sea urchin egg. *Dev. Biol.* **65,** 285–295.

6

Cytoplasmic Microtubule-Associated Motors

J. M. SCHOLEY,[*,†] **M. E. PORTER,**[*] **R. J. LYE,**[*] **AND J. R. McINTOSH**[*]

*Department of Molecular, Cellular, and Developmental Biology
University of Colorado at Boulder
Boulder, Colorado 80309

†Department of Molecular and Cellular Biology
National Jewish Center for Immunology and Respiratory Medicine
Denver, Colorado 80206

I. INTRODUCTION[1]

A number of intracellular motile processes are dependent on microtubules (MTs) (Schliwa, 1984; McIntosh, 1984), and many such MT-based movements occur during fertilization in animal cells (Schatten, 1982). In echinoderms, for

[1]Abbreviations: AMPPNP, 5'-Adenylyl imidodiphosphate; ATP-γ-S, adenosine 5'-O-(3-thiotriphosphate); BSA, bovine serum-albumin; DIC, differential interference contrast; DMSO, dimethyl sulfoxide; DTT, dithiothreitol; EDTA, ethylenediaminetetraacetic acid; EGTA, ethylene glycol bis(βaminoethyl ether)-N,N,N',N'-tetraacetic acid; HM_r, high relative molecular weight; 8-azido-ATP, 8-azidoadenosine 5'-triphosphate; MAPs, microtubule-associated proteins; M_r, relative molecular weight; MT, microtubule; NEM, N-ethylmaleimide; p130, 130-kDa polypeptide; PAGE, polyacrylamide gel electrophoresis; P_i, inorganic phosphate; PMSF, phenylmethyl sulfonyl fluoride; RT, room temperature; SBTI, soybean trypsin inhibitor; SDS, sodium dodecyl sulfate; UV, ultraviolet; valap, vaseline : lanolin : paraffin (1 : 1 : 1).

139

example, which are a favorite subject for studying fertilization, cell division, and embryogenesis (Schatten, 1982; Schroeder, 1986), microtubules are believed to drive (1) sperm flagellar motility; (2) pronuclear migration leading to syngamy; (3) movements associated with the mitotic spindle [such as chromosome congression to the metaphase plate, chromosome to pole movement (anaphase A), and spindle elongation (anaphase B)]; plus (4) organelle, vesicle, and particle transport. It is likely that certain proteins that interact with MTs serve as motors to generate forces for these movements, but none, with the exception of dynein in sperm motility (Gibbons, 1981), has yet been unequivocally identified. Certainly, our understanding of the mechanisms of cytoplasmic MT-based movements would be improved if the identity of the mechanochemical factors that generate the relevant motile forces were known.

Here we review our studies, which have been aimed at identifying mechanochemical proteins that might generate force for cytoplasmic MT-based movements. We have worked with cells from two different animal types. Sea urchins have been used because their gametes and early embryos exhibit a number of MT-based movements; they are therefore likely to represent a rich source of MT-based motors. We have also worked with the nematode, *Caenorhabditis elegans,* because this animal does not possess motile cilia, and it therefore represents a promising system for looking for cytoplasmic motors without risk of contamination by flagellar dynein. The strategy we have used to identify MT-based motors is based on our knowledge of the behavior of two better-characterized mechanoenzymes, myosin and dynein (Johnson, 1985). These proteins both interact with elements of the cytoskeleton and use ATP to generate motile force. Myosin and dynein bind to actin or MTs in a nucleotide-sensitive fashion, such that actomyosin and dynein microtubule complexes form in the absence of nucleotide, but both complexes are dissociated by ATP. Their cycles of attachment, force generation, and detachment are driven by ATP hydrolysis (Johnson, 1985). Recently, a novel MT-based mechanoenzyme called kinesin has been identified in squid axoplasm (Vale *et al.,* 1986). This protein also uses ATP to generate force and shows ATP-sensitive binding to MTs, so it fits into the general pattern defined by myosin and dynein. To identify MT-associated motors from sea urchin eggs and from *C. elegans,* we followed the same paradigm and have sought MT-associated proteins that interact with MTs in a nucleotide (or nucleotide analog)-sensitive manner. This approach has allowed us to identify several candidates for cytoplasmic, MT-based motors in these organisms, as described below.

II. PREPARATION OF SEA URCHIN EGG MICROTUBULES

Unfertilized sea urchin eggs (Schroeder, 1986) stockpile large quantities of mitotic spindle proteins, sufficient to support the multiple divisions, which occur during early embryogenesis; furthermore, it is easy to obtain large quantities of these cells (up to 100 ml packed eggs). The unfertilized sea urchin egg therefore represents a promising source of biochemically useful quantities of spindle proteins. In collaboration with E. D. Salmon, we set out to isolate and analyze these proteins by using taxol to prepare MTs (Vallee, 1982) from unfertilized egg cytosolic extracts (Scholey *et al.*, 1984, 1985; Dinenberg *et al.*, 1986; see also Vallee and Bloom, 1983; Bloom *et al.*, 1985). Briefly, unfertilized sea urchin eggs collected from 25–50 animals are pooled, dejellied, and washed, and then homogenized in "extraction buffer." The homogenate is centrifuged to yield a clear extract supernatant. Addition of 20 μM taxol plus 1 mM GTP to this solution promotes MT assembly, and the resulting polymer is collected by centrifugation into a pellet, then washed with a further centrifugation in "extraction buffer" containing taxol and GTP (see Fig. 1 for details). We consistently obtain 2–5 mg MT protein from 10 ml packed, dejellied eggs. Microtubules have been prepared using taxol from a variety of species, including *Strongylocentrotus purpuratus*, *Strongylocentrotus droebachiensis*, *Strongylocentrotus franciscanus*, *Lytechinus pictus* and *Lytechinus variegatus* (Scholey *et al.*, 1984; J. M. Scholey and R. J. Leslie, unpublished; Vallee and Bloom, 1983; Bloom *et al.*, 1985). When viewed in the electron microscope, negatively stained sea urchin egg cytoplasmic MTs are covered in projections, which may correspond to microtubule-associated proteins (MAPs) (Vallee and Bloom, 1983; Scholey *et al.*, 1984). SDS–PAGE reveals that a number of polypeptides, in addition to tubulin, cosediment with the MTs, and are, therefore, candidates for being MAPs (Fig. 2) (Vallee and Bloom, 1983;, 1985; Scholey *et al.*, 1984,1985). In addition, antibodies raised against a number of polypeptide components of these MT preparations, including polypeptides of $M_r \simeq 235,000$, 205,000, 150,000, 37,000 (Vallee and Bloom, 1983), 77,000 (Bloom *et al.*, 1985), and the 130,000-M_r subunit of sea urchin egg kinesin (Scholey *et al.*, 1985), all stain the mitotic spindle of fixed, dividing sea urchin eggs. These results support the hypothesis that taxol-assembled egg MTs represent useful "affinity ligands" for isolating and characterizing proteins of the mitotic spindle.

III. SEA URCHIN EGG MICROTUBULE-ASSOCIATED ATPase ACTIVITY

An ATPase that is MT associated would represent a plausible candidate for a cytoplasmic or mitotic MT-based motor. Therefore, we analyzed our MT preparations for ATPase activity (Scholey *et al.*, 1984, 1985; Dinenberg *et al.*, 1986; Porter *et al.*, 1988) and found that they hydrolyze ATP with a specific activity approximately 10- to 30-fold higher than the corresponding egg cytoplasmic extract supernatant (Table I). In the absence of assembled

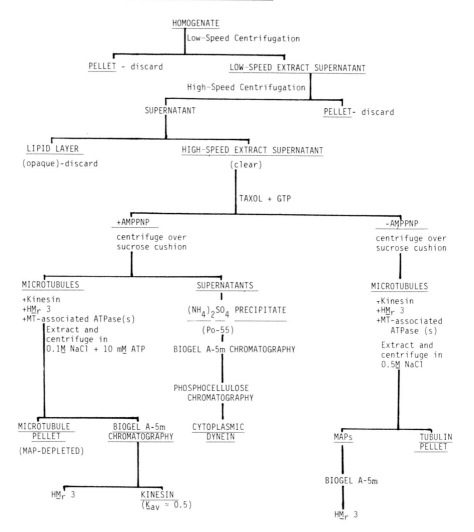

←——————————————————————————————————

Fig. 1. Scheme for Fractionating Sea Urchin Eggs. (1) Collect eggs. (a) Inject 0.56 M KCl into coelomic cavity and collect eggs in tripour beakers filled with seawater (in a cold room). Collect for ~1 hr. (b) Pool eggs and pour 7 times through 150-μm nitex screen. (c) Gently pellet eggs in 50-ml plastic tubes, using tabletop IEC centrifuge (setting 1 for approximately 10–30 sec/spin). Aspirate off supernatant. Resuspend eggs in CA^{2+}-free seawater using plastic dispo pipets. Spin two more times until all eggs are sedimented and jelly coat disappears. Rinse and spin once with small amount of extraction buffer. (2) Prepare microtubules and kinesin. (a) Resuspend eggs into 2 volumes of extraction buffer [0.1 M pipes, 2.5 mM Mg(CH$_3$COO)$_2$, 5 mM EGTA, 0.1 mM EDTA, 0.9 M glycerol, 0.1 mM PMSF, 1 μg/ml pepstatin, 1 μg/ml leupeptin, 10 μg/ml aprotinin, 0.5. mM DTT, pH 6.9]. Homogenize eggs on ice in 45 ml Wheaton homogenizer, pestle A, with six full strokes (down and up equals one stroke). Color should change from a deep orange to a milky "Dreamsicle" orange as cells are lysed. (b) Pour homogenate into Beckman Ti50 tubes, balance them, and spin in Ti50 rotor using the Beckman ultracentrifuge (25,000 rpm for 30 min at 4°C). (c) Remove middle, clear layer with a Pasteur pipet and place into new set of Ti50 tubes. Balance tubes and spin in Ti50 rotor using the Beckman ultracentrifuge (45,000 rpm for 60 min at 4°C). (d) Collect middle clear extract carefully. (If necessary, can freeze this extract in liquid N$_2$ and store in $-70°$C Revco.) Add 2 μl/ml taxol (10 mM stock solution in DMSO) (to 20 μM) and 10 μl/ml GTP (0.1 M stock) (to 1 mM) to extract (in order to assemble MTs) and incubate at room temperature for 20 min on a rocker. (e) Add 25 μl/ml AMPPNP (0.1 M stock) (to 2.5 mM) to MT extract and incubate for 10 min at room temperature on a rocker. (f) Place extract in new Ti50 centrifuge tubes over a cushion of 15% sucrose containing 20 μM taxol, 1 mM GTP, and 2.5 mM AMPPNP. Spin AMPPNP MT extract in Sorvall RC-5, using HB4 Swinging bucket rotor to pellet kinesin/MTs (12,000 rpm for 45 min at 4°C). (g) Resuspend and wash MT pellet with 5 ml extraction buffer containing 20 μM taxol, 0.1 mM GTP in the Beckman ultracentrifuge (spin at 27,000 rpm for 20 min at 4°C) Ti50 rotor. (h) Decant supernatant and remove all supernatant possible with a cotton swab. Resuspend pellet in a 2 ml Wheaton homogenizer on ice with 1 ml extraction buffer containing 0.1 M NaCl, 10 mM MgSO$_4$, and 10 mM ATP, 20 μM taxol, 1 mM GTP (RT or 4°C overnight). Place homogenate in Ti50 centrifuge tubes and spin in Ti50 rotor using Beckman ultracentrifuge (25,000 rpm for 20 min at 4°C). (i) Collect supernatant and run over BioGel A-5m column which has been preequilibrated with extraction buffer containing 1 mM ATP. After loading the sample, elute column with extraction buffer. (j) Collect fractions and assay for MT-gliding activity by video microscopy. Kinesin is expected at $K_{av} \simeq 0.5$ and HM_r 3 nearer the void volume. (3) Isolation of soluble cytoplasmic dynein from unfertilized egg extract depleted of taxol-assembled MTs. (a) A 55% saturated $(NH_4)_2SO_4$ fraction from the supernatant, which remained after pelleting taxol-assembled MTs from extract (plus or minus AMPPNP), is chromatographed on a 2.5 × 43-cm BioGel A-5m column preequilibrated and eluted in the "KCl–Tris" buffer of Pratt *et al.* (1984), and fractions are analyzed for protein concentration and ATPase activity. Fractions, which possess an ATPase activity of approximately 20 nmol of Pi released/min/ml are pooled for subsequent phosphocellulose chromatography. (b) The pooled ATPase fractions are loaded onto a phosphocellulose column preequilibrated in the same KCl–Tris buffer. The ATPase is eluted by a 0.5 M NaCl buffer step and is analyzed for protein concentration and ATPase activity. (See also Fig. 2.)

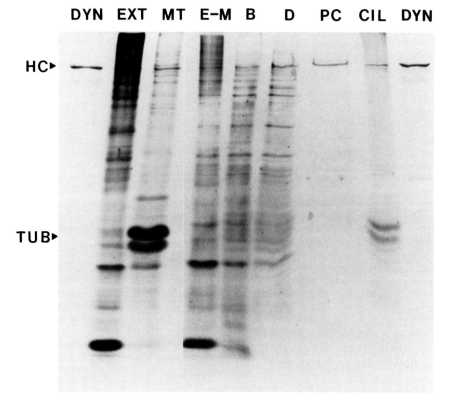

Fig. 2. SDS–polyacrylamide gel analysis of fractions obtained during fractionation of sea urchin egg. Microtubules were prepared from an unfertilized sea urchin egg high-speed extract supernatant (EXT) without AMPPNP treatment as described in detail in Fig. 1. Protein present in the extract minus MTs (E − M) was precipitated in 55% $(NH_4)_2SO_4$, and collected by centrifugation. The resuspended precipitate was loaded onto a 2.5 × 43-cm BioGel A-5m column, preequilibrated, and eluted with KCl–Tris buffer (Pratt *et al.*, 1984). The ATPase peak from the BioGel column (B) was loaded onto a 1.3 × 11-cm DEAE-Sephacel column, preequilibrated, and washed with KCl–Tris buffer. The bound ATPase was eluted using a linear gradient of 0.05–0.30 *M* NaCl buffer. The ATPase peak from the DEAE-Sephacel column (D) was dialyzed against an extraction buffer, then applied to a 1 × 4-cm column of phosphocellulose, preequilibrated, and washed with the extraction buffer. The bound ATPase was eluted with a 0.5 *M* NaCl buffer step (PC). More recently, we have observed that preparations of similar purity and activity can be obtained without using the DEAE-Sephacel column (Porter *et al.*, 1988). DYN, Flagellar dynein; CIL, embryonic cilia; HC, Dynein heavy chain; TUB, Tubulin. This gel was silver stained.

MTs, this ATPase remains soluble when egg extracts are centrifuged (Scholey *et al.*, 1984; Dinenberg *et al.*, 1986). When MTs are formed in egg extracts using taxol, the amount of ATPase that cosediments with the MTs upon subsequent centrifugation depends on the solvent conditions. For example, MTs sedimented from egg extracts supplemented with 0.5 *M* NaCl contain greatly reduced ATPase activity, whereas addition of 1–10 m*M* ATP or AMPPNP (Yount *et al.*, 1971) to the extract had little effect on the specific ATPase activity of the corresponding MT preparations (Table I). Microtubules prepared from egg extracts treated with apyrase (Meyerhof, 1945) to reduce endogenous nucleotide concentrations consistently possessed higher ATPase activity, but we do not know whether the apyrase is itself contributing to the measured activity.

It was originally proposed (Scholey *et al.*, 1984) that the ATPase activity of sea urchin egg MTs was due to the presence of cytoplasmic dynein (Pratt, 1984). This hypothesis was supported by the observation that our MTs contain polypeptides that coelectrophorese with sea urchin sperm flagella dynein A-bands and contain polypeptides that cross-react on immunoblots with blot-affinity-purified antibodies to flagellar dynein heavy chains (Scholey *et al.*, 1984, 1985; Dinenberg *et al.*, 1986). More recently, however, we have found that the amount of the dynein-related ATPase activity in MT preparations is relatively low [less than 25% of the total MT-associated ATPase (Porter *et al.*, 1988)].

Furthermore, we have measured the MT-associated ATPase activity under various conditions (Table I). Several characteristics of this activity differ from those of dynein. For example, the MT-associated ATPase possesses lower substrate specificity than dynein (100% activity in ATP, 71% in GTP), and a higher K_m for ATP (146 μ*M*). In addition, its response to vanadate and changes in the NaCl concentrations is different from dynein (see Dinenberg *et al.*, 1986; Porter *et al.*, 1988). These results suggested that our MT preparations also contain an ATPase activity that differs from dynein (Dinenberg *et al.*, 1986; Porter *et al.*, 1988).

Recently, Collins and Vallee (1986a,b) have described the presence in sea urchin egg cytosolic extracts of an ATPase that possesses a sedimentation coefficient of 10 S and requires the presence of assembled MTs for activity. The ATPase activity of this 10 S fraction requires divalent cations (Mg^{2+} or Ca^{2+}) and is not inhibited by 100 μ*M* vanadate. It hydrolyzes GTP at about 1.4 times the rate at which it hydrolyzes ATP, and its activity is inhibited about 90% by 2 m*M* *N*-ethylmalemide or 0.25 *M* NaCl. Furthermore, the 10 S ATPase cosediments with MTs (both in the presence and absence of ATP) and is therefore likely to contribute to the ATPase activity of sea urchin egg MT preparations. The polypeptide composition of the ATPase present in this 10 S fraction has not yet been defined, and its localization within cells is

TABLE I

ATPase Activity of Sea Urchin Egg Microtubules[a]

	Specific activity	Total activity
1. Homogenate	10.8	1.0
2. High-speed extract supernatant	1.0	0.3
3. Microtubule pellet	18.0	0.8
4. Microtubule-depleted extract supernatant	1.6	
5. Microtubules polymerized with		
a. No taxol or GTP	2.6	0.03
b. 20 μM taxol, 1 mM GTP + 3 mg tubulin	10.0	0.4
c. 20 μM taxol, 1 mM GTP + 0.5 M NaCl	4.1	
d. 20 μM taxol, 1 mM GTP + 10 mM ATP	23.8	
e. 20 μM taxol, 1 mM GTP + 1.5 mM AMPPNP	22.5	0.26
f. 20 μM taxol, 1 mM GTP + apyrase	37.8	
6. Microtubule polymerized at 22°C	18.0	
Microtubule polymerized at 4°C	19.4	
7. Microtubule prepared from		
a. unfertilized eggs	18.0	
b. fertilized eggs	19.6	
8. Microtubule assayed in		
a. Extraction buffer (EB)	18.0	
b. EB diluted 10-fold	24.5	
c. EB at 30°C	26.4	
d. EB + 1% Triton X-100	17.7	
e. EB + 10 mM AMPPNP	13.7	
f. EB + 5 mM EGTA	15.9	
g. EB + 10 μg/ml oligomycin	16.3	

h.	EB + 20 µg/ml oligomycin	17.4
i.	EB + 0.1 mM NaN$_3$	17.7
j.	EB + 1 mM NaN$_3$	16.1
k.	EB + 10^{-5} ouabain	19.7
l.	EB + 10^{-4} ouabain	16.5
m.	EB at pH	
	5.0	2.7
	6.0	16.3
	7.0	18.0
	8.0	19.6
	9.0	29.8
	10.0	15.2
n.	EB + 0.5 M NaCl	6.8
o.	EB + 50 µM Na$_3$VO$_4$	10.5
p.	EB + 2 mM N-ethylmaleimide	14.0 (*S. purpuratus*)
		4.86 (*L. pictus*)

9. K_m for ATP $\simeq$ 146 ± 48 µM ($n = 5$)
10. ATPase : GTPase ratio $\simeq$ 1.4

[a]ATPase activity was measured in "extraction buffer" usually containing 2 mM [γ-^{32}P]ATP at 22°C, unless otherwise indicated, and assayed using a radioisotopic assay (see Cohn *et al.*, 1987, for details). Comparisons of MTs prepared in AMPPNP, apyrase, and ATP were usually done in higher ATP concentrations ($\leq$10 mM). The values presented for MT pellet and high-speed extract supernatant represent the mean of 22 assays all normalized relative to the latter fraction (taken as unity; true specific activity of extract was usually in the range 0.5–3.0 nmol/min/mg protein). The other values were obtained with different preparations and, in all cases, were measured relative to a control MT pellet assayed under standard conditions. For ease of tabulation, the actual activities measured for different preparations were adjusted so that control MT-specific activity equals 18.0.

unknown, but the enzyme is a promising candidate for a cytoplasmic MT-based motor (Collins and Vallee, 1986a,b).

The simplest interpretation of the currently available data is that both cytoplasmic dynein-like ATPases and the 10 S MT-activated ATPase activity contribute to sea urchin egg MT-associated ATPase activity. For example, the vanadate sensitivity (30–50%) and GTPase : ATPase ratio ($\approx$0.7) of egg MTs can be considered to result from a combination of dynein-like activities (vanadate sensitivity approaching 100%; GTPase : ATPase approaching 0) and 10 S ATPase activities (vanadate sensitivity $\simeq$ 0%; GTPase : ATPase $\simeq$ 1.4). Whether other enzymes, such as kinesin (see below), also contribute to ATP hydrolysis by sea urchin egg MTs is not yet known. Microtubule-associated ATPases have been described in other organisms and cell types (reviewed in Pratt, 1984; Scholey *et al.*, 1984; Vale *et al.*, 1986), but their relationship to the sea urchin egg MT-associated ATPase is unclear.

IV. SEA URCHIN EGG HM_r 3

In our initial characterization of sea urchin egg microtubule preparations (Scholey *et al.*, 1984), we described a group of high-molecular-weight (HM_r) polypeptides ($M_r \simeq$ 350,000) which cosedimented with microtubules. We described three polypeptides, HM_r 1, HM_r 2, and HM_r 3 in order of decreasing apparent size on SDS–polyacrylamide gels. The exact composition of these HM_r polypeptides varies in different MT preparations, but the faster migrating of these, HM_r 3, is consistently present. By gel densitometry, there are approximately 0.5 mol HM_r 3/100 mol tubulin dimer in our microtubule preparations (Scholey *et al.*, 1985). On 5–7.5% SDS–polyacrylamide gels (Laemelli, 1970) (using acrylamide dimer : monomer ratio = 2.7 C : 30 T or 1 C : 25 T), HM_r 3 co-electrophoreses with the A-band of sea urchin sperm flagellar 21 S dynein (Fig. 2) (Scholey *et al.*, 1984, 1985; Dinenberg *et al.*, 1986).

HM_r 3 can be partially extracted from microtubules by differential centrifugation in 5–10 mM MgATP (Scholey *et al.*, 1984, 1985; Porter *et al.*, 1988) and essentially totally extracted from microtubules with (1) 0.5 M NaCl (Scholey *et al.*, 1984, 1985) or (2) 0.1 M NaCl, 7.5–10 mM MgATP (Scholey *et al.*, 1985). When sea urchin egg high-speed-extract supernatants containing assembled MTs are treated with 0.5 M NaCl or 0.1 M NaCl plus 10 mM MgATP, HM_r 3 does not cosediment with microtubules (J. M. Scholey and M. E. Porter, unpublished). In contrast, treatment of egg extracts with AMPPNP does not greatly affect the amount of HM_r 3 that cosediments with MTs (Scholey *et al.*, 1985). The ATP sensitivity of microtubule binding by HM_r 3 is consistent with its having a mechanochemical function; myosin, dynein, and kinesin are all dissociated from their cytoskeletal cofactor by ATP.

Our early experiments suggest that HM_r 3 possessed an ATPase like that of cytoplasmic dynein (Scholey et al., 1984). For example, we reported (Scholey et al., 1984) that during fractionation of MT by (1) high salt extraction, followed by gel filtration or sucrose density gradient fractionation, or (2) ATP extraction, a dynein-like MgATPase activity copurified with HM_r 3. Our more recent experiments indicate, however, that these extracts contained a mixture of the soluble axonemal dyneins as well as the MT-associated HM_r 3 polypeptide (Porter et al., 1988).

Our early studies indicated that HM_r 3 was related to flagellar dynein (Dinenberg et al., 1986). Antisera were raised in rats immunized with partially purified flagellar dynein. Dynein heavy chain antibodies were purified from the crude antiserum using blots of electrophoretically purified flagellar dynein A-band as affinity ligands. These monospecific heavy chain antibodies cross-reacted with polypeptides in the HM_r 3 region on 6% (30 T : 2.7 C) SDS gels (Dinenberg et al., 1986). These data suggested that HM_r 3 is immunologically related to flagellar dynein A-bands. However, on high-resolution SDS gels containing 3.2% polyacrylamide and a 0–8 M urea gradient (Piperno and Luck, 1979), the HM_r 3 region is resolved into at least three polypeptides. Two of these coelectrophorese with dynein Aα and dynein Aβ (Dinenberg et al., 1986), but the major band is a lower molecular-weight polypeptide (Porter et al., 1988). The amount of the Aα- and Aβ-bands relative to the third polypeptide is variable from preparation to preparation and species to species. Our more recent experiments have suggested that the apparent cross-reactivity of the HM_r 3 region with flagellar dynein antibodies is due to the presence of the soluble cytoplasmic dynein heavy chains that cosediment with some of the MT preparations and not with the true MT-associated HM_r 3 polypeptide itself (Porter et al., 1988). However, very recent work indicates that the HM_r 3 polypeptide may be a dynein isoform that is distinct from the soluble dynein, as evidenced by its susceptibility to vanadate-induced photocleavage (Porter et al., 1988).

To summarize, we initially thought that HM_r 3 in our MT preparations was cytoplasmic dynein and that this enzyme could account for the sea urchin egg MT-associated ATPase activity (Scholey et al., 1984). Subsequent analysis has revealed that this interpretation was oversimplified. Some cytoplasmic dynein is present in our MT preparations, accounting for (1) the apparent cross-reactivity between HM_r 3 and the flagellar dynein heavy chain antibodies; (2) the presence of Aα- and Aβ-chains on high-resolution SDS gels; and (3) the dynein-like MgATPase activity that copurifies with HM_r 3, following 0.5 M NaCl extraction of S. purpuratus or S. droebachiensis MTs and gel filtration or sucrose density gradient fractionation. It is now clear, however, that the

MT-associated ATPase activity is not just due to dynein (see above). Further, HM_r 3 contains a polypeptide that does not coelectrophorese with the dynein heavy chains, nor does it cross-react with antiflagellar dynein antibodies. This polypeptide seems to be a dyneinlike MAP that binds to MTs in an ATP-sensitive manner. A further characterization of its effects on MT motility, polymerization, and its localization in cells would be valuable.

V. "SOLUBLE" SEA URCHIN EGG CYTOPLASMIC DYNEIN

Sea urchin eggs obviously contain a large number of different ATPases. Egg homogenates display much higher specific ATPase activities than our high-speed supernatants of egg extract (Table I), suggesting that a good deal of ATPase is not extracted from the particulate matter in the eggs under our conditions. High-speed extract supernatants prepared under similar conditions to ours (Collins and Vallee, 1986a,b) yield ATPase activities sedimenting on sucrose gradients with sedimentation coefficients of (1) 6 S, (2) 12 and 20 S (cytoplasmic dynein), and (3) 10 S (MT-activated ATPase activity). We find that 10–50% of the total MgATPase activity present in our egg extract high-speed supernatants cosediments with MTs, the remainder remains soluble, even when additional MTs are added to the extract (Dinenberg *et al.*, 1986).

We have purified a dynein-like MgATPase activity from this MT-depleted extract supernatant by column chromatography. In our early work, we used $(NH_4)_2SO_4$ fractionation, BioGel A-5m chromatography, and hydroxyapatite chromatography (Scholey *et al.*, 1984). However, we have recently modified the procedure by using (1) DEAE-Sephacel and phosphocellulose chromatography or (2) phosphocellulose chromatography alone in place of the hydroxyapatite column (Pratt *et al.*, 1984; Porter *et al.*, 1988) (Fig. 2 and Table II). We have characterized this purified protein in some detail and find that it possesses properties very similar to those of flagellar dynein (Porter *et al.*, 1988). (1) On high-resolution SDS gels (Piperno and Luck, 1979), one sees polypeptides that coelectrophorese with flagellar dynein Aα- and Aβ-polypeptides (Dinenberg *et al.*, 1986; Porter *et al.*, 1988). This is consistent with the pattern of polypeptides seen in preparations of embryonic cilia. (2) On immunoblots of 5–7.5% Laemelli gels, the purified protein binds antibodies raised against sea urchin sperm flagellar 21 S dynein A-bands and blot-affinity purified against dynein A (Dinenberg *et al.*, 1986; Porter *et al.*, 1988). It is therefore immunologically related to sperm flagellar dynein heavy chains. (3) The purified enzyme sediments at 20 S on sucrose density gradients and is converted to a 10 S particle in the presence of 0.1% Triton X-100 (Porter *et al.*, 1988). Collins and Vallee (1986a) also report that the relative amounts of

TABLE II

Preparation of "Soluble" Cytoplasmic Dynein from MT-Depleted Egg Extract[a]

		Protein		ATPase	
Sample	Volume (ml)	Protein (mg/ml)	Total protein (mg)	Specific activity (nmol P_i/ min/mg)	Total activity (nmol P_i/min)
BioGel A-5m	18.0	0.63	11.3	49.8	560.0
DEAE-Sephacel	12.0	0.15	1.8	97.8	176.0
Phosphocellulose	8.0	0.06	0.5	306.0	136.0

[a]Summary of protein yields and specific activities of peak ATPase fractions obtained during purification of soluble ATPase. BioGel A-5m, peak ATPase fraction from BioGel A-5m column; DEAE-Sephacel, peak ATPase fraction from DEAE Sephacel column; phosphocellulose, peak ATPase fraction from phosphocellulose column. Protein concentration was determined using Bio-Rad protein assay dye reagent concentrate (Bradford, 1976). ATPase activity was determined by colorimetric assay for inorganic phosphate (Ames, 1966; Waxman and Goldberg, 1982).

12 and 20 S cytoplasmic dynein depend on buffer conditions. The enzymatic properties of the soluble ATPase are similar to those of flagellar dynein. Under conditions in which flagellar dynein has a specific activity of 1.0–2.5 μmol/ min/mg, the column ATPase has specific activity of 150–600 nmol/min/mg. The effects of various ATPase inhibitors and changes in pH, monovalent cation concentration, divalent cation composition, or nucleotide substrate on the activity of the two enzymes was similar (Dinenberg *et al.*, 1986; Porter *et al.*, 1988). Therefore, on the basis of its size, polypeptide composition, cross-reactivity with flagellar dynein antibodies, and enzymatic properties, we conclude that the ATPase we have purified from MT-depleted high-speed supernatants of egg extracts is a cytoplasmic dynein (Pratt, 1984; Hisanaga and Sakai, 1983; Dinenberg *et al.*, 1986; Porter *et al.*, 1988).

The soluble cytoplasmic dynein binds weakly to MTs in high-speed egg extracts (Scholey *et al.*, 1984; Porter *et al.*, 1988). Based on recovery of ATPase activity and flagellar dynein antigens, we estimate that less than 10% of soluble egg dynein is recovered in taxol-assembled MTs (Porter *et al.*, 1988).

VI. SEA URCHIN EGG KINESIN

We began looking for a MT translocator on the basis of reports that extracts of squid axoplasm would support MT-based, ATP-dependent movement of vesicles, driven by a translocator that existed primarily in soluble form, with a fraction serving to cross-link MTs to the moving vesicles (see Vale *et al.*,

1986, for a review of this field). We were struck by the similarity between this translocator and cytoplasmic dynein, which also exists predominantly in a soluble form. The key observation by Lasek and Brady (1985) that the non-hydrolyzable ATP analog adenylyl imidodiphosphate (AMPPNP) (Yount et al., 1971) "freezes" axoplasmic vesicle motion on MTs suggested that the analog could induce strong binding between the translocators and MTs. We reasoned that AMPPNP might enhance MT binding by the dynein present in sea urchin egg cytoplasmic extracts (Scholey et al., 1985).

Taxol-assembled MTs were, therefore, prepared from cytoplasmic extracts of unfertilized sea urchin eggs in the presence and absence of AMPPNP; the amount of various MAPs (MT-associated proteins) cosedimenting with tubulin was examined on SDS gels. Surprisingly, addition of AMPPNP did not affect the amount of dynein heavy chain or MT-associated ATPase activity appearing in the pellet, but it greatly enhanced the amount of a 130,000 M_r (130K) poly-peptide ($M_r = 129,000 \pm 4,000$, $n = 11$) cosedimenting with the MTs. This 130K polypeptide could be released from the MTs by differential centrifugation in ATP, then partially purified by gel filtration chromatography (Scholey et al., 1985; Porter et al., 1987).

In several respects this 130K protein resembled kinesin, the protein believed to translocate vesicles along squid axonal MTs. (1) Kinesin too was identified and isolated by its affinity for MTs in the presence of AMPPNP (Vale et al., 1985b). (2) A video microscopy assay showed that fractions containing the 130K polypeptide caused beads to move along MTs and MTs to move over a glass coverslip (Figs. 3 and 4) in a manner indistinguishable from that de-scribed for squid axoplasmic kinesin. (3) Specific antibodies raised against electrophoretically purified 130K polypeptide exhibited a clear cross-reaction with the major polypeptide component (110K) of squid kinesin, supporting the hypothesis that the sea urchin egg 130K polypeptide represents kinesin. (4) The 130K protein sediments with an S-value of 9.5 S on sucrose density gradients, and it elutes from BioGel A-5m columns with a partition coefficient

———→

Fig. 3. Cosedimentation of p130 kinesin activity in crude extracts of sea urchin eggs. The following fractions were run on a 5–15% polyacrylamide gel (b), blotted to nitrocellulose (a), and probed with a blot-affinity-purified antibody to p130 (Scholey et al., 1985). ATP-P, Pellet of MT following addition of 10 mM MgATP to the cell extract. ATP-S, ATP su-pernatant containing motility activity, which was loaded onto a 5–15% sucrose density gra-dient. 12–20, Even numbered fractions from the sucrose gradient of the ATP supernatant. Motility activity peaked at 9.5 S in fractions 13–17, which also contained the p130 antigen. AMPPNP-S, Supernatant from cell extract, which was treated with 10 mM MgAMPPNP and centrifuged to remove MTs. AMPPNP-P, Pellet of AMPPNP-treated MTs containing the p130 antigen. Identical results were obtained with the antibody to the squid 110-kDa kinesin polypeptide. (See Porter et al., 1987.)

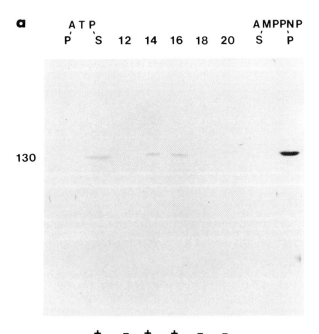

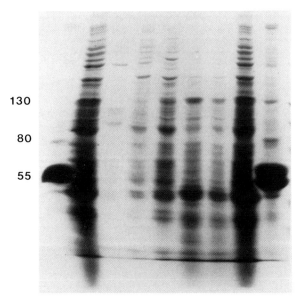

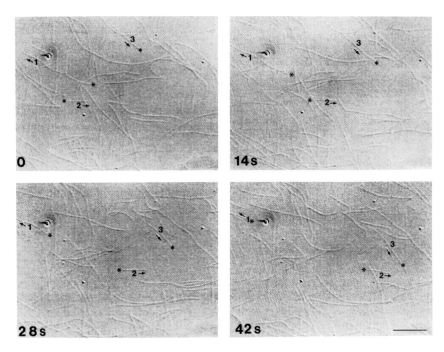

Fig. 4.   Sea urchin egg kinesin induces MT gliding over glass coverslips. Partially purified sea urchin egg kinesin was allowed to adhere to a glass coverslip, mixed with taxol-stabilized brain microtubules and 50 μ*M* ATP, and then sealed to a glass slide. Microtubule gliding on the coverslip surface was viewed by video-enhanced DIC microscopy. Shown here are four frames taken from the video monitor at 14-sec intervals. The ends of three numbered microtubules are indicated (∗), and their direction of movement is shown through the sequence (→). These microtubules are gliding at a rate of 0.39 ± 0.08 μm/sec. Bar, 5 μm. [From Porter *et al.* 1987).]

$K_{av} \simeq 0.5$, suggesting that it resembles squid kinesin in containing multiple subunits. Whether the native sea urchin protein contains subunits in addition to the 130K polypeptide has not yet been determined (Scholey *et al.*, 1985; Porter *et al.*, 1987; see Vale *et al.*, 1986, for review). Nonetheless, the similarities between the 130K protein from sea urchins and the microtubule translocator from squid axoplasm are strong enough to warrant our calling the sea urchin protein kinesin.

   We have now used an *in vitro* video microscopic assay (Allen *et al.*, 1981; Inoué, 1981; Vale *et al.*, 1985a,b,c; Schnapp *et al.*, 1985) to characterize some of the motile properties of sea urchin egg kinesin (Porter *et al.*, 1987; Cohn *et al.*, 1987). Partially purified kinesin is adsorbed to a glass coverslip, mixed with taxol-stabilized MTs and ATP, and viewed by video-enhanced differential interference contrast microscopy. The MT-translocating activity of the purified egg kinesin is qualitatively similar to the analogous activity observed in crude extracts of sea urchin eggs (Fig. 4) (Pryer *et al.*, 1986) and resembles the activity of neuronal kinesin with respect to the maximal rate of MT movement (>0.5 μm/sec). Kinesin also induced gliding of flagellar axonemes stripped of their dynein arms. Axonemes glide at rates comparable to those of singlet MTs. Latex beads coated with sea urchin egg kinesin move toward the plus end of MTs attached to isolated centrosomes. In addition, axonemes, which were seeded with brain tubulin, were observed to glide with their "minus ends" leading. Thus, the polarity of MT-based movement induced by sea urchin egg kinesin is the same as that induced by neuronal kinesin. The motility–activity of sea urchin egg kinesin requires nucleotide turnover. For example, the nucleotide analogs ATP-γ-S and AMPPNP immediately inhibit MT gliding in the presence of equimolar ATP and cause MTs to bind "immotile" to kinesin on the glass surface. The photoaffinity analog of ATP, 8-azido-ATP, inhibits sea urchin egg kinesin-induced motility in a UV-dependent, ATP-sensitive fashion. Mg^{2+} and nucleotides are both required for activity; ATP depletion, by dialysis or apyrase, and chelation of Mg^{2+}, using EDTA, block motility and induce a "rigorlike" state. The activity of sea urchin egg kinesin is inhibited by high concentrations of [NEM] (>3–5 mM) or vanadate (>50 μM). The nucleotide requirement of sea urchin egg kinesin is fairly broad (ATP > GTP > ITP, CTP, and UTP), and the rate of MT movement increases in a saturable fashion with the [ATP]. When the reciprocal of the rate of MT gliding versus the reciprocal of the ATP concentration are plotted (Lineweaver–Burke), we obtain a $V_{max} \simeq 0.5$–0.7 μm/sec and an apparent K_m that varies around 60 μM ATP. We conclude that the motile activity of egg kinesin is indistinguishable from that of neuronal kinesin (Porter *et al.*, 1987; Cohn *et al.*, 1987; Scholey *et al.*, 1988).

The observation that the MT-translocating activity of sea urchin egg kinesin is ATP dependent suggests that the protein might be an ATPase. Other mechanochemical proteins (dynein and myosin) are ATPases, and the rate at which they hydrolyze ATP is coupled to motility and consequently activated by MTs and actin filaments, respectively. Since kinesin moves MTs at a velocity of 500 nm/sec, and assuming each kinesin cross-bridge moves the MT a distance of about 10 nm (1–2 subunits) and assuming 1 molecule of ATP is hydrolyzed per cross-bridge cycle, we might expect a rate of ATP hydrolysis of 500/10 = 50 ATP/sec for active kinesin molecules. Recently, Kuznetsov and Gelfand (1986) succeeded in obtaining preparations of bovine brain kinesin which, in the presence of MTs, hydrolyze ATP at rates expected for active kinesin. Their purification protocol involves chromatographic fractionation of a brain extract, followed by affinity purification of kinesin via MT binding in inorganic tripolyphosphate instead of AMPPNP. The bound kinesin is released from MTs with ATP, then subjected to gel filtration and ion-exchange chromatography. The purified material hydrolyzes ATP at a rate of 60–80 nmol/min/mg, but the activity is stimulated severalfold by MTs to a $V_{max} \simeq$ 4.6 μmol/min/mg.

Recently, Cohn *et al.* (1987) reported that sea urchin egg kinesin is also an MT-activated Mg-ATPase, and they reported that the inhibitors AMPPNP, vanadate, EDTA, and Mg-free ATP cause a dose-dependent inhibition in both the velocity of kinesin-driven MT gliding and the MT-activated Mg-ATPase activity. Furthermore, on the basis of observations that three properties are shared by myosin, dynein, and kinesin, namely, (1) formation of a bound complex with actin or MTs in the absence of Mg-ATP, (2) dissociation of the complex by Mg-ATP, and (3) activation of ATPase activity by actin or MTs, these workers propose that the mechanochemical cycles of kinesin, myosin, and dynein may have features in common (Scholey *et al.*, 1988).

One important question remaining concerns the function of kinesin within cells. As discussed in a recent review (Vale *et al.*, 1986), kinesin is a good candidate for both an organelle translocator and a mitotic motor. Indeed, the observation that kinesin is present in sea urchin eggs, a cell type where mitosis is a primary function of MTs, supports the hypothesis that kinesin may be a mechanochemical factor for some form of motility associated with the mitotic spindle. In addition, the observation that monospecific antibodies to kinesin stain the mitotic spindle in fixed, dividing sea urchin eggs adds further support for the notion that kinesin might be important for chromosome or vesicle movement during mitosis (Scholey *et al.*, 1985), as discussed in more detail in recent reviews (Vale *et al.*, 1986; Mitchison, 1986). However, it is important to emphasize that the localization of kinesin in the spindle does not address the question of its function there, and future work will be aimed at probing the function of kinesin in living cells (see Vale *et al.*, 1986, for discussion).

VII. A MICROTUBULE-ASSOCIATED MOTOR FROM *C. ELEGANS*

The evidence that sea urchin egg dyneins play a role in mitotic motion or intracellular transport is equivocal. These dyneins may be ciliary precursors that are stockpiled in the egg and may have no role in cytoplasmic MT-based transport. Identifying a cytoplasmic dynein in an organism that does not make cilia or flagella would greatly simplify the study of its role(s).

The free-living soil nematode *C. elegans* does not form motile cilia or flagella. The sperm of this organism are amoeboid (Nelson *et al.*, 1982); neither the embryos, the gonads, nor the intestine are ciliated (Chitwood and Chitwood, 1974); and although there are doublet MTs in the six mechanosensory neurons (Ward *et al.*, 1975), these appear to be passively involved in force transduction rather than motility. The fine structure of these neurons reveals no dynein arms on the doublet MTs (Chalfie and Thompson, 1982). We have chosen *C. elegans* as a suitable species for the study of cytoplasmic dyneins in cellular transport because of its lack of contaminating axonemal dyenins and its suitability for genetic, molecular genetic, and detailed morphological analysis.

We have identified a protein in *C. elegans* that shares many properties with axonemal dynein (Lye *et al.*, 1987). We have again used taxol (Vallee, 1982) to stimulate the polymerization of MTs in cold extracts of gravid adult nematodes. Several polypeptides pellet in the presence, but not in the absence, of assembled MTs. One of these associated polypeptides exhibits ATP-sensitive binding to the polymer. If the ATP concentration in the extract is reduced using either apyrase or hexokinase and glucose, the amount of the HM_r polypeptide (M_r, ~400K) that binds to the MTs is significantly enhanced. This polypeptide can subsequently be extracted from the MTs with 10 mM MgATP plus 100 mM NaCl. The HM_r polypeptide and a MgATPase activity cosediment on sucrose gradients at approximately 20 S (Fig. 5). This ATPase activity is more than 50% inhibited by either 10 μM vanadate, 1 mM N-ethylmaleimide, or 5 mM AMPPNP. The ATPase is enhanced 50% by 0.2% Triton X-100. Like axonemal dynein, the *C. elegans* 400K polypeptide displays vanadate-sensitive photocleavage. These properties are all dynein-like.

When the 20 S protein is adsorbed to a glass coverslip and either MTs or flagellar axonemes are added, it promotes a nucleotide-dependent MT translocation. In our early work, axonemes that have been elongated at their "plus" end with phosphocellulose-purified tubulin glided with their minus ends leading, suggesting that the *C. elegans* motor displayed similar polarity of motion to that of kinesin. Subsequent work, however, using demembraned sperm has shown that the *C. elegans* motor moves objects toward the minus end of axonemal MTs (Lye *et al.*, 1988). This is the same direction of motility that is found for axonemal and bovine brain dyneins (Fox and Sale, 1987; Paschal and Vallee, 1987), supporting the hypothesis that the *C. elegans* motor is a cytoplasmic dynein. Microtubules move at about 1.5 ± 0.53 μm/sec, significantly faster than with kinesin under conditions of saturating ATP, and the motion is ATP specific and is blocked by 10 μM vanadate, 1 mM N-ethylmaleimide, or by 0.5 mM ATP-γ-S. Motility is slowed, but not blocked, by [AMPPNP] = [ATP]. These characteristics differ from the properties described for sea urchin egg kinesin. The roughly parallel inhibition of the ATPase and the motility by several reagents suggests that the enzyme activity and the motility are caused by the same protein or protein complex, but its role in cytoplasmic or mitotic MT-based motility is presently unknown (Lye *et al.*, 1987).

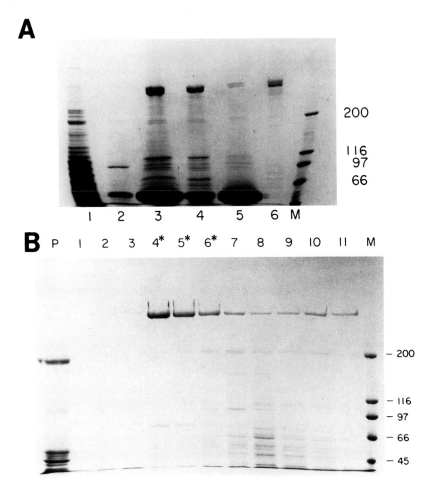

Fig. 5. Purification of the nematode motility factor from ATP depleted extracts of gravid adult *C. elegans*. (A) A Coomassie-stained 3–8% polyacrylamide gradient gel showing the binding of the 400K polypeptide to assembled MTs. Lane 1 contains the extract to which GTP and taxol will be added. Lane 2 is the pellet that results when MTs are prepared in the presence of nocodazole rather than taxol. Lane 3 contains the taxol MT pellet. Lane 4 contains the material that remains in the supernatant when taxol MTs are incubated in ATP-containing buffer, and lane 5 contains the MTs that pellet in the presence of ATP. Lane 6 contains *Tetrahymena* flagellar dynein as a marker, and lane M contains molecular-weight standards ($\times 10^3$). (B) A Coomassie-stained 3–8% polyacrylamide gradient gel showing the protein profile of a 5–20% sucrose density gradient. Lane P is the material that sediments through the gradient into a pellet. Lanes 1–11 are fractions from the gradient. Lane M contains molecular-weight standards ($\times 10^3$). The fractions that are indicated by an asterisk are the only ones that contain *in vitro* motility-inducing activity. They also cosediment with a peak of ATPase activity at approximately 20 S.

VIII. CONCLUDING REMARKS

A variety of putative MT-based motors have been described in recent years (Table III). These proteins can be considered as candidates for performing a number of motile functions within cells. For the purposes of discussion, we have divided these proteins into three classes. Class I, the proteins that perform a known, MT-based, mechanochemical function in living cells, contain only a single member (flagellar dynein, the ATPase that generates force for MT-sliding in axonemes). It should be noted, however, that current evidence suggests that inner and outer arm dynein differ, and it is likely that there are at least two flagellar dyneins. Future work will be aimed at understanding in detailed molecular terms the precise mechanisms by which these enzymes convert the chemical energy stored in ATP into MT sliding (Johnson, 1985).

In Class II, we have grouped together those proteins that have been demonstrated to induce MT-based movements (bead translocation and microtubule gliding) *in vitro*. These are attractive candidates for performing mechanochemical functions in living cells, although there is as yet no direct evidence that this is the case (Vale *et al.*, 1986). Kinesin, for example, is thought to

TABLE III

Summary of Putative Cytoplasmic Microtubule-Associated Motors

Class	Putative translocator	Source
I. Proteins known to cause MT-movement *in vivo*	Flagellar dynein	Eukaryotic cilia and flagella
II. Proteins known to cause MT-movement *in vitro*	Kinesin	Various cell types
	Retrograde translocator = cytoplasmic dynein	Squid axoplasm, *C. elegans*, bovine brain, and HeLa cells
	C. elegans "motor"	*C. elegans*
III. Proteins not yet demonstrated to cause MT-movement *in vivo* or *in vitro*	STOPs (stable tubule only proteins)	Vertebrate brain
	HM_r 3	Sea urchin eggs
	Soluble, axonemal, and cytoplasmic dynein	Sea urchin eggs
	Microtubule-associated ATPases	Various cell types
	10 S MT-activated ATPase	Sea urchin eggs

generate force for moving axoplasmic vesicles in the anterograde direction (from cell body to synapse) while cytoplasmic dyneins may serve as "retrograde translocators" (Vale *et al.*, 1985c; Paschal and Vallee, 1987; Lye *et al.*, 1988), which are thought to move vesicles in the reverse direction. Sea urchin egg kinesin is a candidate for playing some role in chromosome, vesicle, organelle, or pronuclear translocations. Speculation about the function of the *C. elegans* dynein-like motor must await determination of its cellular localization. Future research on these proteins will be aimed at (1) elucidating their functions in living cells, for example, by antibody or antisense RNA microinjection (Mabuchi and Okuno, 1977; Izant and Weintraub, 1984), or by gene disruption (Neff *et al.*, 1983) and (2) identifying the cellular components with which these proteins interact to generate motile force.

Class III includes proteins for which only circumstantial evidence suggests a motility-related function, but at present there is no direct evidence that these proteins can generate motile force either *in vivo* or *in vitro*. These include the MT-associated ATPases found in a variety of cell types, whose roles in MT-based motility are unknown: the MT-associated HM_r 3 and the soluble axonemal-like cytoplasmic dynein, which have been identified in sea urchin egg cytoplasm. Also in this class are the STOP (stable-tubule only proteins), which confer stability to brain MTs and are reported to "slide" along MTs. Whether these proteins can generate motile force is unknown (Margolis *et al.*, 1986). No doubt future studies will be aimed at improving the characterization of these proteins and probing their motility-inducing properties.

It seems likely that the list of putative MT-associated motors (Table III) will change very rapidly in the near future. The active research in this area may be expected to increase the catalog of putative MT-based motors and clarify our understanding of those already identified.

ACKNOWLEDGMENT

J. M. S. is grateful to Dr. Ted Salmon for introducing him to the sea urchin as a system for studying microtubule motors and to Shirley Downs for typing the manuscript.

REFERENCES

Allen, R. D., Allen, N. S., and Travis, J. L. (1981). Video-enhanced differential interference contrast A.V.E.C. (D.I.C.) microscopy; a new method capable of analyzing microtubule related movement in the reticulopodial network of *Allogromia laticollaris*. *Cell Motil.* **1,** 191–302.

Ames, B. N. (1966). Assay of inorganic phosphate, total phosphate and phosphatases. *In*

"Methods in Enzymology" (E. F. Neufeld and V. Ginsburg, eds.), Vol. 8, pp. 115–118. Academic Press, New York.

Bloom, G. S., Luca, F. C., Collins, C. A., and Vallee, R. B. (1985). Use of multiple monoclonal antibodies to characterize the major MAP in sea urchin eggs. *Cell Motil.* **5,** 431–446.

Bradford, M. M. (1976). A rapid and sensitive method for the quantitation of microgram quantities of protein using the principle of protein dye binding. *Anal. Biochem.* **72,** 248–254.

Chalfie, M., and Thompson, J. N. (1982). Structural and functional diversity in the neuronal microtubules of *Caenorhabditis elegans. J. Cell Biol.* **93,** 15–23.

Chitwood, B. G., and Chitwood, M. B. (1974). "Introduction to Nematology." University Park Press, Baltimore, Maryland.

Cohn, S. A., Ingold, A. L., and Scholey, J. M. (1987). Correlation between ATPase and microtubule-translocating activity of sea urchin egg kinesin. *Nature (London)* **328,** 160–163.

Collins, C. A., and Vallee, R. B. (1986a). A microtubule activated ATPase from sea urchin eggs distinct from cytoplasmic dynein and kinesin. *Proc. Natl. Acad. Sci. U.S.A.* **83,** 4799–4803.

Collins, C. A., and Vallee, R. B. (1986b). Characterization of the sea urchin egg MT-activated ATPase. *J. Cell Sci., Suppl.* **5,** 197–204.

Dinenberg, A. S., McIntosh, J. R., and Scholey, J. M. (1986). Studies on sea urchin egg cytoplasmic ATPases of possible significance for microtubule functions. *Ann. N.Y. Acad. Sci.* **466,** 431–435.

Fox, L. A., and Sale, W. S. (1987). Direction of force generated by the inner row of dynein arms on flagellar microtubules. *J. Cell Biol.* **105,** 1781–1788.

Gibbons, I. R. (1981). Cilia and flagella of eukaryotes. *J. Cell Biol.* **91,** 107–124.

Hisanaga, S. I., and Sakai, H. (1983). Cytoplasmic dynein of the sea urchin egg. II. *J. Biochem. (Tokyo)* **93,** 87–98.

Inoué, S. (1981). Video image processing greatly enhances contrast, quality and speed in polarization-based microscopy. *J. Cell Biol.* **89,** 346–356.

Izant, J. A., and Weintraub, M. (1984). Inhibition of thymidine kinase gene expression by anti-sense RNA: A molecular approach to genetic analysis. *Cell (Cambridge, Mass.)* **36,** 1007–1015.

Johnson, K. A. (1985). Pathway of the microtubule dynein ATPase and the structure of dynein. *Annu. Rev. Biophys. Biophys. Chem.* **14,** 161–188.

Kuznetsov, S., and Gelfand, V. I. (1986). Bovine brain kinesin is a MT-activated ATPase. *Proc. Natl. Acad. Sci. U.S.A.* **83,** 8530–8534.

Laemelli, U.K. (1970). Cleavage of structural proteins during the assembly of bacteriophage T4. *Nature (London)* **227,** 680–685.

Lasek, R., and Brady, S. (1985). Attachment of transported vesicles to microtubules is facilitated by AMPPNP. *Nature (London)* **316,** 645–647.

Lye, R. J., Porter, M. E., Scholey, J. M., and McIntosh, J. R. (1987). Identification of a microtubule-based cytoplasmic motor in the nematode *C. elegans. Cell (Cambridge, Mass.)* **51,** 309–318.

Lye, R. J., Pfarr, C. M., and Porter, M. E. (1988). Cytoplasmic dynein and microtubule translocations. *In* "Cell Movement" (F. D. Warner and J. R. McIntosh, eds.) Vol. 2. Liss, New York (in press).

Mabuchi, I., and Okuno, M. (1977). The effect of myosin antibodies on the division of starfish blastomeres. *J. Cell Biol.* **74,** 251–263.

McIntosh, J. R. (1984). Mechanisms of mitosis. *Trends Biochem. Sci.* **100,** 195–198.

Margolis, R. L., Rauch, C. T., and Job, D. (1986). *Proc. Natl. Acad. Sci. U.S.A.* **83**, 639–643.

Meyerhof, O. (1945). *J. Biol. Chem.* **157**, 105.

Mitchison, T. J. (1986). The role of microtubule polarity in the movement of kinesin and kinetochores. *J. Cell Sci., Suppl.* **5**, 121–128.

Neff, N. F., Thomas, J. H., Grisafi, P., and Botstein, D. (1983). Isolation of the β-tubulin gene from yeast, and demonstration of its essential function "in vivo." *Cell (Cambridge, Mass.)* **33**, 211–219.

Nelson, G. A., Roberts, T. M., and Ward, S. (1982). *C. elegans* spermatozoan locomotion: Ameboid movement with almost no actin. *J. Cell Biol.* **92**, 121–131.

Olmsted, J. B. (1981). Affinity purification of antibodies from diazotined paper blots of heterogeneous protein samples. *J. Biol. Chem.* **256**, 11955–11957.

Paschal, B. M., and Vallee, R. B. (1987). Retrograde transport by the microtubule-associated protein, Map 1C. *Nature (London)* **330**, 181–183.

Piperno, G., and Luck, D. J. L. (1979). Axonemal ATPases from flagella of *Chlamydomonas reinhardtii.* *J. Biol. Chem.* **254**, 3084–3090.

Porter, M. E., Scholey, J. M., Grissom, P. M., Salmon, E. D., and McIntosh, J. R. (1988). Dynein isoforms in sea urchin eggs. *J. Biol. Chem.* **263**, 6759–6771.

Porter, M. E., Scholey, J. M., Stemple, D. L., Vigers, G., Vale, R. D., Sheetz, M. P., and McIntosh, J. R. (1987). Characterisation of the MT-movement produced by sea urchin egg kinesin. *J. Biol. Chem.* **262**, 2794–2802.

Pratt, M. M. (1984). ATPases in mitotic spindles. *Int. Rev. Cytol.* **87**, 83–105.

Pratt, M. M., Hisanaga, S., and Begg, D. A. (1984). An improved purification method for cytoplasmic dynein. *J. Cell Biochem.* **26**, 19–34.

Pryer, N. K., Wadsworth, P., and Salmon, E. D. (1986). Microtubule, bead, and organelle movements driven by soluble factors in sea urchin eggs. *Cell Motil.* **6**, 537–548.

Schatten, G. (1982). Motility during fertilization. *Int. Rev. Cytol.* **79**, 59–163.

Schliwa, M. (1984). Mechanisms of intracellular organelle transport. *Cell Muscle Motil.* **5**, 1–82.

Schnapp, B. J., Vale, R. D., Sheetz, M. P., and Reese, T. S. (1985). Single MTs from squid axoplasm support bidirectional movement of organelles. *Cell (Cambridge, Mass.)* **40**, 455–462.

Scholey, J. M., Neighbors, B., McIntosh, J. R., and Salmon, E. D. (1984). Isolation of MTs and a dynein-like MgATPase from unfertilised sea urchin eggs. *J. Biol. Chem.* **259**, 6516–6525.

Scholey, J. M., Porter, M. E., Grissom, P. M., and McIntosh, J. R. (1985). Identification of kinesin in sea urchin eggs and evidence for its localization in the mitotic spindle. *Nature (London)* **315**, 483–486.

Scholey, J. M., Ingold, A. L., and Cohn, S. A. (1988). Biochemical and motile properties of sea urchin egg kinesin. *In* "Cell Movement" (F. D. Warner and J. R. McIntosh, eds.) Vol. 2, Liss, New York (in press).

Schroeder, T. E., ed. (1986). "Methods in Cell Biology," Vol. 27. Academic Press, Orlando, Florida.

Vale, R. D., Schnapp, B. J., Reese, T. S., and Sheetz, M. P. (1985a). Movement of organelles along filaments dissociated from axoplasm of squid giant axons. *Cell (Cambridge, Mass.)* **40**, 449–454.

Vale, R. D., Reese, T. S., and Sheetz, M. P. (1985b). Identification of a novel force-generating protein (kinesin) involved in MT-based motility. *Cell (Cambridge, Mass.)* **42**, 39–50.

Vale, R. D., Schnapp, B. J., Mitchison, T., Steuer, E., Reese, T. S., and Sheetz, M. P. (1985c). Different axoplasmic proteins generate movement in opposite directions along MTs in vitro. *Cell (Cambridge, Mass.)* **43**, 623–632.

Vale, R. D., Scholey, J. M., and Sheetz, M. P. (1986). Kinesin; possible biological roles for a new MT-motor. *Trends Biochem. Sci.* **11**(11), 464–468.

Vallee, R. B. (1982). A taxol dependent procedure for the isolation of MTs and MAPs. *J. Cell Biol.* **92**, 435–442.

Vallee, R. B., and Bloom, G. S. (1983). Isolation of sea urchin egg microtubules with taxol and identification of mitotic spindle MAPs with monoclonal antibodies. *Proc. Natl. Acad. Sci. U.S.A.* **80**, 6259–6263.

Ward, G. E., Thompson, S. N., White, J. G., and Brenner, S. (1975). Electron microscopical reconstruction of the anterior anatomy of the nematode *Caenorhabditis elegans. J. Comp. Neurol.* **160**, 313–318.

Waxman, L., and Goldberg, A. (1982). Protease La from *E. coli* hydrolyses ATP and proteins in a linked fashion. *Proc. Natl. Acad. Sci. U.S.A.* **79**, 4883–4887.

Yount, R. A., Babcock, P., Ballantyne, W., and Ojala, D. (1971). Adenylyl-imidodiphosphate (AMPPNP), an ATP analogue containing a P-N-P linkage. *Biochemistry* **10**, 2484–2489.

7

The Fine Structure of the Formation of Mitotic Poles in Fertilized Eggs

NEIDHARD PAWELETZ[*] AND DANIEL MAZIA[†]

[*]Institute of Cell and Tumor Biology
German Cancer Research Center
D-6900 Heidelberg 1, Federal Republic of Germany

[†]Department of Biological Sciences
Stanford University
Hopkins Marine Station
Pacific Grove, California 93950

I. INTRODUCTION

The broad problems of fertilization were defined by the opening years of the present century. They were the activation of the egg, the intraspecific fusion of gametes leading to a union of their nuclei, and the fates of the centrosomes of the respective gametes.

The study of the activation of the egg has thrived for reasons first noted by Loeb (1909); it represents an opportunity for a deeper study of the nature of "stimulation." Loeb's perception has been justified by modern develop-

165

ments, some of which are discussed in other chapters in this book. However, one detail has evaded us so far; we do not yet know how to think of a spermatozoon as an activator.

Clearly, we have learned more about the fusion of gametes and nuclei since Boveri (1907) proved the chromosomal basis of inheritance by his experiments on dispermic eggs (experiments that were so simple technically and so prodigious intellectually). The binding of spermatozoa to eggs has been studied at a quite deep molecular level, and the approach of pronuclei is understood in terms of the behavior of microtubules.

The role of centrosomes in fertilization was once the subject of lively literature, beginning with Boveri's theory of fertilization (Boveri, 1891), according to which unfertilized eggs derive their centrosome from the male gametes. These centrosomes divide and separate to form the poles for the first mitosis; they are the ancestors of the centrosomes in all the cells in the developing organism. The maternal centrosomes, which had been operative during the meiotic divisions, become inactive in some way. Boveri based his theory on sound evidence. One cannot dispute it in the cases he studied (*Ascaris* eggs and sea urchin eggs); there centrosomes could be resolved as particles. Modern techniques employing anticentrosome antibodies confirm that the hematoxylin-stained particles were indeed centrosomes. The poles of the first mitotic spindle were formed at the male pronucleus; that fact has been confirmed by recent tests. In polyspermic eggs, the number of poles at the first mitosis was twice the number of sperm that entered the egg. Thus, the evidence was overwhelming—but could the conclusion be generalized? That remains a problem, but it is only one of the sources of later confusion.

Boveri's theory of fertilization is almost certainly valid for many cases and it makes biological sense. It is logical, because it is one way of assuring that reproduction will require two parents, thus combining two sets of genes. As we shall see (Figs. 1–5), the newest evidence (Schatten *et al.*, 1986; Paweletz *et al.*, 1987a,b) confirms the observations of Boveri and others (e.g., Wilson and Leaming, 1895) on sea urchin eggs. In the normal case, the poles of the first cleavage spindle can be traced back to the sperm centrosome. The source of confusion was the proposition that the maternal centrosome becomes inactive following meiosis. That proposition disturbed the otherwise defensible thesis of the genetic continuity of centrosomes. It was soon challenged by the successes of experiments on artificial parthenogenesis.

The subsequent debate, the outcome of which was general indifference to the problem itself, would not have been as unproductive if the contestants had remembered that artificial parthenogenesis was indeed artificial. It might then have been recognized that the strenuous methods of parthenogenesis were the means of restoring the activity of the maternal centrosome.

II. THE CENTROSOME: STRUCTURE AND FUNCTION

Fertilization of the sea urchin egg is a highly complex process (see Longo, 1973; Guidice, 1973; and this volume): the encounter of the spermatozoon and the egg sets a cascade of metabolic events into motion, which are accompanied by structural alterations starting at the egg surface and moving on toward the center of the egg. The spermatozoon is engulfed by microvilli and then the cytoplasm of the egg and soon the sperm head starts to transform into the male pronucleus. Movements of the male and female pronuclei, which are mediated by cytoskeletal elements of the egg cytoplasm, bring them near to each other to enable fusion (see Schatten, 1981a,b,c; Schatten and Schatten, 1981). After fusion, the male chromatin continues its decondensation and becomes mixed with the female chromatin, until condensation of the total chromatin of the zygote nucleus prepares the cell for mitosis. During embryogenesis, particularly in its early stages, cell division is the most evident and most important process. After fusion of the pronuclei, two microtubular asters can be seen in close association with the zygote nucleus, but in opposite positions (see Schatten et al., 1986; Paweletz et al., 1987a,b). They represent two poles of the mitotic spindle, which soon become active in the transport of the chromosomes to the two poles. After cytokinesis, the daughter cells do not rest for a long period, but immediately prepare for the next division. A chain of divisions has now begun, which does not end until the organism is complete. During all these divisions, the formation of two mitotic poles out of one center (bipolarization; Mazia, 1961, 1978) is the most important part, since only a bipolar spindle can guarantee an ordered distribution of the chromosomes to the daughter cells. By this bipolarization process, the long axis of the spindle is determined, and the destination of the separating chromosomes is defined. The plane of cleavage always is orientated perpendicular to the long axis of the spindle; thus, the direction of cytokinesis is determined also by the location of the two poles. The physical embodiment of the poles is represented by the centrosomes (Boveri, 1900; Mazia, 1984, 1987; Paweletz et al., 1984, 1987a,b; Schatten et al., 1986). Centrosomes are definite corpuscles; their form is altered in a cyclical way in the mitotic cycle, and during these structural changes, the two poles for the next mitosis arise.

In detailed review articles, Mazia (1984, 1987) described the role of the centrosomes and the function of mitotic poles: the centrosome must be considered as a flexible structure. During division, centrosomes are transferred to the daughter cells, they are reproduced there, and by their separation from each other, the two poles for the next mitosis are formed. Centrosomes can have different forms in different cell types, which can alter during the life of the cell, but the form of the centrosomes during mitosis finally determines the shape of the spindle.

During the early investigations of the problems of fertilization and cell reproduction (see Wilson, 1925; Wassermann, 1929), in order to explain the process of fertilization the reality of centrosomes as physical bodies was discovered (Boveri, 1900). The egg loses its centrosomes (in their functioning form) during oogenesis and, therefore, must "wait" for the sperm to bring in functioning centrosomes, which have been preserved in the sperm during spermatogenesis. These incorporated centrosomes act as microtubule-organizing centers (MTOCs) to form the mitotic apparatus in the fertilized egg. These microtubules are responsible for the connection between chromosomes and poles, but to realize this engagement of the chromosomes to the poles, an intensive cooperation between centrosome (pole) and kinetochores is necessary (Mazia *et al.*, 1981).

Though centrosomes (for terminology in the old literature see Wilson, 1925; Wassermann, 1929) have been known for nearly a hundred years (see Boveri, 1900), electron microscopic studies have added more information to the understanding of their structure and function by studying centrioles, which must be considered as indicators for centrosomes (see Wheatley, 1982). Detailed, fine structural observations, however, have revealed that microtubules of the mitotic apparatus do not originate from centrioles proper, but from the osmiophilic cloud around the centrioles (pericentriolar material, osmiophilic material; see Peterson and Berns, 1980; Wheatley, 1982). In recent years, it has become more evident that centrosomes comprising osmiophilic material, short microtubules, and in a number of cell types, centrioles are real MTOCs (Gould and Borisy, 1977). Centrioles in this respect only act as advertisements of centrosomes (Mazia, 1984). There is now a host of literature (Bajer and Molé-Bajer, 1972; Pickett-Heaps and Tippit, 1978; Wheatley, 1982) demonstrating a regular development of mitotic spindles without centrioles, but centrosomes must be present to organize and orientate microtubules to form a bipolar spindle. Boveri (1900) has described "das Centrosom als cyklisches Gebilde" (see also Mazia, 1984, 1987; Paweletz *et al.*, 1984; Schatten *et al.*, 1986). He could show that the centrosome in the sea urchin egg undergoes a cyclical change in shape. Since then the centrosome cycle has been nearly forgotten and neglected. Only recently could these data be confirmed and extended by electron microscopic studies on "the centrosome cycle in the mitotic cycle" (Paweletz *et al.*, 1984, 1987a,b). We shall see in the next part of this chapter that the centrosome transforms from a spherical structure to a flat, elongated platelike formation, which subdivides into two club-shaped bodies, which in turn go over into spheres again. These alterations in shape do not only influence the shape of the spindle but also the arrangements of chromosomes, implicating strong lateral interactions of the microtubules in the half-spindle.

Centrosomes act as mitotic poles: from the work on monopolar mitotic apparatus (Boveri, 1903; Bajer *et al.*, 1980; Bajer and Molé-Bajer, 1981; Mazia *et al.*, 1981), it is well known that a single centrosome is capable of cooperating with kinetochores to form a complete and normally functioning half-spindle. In these monopolar spindles, the chromosomes are arranged in a metaphase-like plate, the kinetochore facing the pole (centrosome) is engaged to the half-spindle by microtubules and, thus, is connected to the pole, while the sister kinetochore pointing away from the pole is completely void of microtubules (Mazia *et al.*, 1981). It can be demonstrated that the chromosomes in this half-spindle can approach the pole and form a new nucleus, but the cell fails to undergo cleavage. These data indicate that centrosomes are responsible for the orderly distribution of the chromosomes and can, therefore, be considered as "reproductive organs" of the cell (Mazia, 1984, 1987). As was already briefly mentioned in eggs, functioning centrosomes are lost during oogenesis: it must be assumed that, at the end of the meiotic divisions during which centrosomes still exist in the maturing egg, the functioning centrosomes are disintegrated, and the remnants, which remain in the mature, unfertilized egg, are obviously not capable of functioning like intact centrosomes. This can clearly be seen during the development of the nonpolar mitotic apparatus (Mazia, 1974; Paweletz and Mazia, 1978, 1979): the unfertilized sea urchin egg can be activated by various different means, e.g., ammoniacal sea water, (see Mazia *et al.*, 1975) to enter the cell cycle. It is evident from the replication of the chromosomes, which then condense, and the subsequent breakdown of the nuclear envelope, that a defective mitotic apparatus is formed; the chromosomes are arranged at the periphery of an asterlike formation of microtubules. Neither centrioles nor the typical compact centrosomes are present in the nonpolar mitotic apparatus, but microtubules originate from small foci of osmiophilic material, which is very similar to that found in normal centrosomes. These osmiophilic foci must be remnants of the functioning centrosome. We have to assume that such a denatured centrosome is responsible for the formation of the nonpolar mitotic apparatus, but cannot develop a bipolar spindle. In this relation, we have to ask what occurs during artificial parthenogenesis in which bipolar mitotic apparatuses are formed in spite of the lack of the sperm and cleavage takes place (see Mazia, 1978). The question can be answered by assuming that, during these experimental treatments, the remnants of the disintegrated centrosome in the unfertilized egg are renatured to form a functioning centrosome, which can then act similarly to a normal one (see Kallenbach and Mazia, 1982). These renaturing processes, however, obviously need very special prerequisites, because they occur only infrequently; parthenogenesis is a rare event (Mazia, 1984).

Examples in the literature describing various different types of spindles with different types of bodies in the polar region, e.g., in lower plant and animal cells (Fuge, 1976), and the existence of, e.g., barrel-shaped spindles in higher plants, apparently with no polar structures (see Bajer and Molé-Bajer, 1972) clearly prove that centrosomes can assume different shapes and forms and that advertising structures (centrioles and polar bodies) can either be absent or can vary from cell type to cell type. In all cases, however, the centrosomes are capable of forming a bipolar mitotic apparatus. By means of anticentrosome antibodies, Wick (1985) could clearly show centrosomes in the polar regions of the barrel-shaped spindle of higher plants.

It was mentioned earlier that centrosomes are characterized by their capacity for organizing and orientating microtubules (acting as MTOCs); it could, however, be shown that they obviously can also be considered as membrane-organizing or -assembling centers (see Mazia, 1984). During the formation of the mitotic apparatus in the center of the fertilized sea urchin egg, for example, where later on the spindle is found, a clear zone can be identified which is free of obscuring structures such as yolk granules or mitochondria. This clear zone is filled with densely packed membranous structures which surround the chromosomes and the microtubules (Harris, 1975) and seem to originate from the centrosomal region. In recent years, a host of articles have been published (see Paweletz, 1981; Hepler and Wolniak, 1984; Paweletz and Schroeter, 1986) about the existence of membranes in a definite pattern in the mitotic apparatus of a variety of cells. The ordered arrangement of such structures is asking for an organizing principle; the best candidate can be seen in the centrosome.

Taking all the before-mentioned data into account, Mazia (1984) proposed the following nature of the centrosome. (1) Centrosomes can assume different shapes in different cells and can change their shapes in different physiological stages of the cell type. (2) Centrosomes are composed of subunits. (3) Each centrosomal unit initiating microtubules defines its starting point and its direction of growth. (4) The arrangement of the initiating units and of their orientation determines the form of the structures (e.g., mitotic apparatus), which is built up of microtubules.

These propositions on the nature of centrosomes are supported by experimental data, which will be presented in the next part of this chapter. Another idea, however, should briefly be mentioned here although the available experimental data have not yet been interpreted in this way: when studying the formation of the polar bodies during the maturation of the starfish egg (N. Paweletz and D. Mazia, unpublished observations), an intimate association of one of the centrosomes (i.e., poles of the meiotic spindle) with the cortex of the egg becomes evident. The shape of this centrosome depends on the behavior of the cortical region, which, later on, forms the envelope of the polar bodies. It can be supposed that the centrosome of the meiotic spindle in the neighborhood of the cortex determines the size of the cortical patch, which becomes the envelope of the polar bodies, and the arrangement of the contractile elements to pinch it off. If we transfer these results and suppositions to the problems of cleavage after nuclear division, we can argue that the centrosomes might also determine the part of the cortex which forms the cleavage furrow and its direction in sending messages or even centrosomal units to the equator of the egg using microtubules, which radiate out from both asters and meet and cross in the equatorial region as rails for the transport.

This hypothesis could be experimentally tested by studying carefully the shape and behavior of the centrosomes in cells in which cleavage is only attempted and does not take place in the form of a contractile ring. For now the question of how the cleavage furrow is determined in an anastral apparatus has to remain open.

This section has tried to elucidate the structure and the function of the centrosome as the embodiment of the mitotic pole, but only little has been said about the bipolarization process during which the two mitotic poles of a bipolar spindle are formed; the bipolarization, however, is one of the most important processes during cell division (Mazia, 1978). It has been described before that during maturation, the sea urchin egg, as in many other oocytes, loses its functioning centrosome by disintegration. To guarantee biparental inheritance, the sperm must, therefore, bring in functioning centrosomes, which have to take over the necessary functions during all subsequent mitoses.

In the following sections, we will describe the behavior of the centrosomes before and during the first mitosis in the fertilized sea urchin egg and try to explain the bipolarization process (see also Paweletz *et al.*, 1984, 1987a,b).

III. THE FIRST BIPOLARIZATION IN THE FERTILIZED EGG

Even careful ultrastructural analyses of the unfertilized sea urchin egg by means of sequential sections could not demonstrate normally functioning centrosomes, assuming that they are indicated by the existence of centrioles or any other organized structure; centrioles or similar structures are neither present somewhere in the cytoplasm nor in the neighborhood of the female pronucleus. Kallenbach and Mazia (1982) have shown in sea urchin eggs that artificial means such as hypertonic treatment can induce the formation of centrioles, which, in turn, indicate the development of centrosomes. This occurs always in the neighborhood of the female pronucleus. Their data show that the potency to form centrosomes is present in the unfertilized egg, that it can be renatured by special treatment, but that it does not represent a normal, functioning centrosome. It looks as if a functioning centrosome in interphase or an analog must somehow be related to or even associated with the nucleus. The type of association is not yet known (for discussion, see Bornens, 1977).

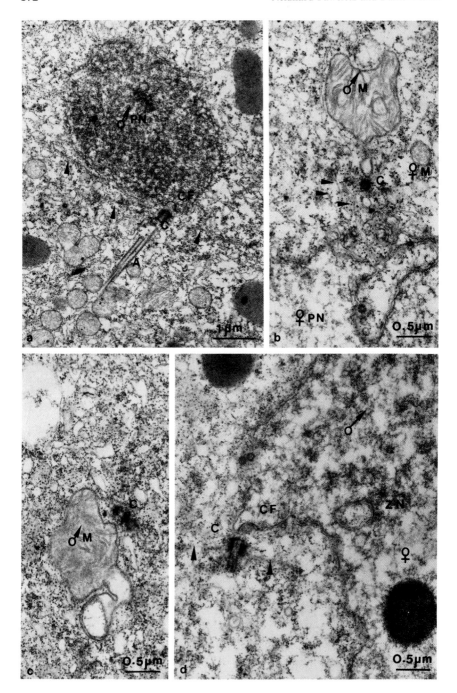

This is completely different in the spermatozoon: Longo and Anderson (1968, 1969) and Longo (1973) have described two centrioles at the base of the sperm head. The proximal centriole is located near a bowl-shaped depression of the sperm nucleus, the centriolar fossa, while the distal centriole can be found in its neighborhood perpendicular to the proximal one (Longo and Anderson, 1969). After the sperm has entered the egg, the following alterations can be seen: the sperm head transforms into the male pronucleus by decondensing its chromatin, and the bowl-shaped depression of the nuclear envelope (centriolar fossa equals centrosomal fossa) remains visible at the base. At the bottom of the fossa, the inner nuclear membrane is strongly osmiophilic and somewhat thicker than the rest of the nuclear membranes, obviously by apposition of osmiophilic material. The fossa is filled with fine, fibrillar material with a rather high osmiophilia. These fibrils connect one centriole (Longo's and Anderson's proximal centriole) with the nuclear envelope of the male pronucleus. This centriole is surrounded by osmiophilic material and is often so densely enwrapped that its tubular composition is partly hidden (Fig. 1d; see also Fais et al., 1986). In spite of this obscuring material, it is evident that only one centriole is present. In some cases, this centriole still acts as a basal body for the axoneme, the former sperm tail (Fig. 1a). Later on, the sperm tail detaches from the sperm head, can be found in its neighborhood (Fig. 2a), and soon becomes resorbed. Originating from the osmiophilic material, microtubules radiate out especially toward both flanks of the centriole (Fig. 1a), so that this structure must be considered as a MTOC. Its function as a MTOC and the osmiophilic material surrounding the centriole clearly identify this structure as a centrosome. In the following text, we will call this assembly the sperm-head centrosome. The aster of microtubules around this centrosome must be considered as the sperm aster, which can easily be identified by immunofluorescence by means of antibodies against tubulin (see Schatten, 1981a,b,c).

Fig. 1. (a) The chromatin of the male pronucleus ($\male$ PN) is slightly decondensed. The sperm-head centrosome (C) is in connection with the centrosomal fossa (CF); the axoneme (A) of the sperm tail is still present. Arrowheads indicate microtubules. (b) The male mitochondrion ($\male$ M) is clearly distinguishable from the female mitochondria ($\female$ M) by its size and the arrangement of the cristae mitochondriales. The mitochondrial centrosome (C) is partly separated from the mitochondrial protrusion. $\female$ PN is the female pronucleus; arrowheads indicate microtubules. (c) The mitochondrial centrosome (C) attached to the male mitochondrion ($\male$ M) has started to reduplicate, though its centriole is still immature. (d) Zygote nucleus (ZN) just after complete fusion of the two pronuclei. The chromatin of the pronuclei ($\female$ and $\male$) have not yet mixed completely. The sperm-head centrosome (C) is now attached to the envelope of the zygote nucleus. Microtubules (arrowheads) are originating from the centrosome. The centrosomal fossa (CF) is still recognizable.

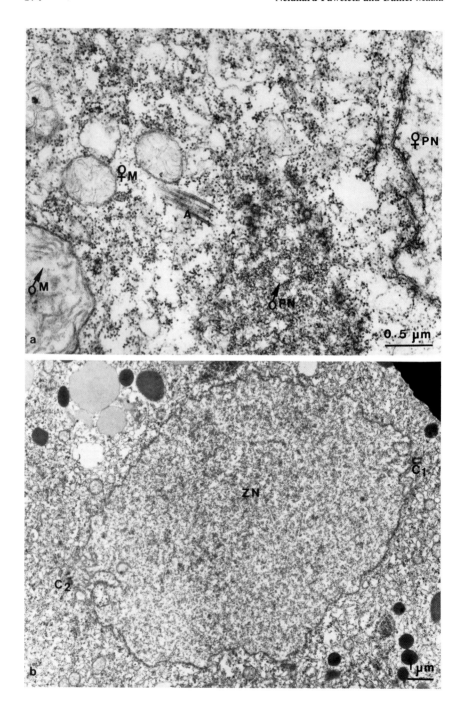

The fact that the sperm tail (axoneme) remains attached to the sperm head in the early stages after incorporation of the sperm leads to the assumption that the midpiece, which mainly consists of one large mitochondrion (Longo and Anderson, 1969), is also incorporated together with the sperm head. Careful fine structural analyses (Paweletz *et al.*, 1987a) of the perinuclear region about 20 min after the entrance of the sperm, but prior to the fusion of the two pronuclei, have shown the sperm mitochondrion detached from the sperm head and tail, but remained in the neighborhood of the female pronucleus (Fig. 2a and b). This mitochondrion is clearly distinguishable from the egg mitochondria by its larger size and a different arrangement of the cristae mitochondriales. Very often, a noselike protusion of the mitochondrion is formed, which is crowned by a small, densely packed accumulation of osmiophilic material (Fig. 1b). This accumulation obviously is tightly attached to the mitochondrial protrusion; within this accumulation, microtubules cannot be identified, but it acts as a center of a small aster of microtubules, which radiate out from the periphery of the accumulation (Fig. 1b). Due to the strong osmiophilia and the nature of this structure as a MTOC, we assume that it is the second centrosome, which comes in with the sperm. In the following text, we will call this the mitochondrial centrosome. It is identical with the distal centriole described by Longo and Anderson (1968, 1969).

Immunocytochemical studies with antibodies against centrosomes (Schatten *et al.*, 1986) often reveal two bright spots at the base of the sperm head, thus demonstrating two centrosomes per spermatozoon. Shortly after incorporation of the sperm into the egg, one sperm aster can be identified (Schatten, 1981a). The electron microscopic examination, however, shows two rather small asters around each of the two centrosomes. How can this apparent contradiction be explained? At the beginning of the migration of the two pronuclei toward each other, the two asters at the sperm head are so near to each other that the relatively low resolution of the immunofluorescent image does not resolve two centers. Later on when already two asters can be seen, the two centrosomes have separated far enough from each other to be recognized by light microscopy, but this occurs only after fusion of the two pronuclei (Schatten *et al.*, 1986). Already before fusion when the male pronucleus has approached the female one, the two centrosomes start to separate from each other, but both are in the intimate neighborhood of the female pronucleus (Fig. 2a) and,

---

Fig. 2. (a) The three important elements of fertilization are in close neighborhood to each other. δM is the male mitochondrion, $\female$M is the female mitochondria, A is the axoneme of the sperm tail, δPN is the male pronucleus, and $\female$PN is the female pronucleus. (b) Two centrosomes (C_1 and C_2) are located in opposite position to the zygote nucleus (ZN). C_1 is the stationary (former sperm head), C_2 is the movable (former mitochondrial) centrosome.

as will be described in detail later, in contact with the nuclear membrane by means of microtubules.

The sperm head with the attached centrosome approaches the female pronucleus more and more. In spite of the decondensation of the male chromatin, the form of the male pronucleus still is an ellipsoid (Fig. 1a). The centrosomal fossa is clearly visible and has approached the surface of the female pronucleus. Short protrusions of the envelope of the maternal nucleus are forming toward the male pronucleus. Finally, the two neighboring envelopes fuse, so that a continuum between the two interiors of the pronuclei is present (see Longo and Anderson, 1968; Longo, 1973). The initially narrow, diaphragm-like connection between the two former pronuclei increases in size until the male chromatin, which is still more condensed than the female one, is incorporated completely into the former female nucleus. Both types of chromatin obviously mix with each other, since, a short time after fusion, they can no longer be distinguished from each other.

After fusion, the envelope of the male pronucleus becomes completely part of the envelope of the zygote nucleus; this means that the sperm-head centrosome, which has been attached previously to the nuclear envelope of the male pronucleus by a fine fibrillar network, is now connected with the zygote nucleus (Fig. 1d). Since the centrosomal fossa does not regress rapidly nor does the osmiophilia of the inner nuclear membrane vanish, this centrosome can be recognized clearly until shortly before the breakdown of the nuclear envelope for the first mitotic prophase.

The mitochondrial centrosome also undergoes some changes. The accumulation of osmiophilic material separates from the protrusion of the mitochondrion (Fig. 1b). During this time, the compact accumulation transforms into a hollow cylinder in which short pieces of microtubules become visible. This indicates the "crystallization" of a new centriole within the osmiophilic mass (see Fig. 1c). While the centriole in the sperm-head centrosome exhibits the length, which is generally known for centrioles of higher animals, the cylinder (procentriole) of the mitochondrial centrosome has only half the length of a normal centriole (Figs. 1c and 2b).

In spite of the immature centriole, this centrosome also acts as a MTOC and is the center of a second aster (Fig. 1b; for discussion of the identification of two "sperm" asters, see above), since microtubules radiate out into nearly all directions, except toward the former point of attachment at the mitochondrion. There must be a close relationship between this centrosome and the surface of the female pronucleus or the zygote nucleus, respectively, since the nuclear surface exhibits deep invaginations and short protrusions just opposite to the centrosome (Fig. 1b). This is obviously realized by small bundles of microtubules, which span between nucleus and centrosome. In slightly later stages, the sperm mitochondrion is no longer visible; it is either resorbed completely or disintegrated into smaller units, which can no longer be distinguished from the egg mitochondria. The separation of the mitochondrial centrosome and the sperm mitochondrion from each other occurs only when the complex has reached a definite short distance from the egg or zygote nucleus, since a solitary (mitochondrial) centrosome far away from the nucleus could never be found.

The facts that only one sperm aster can be seen in immunofluorescence images of antitubulin antibodies in the early stages after incorporation of the sperm (see Schatten, 1981a) and that the male mitochondrion remains always in close neighborhood to the sperm head indicate that the separation of the two centrosomes from each other into opposite positions at the zygote nucleus can take place only when the whole complex has reached a definite distance from the egg or zygote nucleus, respectively. The sperm-head centrosome and the mitochondrial centrosome seem to be associated or related to the zygote nucleus at random, but, at the beginning, never in opposite positions (Fig. 2a). How then does the first bipolarization occur (see also Paweletz *et al.*, 1987b)? The sperm-head centrosome stays attached to the nuclear envelope of the zygote nucleus nearly all the time; it is, therefore, conceivable to assume that it remains stationary without being moved, while the mitochondrial centrosome is the movable (motile?) one. The sperm-head centrosome very often can be found at the tip of a protrusion of the nuclear envelope (Fig. 2b). This tip is characterized by the bowl-shaped depression of the centrosomal fossa and the strong osmiophilia of the inner nuclear membrane at the bottom of the bowl. The fossa now has become shallower, but is still clearly recognizable (Fig. 1d). The centriole of the stationary centrosome, which has been formerly the sperm-head centrosome, is characterized by an osmiophilic "bottom," which "closes" the proximal (orientated toward the nuclear surface) end of the cylinder (Fig. 2b). In addition, the proximal half of the microtubular triplets is enwrapped densely in strongly osmiophilic material. This covered part of the sperm-head centriole corresponds exactly to the initial size of the mitochondrial centriole.

Soon after fusion of the two pronuclei (in a few cases also before fusion), both centrioles start to reduplicate (see Fig. 1c). The daughter of the sperm-head centriole is formed in the proximal half exactly perpendicular to the mother centriole. The mitochondrial centriole, which is only half as long as a normal centriole, starts reduplication still in its initial small size; here, the daughter is of the same (reduced) size as the mother (see Fig. 1d).

From the results obtained from observations of the mitochondrial centrosome, one can argue that, to function as a MTOC, a mature (fully sized) centriole is not necessary; the same is true for its reduplication. About 10 min after fusion of the two pronuclei, the two centrosomes, each containing one pair of centrioles, can be found near or at the nuclear surface, but not yet in opposite positions. One pair represents the former sperm-head centrosome; it is still in connection with the nuclear surface, while the second pair belongs to the former mitochondrial centrosome, which is still free (Fig. 2b). As already mentioned before, microtubules are radiating out from both centrosomes in all directions, but mainly toward the nuclear surface. In the neighborhood of both centrosomes, the nuclear surface is deeply invaginated, and protrusions are also formed. Small bundles of microtubules are found within these nuclear folds, sometimes in direct contact with the nuclear envelope at the bottom of the crypt. These bundles originate obviously at the centrosomes and, by elongation, are forcing the nuclear envelope to invaginate. Similar observations could be made in mammalian cells during early prophase, when microtubules induce invaginations of the nuclear surface (Paweletz, 1974). Microtubules cannot only be found within the folds of the nuclear envelope, but also as tangents to the surface of the nucleus. Sometimes these microtubules are surrounded and enwrapped densely by osmiophilic material. From these results, we conclude that the two centrosomes are separated from each other and brought into opposite positions by means of microtubules, which grow between them, thereby pushing them apart. This is very similar to what was described for cells of higher animals by Rattner and Berns (1976a,b) for the separation of centrioles. Together with the outgrowing microtubules, osmiophilic material is spread along one side of the surface of the zygote nucleus, until the two centrosomes form one flat centrosomal plate on top of the nuclear surface.

Such a centrosomal plate should be able to act as a MTOC; therefore, one would expect an asymmetric microtubular structure originating from or near the nuclear surface. When mitotic apparatuses of stages before the first mitotic prophase are isolated, nuclei with an asymmetric microtubular apparatus are found, and the microtubules seem to originate from the surface of the nucleus. Careful light microscopic observations (D. Mazia and N. Paweletz, unpublished) reveal that the microtubules in fact originate at or very near to the nuclear surface, but at the opposite side from where the microtubular apparatus seems to start, so that the microtubules are enwrapping the nucleus like a basket. We have called these apparatuses "beards." Their development can be explained by the assumption that the flat centrosomal plate acts as a MTOC from which the microtubules grow out in only one direction, namely, toward the nucleus and past it. These beards are precursors of the mitotic spindle, which "crystallizes" out as soon as the centrosomal plate continues its development and splits into two half-plates. At the beginning of the spindle formation, the asters appear rather asymmetric with longer microtubules at the side of the previous beard; with proceeding time, however, the astral rays become symmetric. While the beard formation before the first mitosis is expressed only rather weakly, it can best be identified before the second and the third mitosis.

The phenomenon of the beard indicates that the centrosome at this time is a flat cap on top of one side of the zygote nucleus with the pairs of centrioles near or at the ends, but always in the neighborhood of the nucleus (see also Mazia, 1987).

During all these processes, the former sperm-head centrosome still stays attached to the nuclear envelope; it is, therefore, conceivable that it remains stationary, while the free, former mitochondrial centrosome is moved on an arc around one-half of the nucleus during the spreading process of the centrosome. There is no other example known in which one centrosome is stationary, while the second is movable. Though the mechanism of spreading of the centrosome does not seem to be unique, the situation of the two centrosomes is not only unique, but this is the only time in the embryonic development of the sea urchin when one centrosome is attached to the nuclear envelope. It can already be revealed here that, during all following mitoses, both centrosomes are free from the nuclear envelope and that, only before the first mitosis, is this peculiar behavior expressed.

It can be demonstrated that the characteristic features of the former sperm-head centrosome, the bowl-shaped depression and the strong osmiophilia, become weaker with proceeding time. Shortly before the onset of the breakdown of the nuclear envelope, the former sperm-head centrosome, too, can be found detached from the nuclear envelope. Deep invaginations near the centrosomes on both poles of the slightly elongated preprophase nucleus are characteristic for this stage. The previously small-sized mother and daughter centrioles of the former mitochondrial centrosomes now exhibit their full size, which is known from many other cell types and characteristic of the sperm-head centrioles from the beginning. Both pairs of centrioles are enwrapped in a thin layer of osmiophilic material. The two centrosomes have reached their final position, the first bipolarization process of the sea urchin development is finished, and the bipolar mitotic apparatus can be fully established.

IV. THE CENTROSOMAL CYCLE IN THE MITOTIC CYCLE

We described above that the centrosomes, which had been brought in by the sperm, have formed a thin flat plate near one-half of the surface of the preprophase nucleus. With proceeding time during prophase, the centrosomal plate separates into two half-plates; during prometaphase, it starts to condense more around the pair of centrioles to become nearly a sphere (Fig. 3a and b; see Paweletz et al., 1984; Schatten et al., 1986).

In the following text, we will concentrate our attention on only one of the two centrosomes but always keep in mind that the same changes also take place in the second half-spindle.

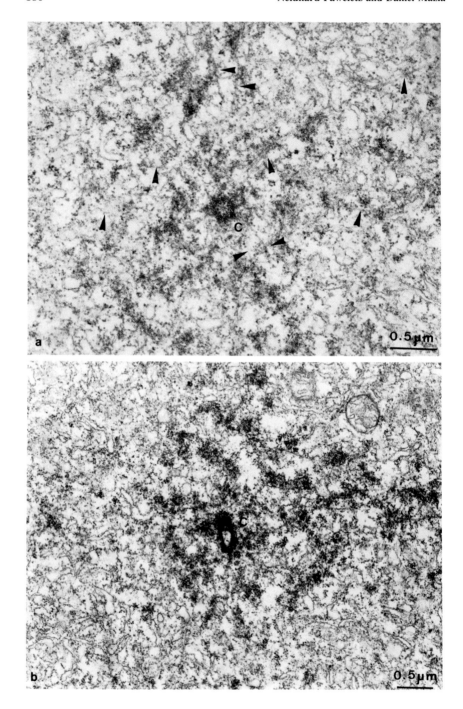

While, during prometaphase, the centrosomal mass concentrates around the pair of centrioles, in metaphase, the centrosome has become a sphere (Fig. 3a). The pair of centrioles is surrounded by strongly osmiophilic material and is located nearly in the center of the sphere. Some microtubules are radiating out from the periphery of the centrosome, and a few of them can be found within the centrosome itself. The centrosome is now the center of the aster; the majority of aster microtubules grow out from a spherical shell, which is larger than the centrosome and located in some distance from it. The zone in between exhibits only a few microtubules, which are not easily identifiable, e.g., by immunological means (see Schatten *et al.*, 1986).

To uncouple the behavior of the chromosomes during mitosis from that of the centrosomes, which normally go hand in hand but make an ultrastructural analysis more complicated, we used a technical trick, which was introduced by Mazia *et al.* (1960) and discussed later on by Harris (1976), Sluder (1978), and Mazia *et al.* (1981): fertilized eggs entering the first mitosis are treated with 2-mercaptoethanol. While the chromosomal cycle is arrested in metaphase, the centrosomal cycle goes on and can, thus, be investigated in detail independent of the mitotic stage (Paweletz *et al.*, 1984).

The spherical centrosome of the metaphase spindle soon starts to elongate, thereby becoming nearly ellipsoid (Fig. 3b). Thin sections through the centrosome reveal alterations in the location of the centrioles. They have separated from each other, but maintain their rectangular orientation (Fig. 4a). The elongation process of the centrosome continues, resulting in a stretched, flattened structure, which now resembles a cigar (Fig. 4a). The flattening process goes on until, in thin sections, only a rather thin osmiophilic line can be seen (Figs. 4b and 5b); sections under various angles through the flat centrosome, however, clearly show that a flat plate, instead of a line, has formed. During the spreading and flattening of the centrosome, the centrioles have separated more; they are now located in the foci of the very flat ellipsoid (Fig. 4a). The final stage of the centrosome is a very elongated flat plate (Figs. 4a and 5a and b) with centrioles near to, but not exactly at the end of, the osmiophilic accumulation. The data clearly indicate that, during the spreading and flattening process of the centrosome, the centrioles have also separated from each other. However, they always remain inside the centrosome. As can be identified in ultrathin sections, they still retain their characteristic rectangular orientation

←———

Fig. 3. (a) A centrosome during early prometaphase. One centriole (C) is found in this section in the middle of the accumulation of osmiophilic material. Microtubules (arrowheads) are visible within the centrosome. (b) A centrosome during late metaphase. The osmiophilic material has concentrated more around one of the centrioles (C). Compare to a.

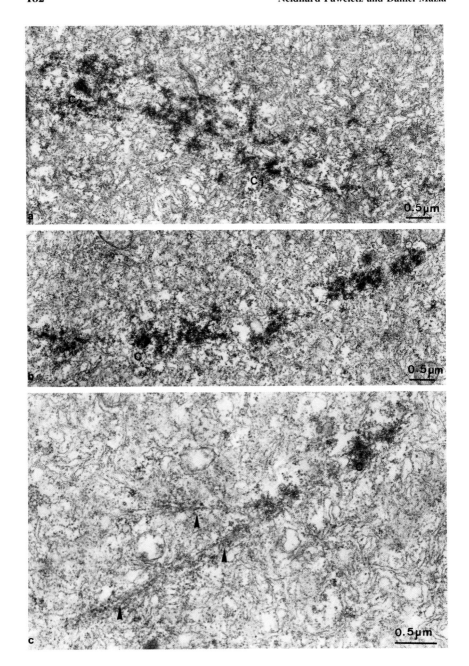

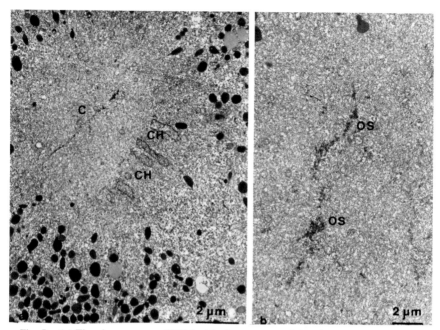

Fig. 5. (a) The chromosomes (CH) in telophase are parallel and move to the flat plate of the centrosome (C), not to the centrioles (not visible in this section). (b) The osmiophilic material (OS) in this thin section formed an extended thin osmiophilic line. This is the form of the centrosome in late telophase. Compare to a.

(Fig. 4a). This fact asks for a system of cytoskeletal elements, which keeps the centrioles in their respective orientations. Within the metaphase centrosome, microtubules have been shown (Fig. 3a). Careful ultrastructural analyses reveal a whole system of microtubules; small bundles of them are extending between the centrioles within the centrosomal plate. While a few microtubules seem to originate in the neighborhood of the centrioles, there are small bundles within the centrosome running nearly perpendicular to the long axis of the spindle (Fig. 4c). The electron microscopic data are confirmed by observations with the polarization microscope (Paweletz *et al.*, 1984; D. Mazia, unpublished)

Fig. 4. (a) The centrosome has started to spread and elongate. The two centrioles (C_1 and C_2) have separated from each other, their rectangular orientation is preserved. (b) The centrosome is visible in the form of an osmiophilic line which corresponds to an elongated flat plate. C is one of the centrioles. (c) Small microtubular bundles (arrowheads) are visible within the centrosome. C is one of the centrioles.

by antitubulin immunofluorescence (G. Schatten and H. Schatten, unpublished), showing a microtubular cytoskeleton within the spread centrosome. These data speak in favor of the assumption that microtubules are not involved only in the separation of the centrioles within the centrosome, but also are responsible for the spreading and flattening of the centrosome itself. In cells of higher animals, it could be shown by Rattner and Berns (1976a,b) that microtubules are separating the centrioles during early prophase. In the old literature (see Wilson, 1925; Wassermann, 1929), the separation of the centrosomes by means of spindle fibers (i.e., bundles of microtubules) is described as centrodesmose. It now becomes evident that the same or a very similar mechanism of spreading of the centrosome during the early and the late mitotic stages is effective.

The flat centrosomal plate thins out in its center, and two club-shaped accumulations are formed, which soon transform into spheres. Thus, one centrosomal cycle comes to its end and the next begins, while the cell goes into its next division.

The results described here for the fertilization of sea urchin eggs might be valid for several other types of sperm and oocytes, while, in other types, still other mechanisms of bipolarization and the centrosomal cycle might occur. In any case, these findings must be seen in a much wider context of the structure and function of centrosomes and their role in fertilization. In detailed reviews, Mazia (1984, 1987) has presented a picture of the centrosome, which reaches far beyond fertilization.

ACKNOWLEDGMENTS

The work of Neidhard Pawletz was supported by grants from the NATO, the Verein zur Förderung der Krebsforschung in Deutschland e.V., and the German Cancer Research Center. The work of Daniel Mazia was supported by grants from the National Science Foundation.

REFERENCES

Bajer, A. S., and Molé-Bajer, J. (1972). Spindle dynamics and chromosome movements. *Int. Rev. Cytol. Suppl.* **3**, 1–271.

Bajer, A. S., and Molé-Bajer, J. (1981). Mitosis: Studies of living cells—A revision of basic concepts. In "Mitosis/Cytokinesis" (A. M. Zimmerman and A. Forer, eds.), pp. 227–299. Academic Press, New York.

Bajer, A. S., De Brabander, M., Molé-Bajer, J., De Mey, J., Geuens, G., and Paulaitis, S. (1980). Aster integration and functional autonomy of monopolar spindle. In "Microtubules and Microtubule Inhibitors" (M. De Brabander and J. De Mey, eds.), pp. 339–425. Elsevier, Amsterdam.

Bornens, M. (1977). Is the centriole bound to the nuclear membrane? *Nature (London)* **270**, 80–82.

Boveri, M. (1903). Über Mitosen bei einseitiger Chromosomenbindung. *Jena. Z. Naturwiss.* **37**, 401–446.

Boveri, T. (1900). Zellen-Studien. IV. Über die Natur der Centrosomen. *Jena. Z. Naturwiss.* **356**, 1–220.

Boveri, Th. (1891). Merkel, F. S., Bonnet, R. Befruchtung. *Anat. Hefte* Abt. 2, Bd. 1.

Boveri, Th. (1907). Zellenstudien. VI. Die Entwicklung dispermer Seeigeleier. Ein Beitrag zur Befruchtungslehre und zur Theorie des Kerns. Gustav Fischer, Jena.

Fais, D. A., Nadezhina, E. S., and Chentsov, Y. S. (1986). The centriolar rim. The structure that maintains the configuration of centrioles and basal bodies in the absence of their microtubules. *Exp. Cell Res.* **164**, 27–34.

Fuge, H. (1976). Ultrastructure of mitotic cells. *In* "Mitosis Facts and Questions" (M. Little, N. Paweletz, C. Petzelt, H. Ponstingl, D. Schroeter, and H. -P. Zimmermann, eds.), pp. 57–68. Springer-Verlag, Berlin.

Gould, R. R., and Borisy, G. G. (1977). The pericentriolar material in Chinese hamster ovary cells nucleates microtubule formation. *J. Cell Biol.* **73**, 601–615.

Guidice, G. (1973). "Developmental Biology of the Sea Urchin Embryo." Academic Press, New York.

Harris, P. (1975). The role of membranes in the organization of the mitotic apparatus. *Exp. Cell Res.* **94**, 409–425.

Harris, P. (1976). Structural effects of mercaptoethanol during mitotic block of sea urchin eggs. *Exp. Cell Res.* **97**, 63–73.

Hepler, P. K., and Wolniak, S. M. (1984). Membranes in the mitotic apparatus: Their structure and function. *Int. Rev. Cytol.* **90**, 169–238.

Kallenbach, R. J., and Mazia, D. (1982). Origin and maturation of centrioles in association with the nuclear envelope in hypertonic-stressed sea urchin eggs. *Eur. J. Cell Biol.* **28**, 68–76.

Lillie, F. R. (1919). "Problems of Fertilization." University of Chicago Press, Chicago.

Loeb, J. (1909). "Die chemische Entwicklungserregung der tierischen Eier (künstliche Parthenogenese)." Springer-Verlag, Berlin and New York.

Longo, F. J. (1973). Fertilization: A comparative ultrastructural review. *Biol. Reprod.* **9**, 149–215.

Longo, F. J., and Anderson, E. (1968). The fine structure of pronuclear development and fusion in the sea urchin, *Arbacia punctulata. J. Cell Biol.* **39**, 339–368.

Longo, F. J., and Anderson, E. (1969). Sperm differentiation in the sea urchins *Arbacia punctulata* and *Strongylocentrotus purpuratus. J. Ultrastruct. Res.* **27**, 486–509.

Mazia, D. (1961). Mitosis and the physiology of cell division. *In* "The Cell" (J. Brachet and A. E. Mirsky, eds.), Vol. III, pp. 80–440. Academic Press, New York.

Mazia, D. (1974). Chromosome cycles turned on in unfertilized sea urchin eggs exposed to NH_4OH. *Proc. Natl. Acad. Sci. U.S.A.* **71**, 690–693.

Mazia, D. (1978). Origin of twoness in cell reproduction. *In* "Cell Reproduction. In Honor of Daniel Mazia" (E. R. Dirksen, D. M. Prescott, and C. F. Fox, eds.), pp. 1–14. Academic Press, New York.

Mazia, D. (1984). Centrosomes and mitotic poles. *Exp. Cell Res.* **153**, 1–15.

Mazia, D. (1987). The chromosome cycle and the centrosome cycle in the mitotic cycle. *Int. Rev. Cytol.* **100**, 49–92.

Mazia, D., Harris, P. J., and Bibring, T. (1960). The multiplicity of the mitotic centers and the time-course of their duplication and separation. *J. Biophys. Biochem. Cytol.* **7**, 1–20.

Mazia, D., Schatten, G., and Steinhardt, R. A., (1975). Turning on of activities in unfertilized sea urchin eggs: Correlation with changes of the surface. *Proc. Natl. Acad. Sci. U.S.A.* **72,** 4469–4473.

Mazia, D., Paweletz, N., Sluder, G., and Finze, E. -M. (1981). Cooperation of kinetochores and pole in the establishment of monopolar mitotic apparatus. *Proc. Natl. Acad. Sci. U.S.A.* **78,** 377–381.

Paweletz, N. (1974). Elektronenmikroskopische Untersuchungen an frühen Stadien der Mitose bei HeLa-Zellen. *Cytobiologie* **9,** 368–390.

Paweletz, N. (1981). Membranes in the mitotic apparatus. *Cell Biol. Int. Rep.* **5,** 323–336.

Paweletz, N., and Mazia, D. (1978). The nuclear mitotic apparatus in sea urchin eggs. *In* "Cell Reproduction. In Honor of Daniel Mazia" (E. R. Dirksen, D. M. Prescott, and C. F. Fox, eds.), pp. 495–503. Academic Press, New York.

Paweletz, N., and Mazia, D. (1979). Fine structure of the mitotic cycle of unfertilized sea urchin eggs acitvated by ammoniacal sea water. *Eur. J. Cell Biol.* **20,** 37–44.

Paweletz. N., and Schroeter, D. (1986). On the fine structure of the mitotic apparatus of mammalian cells. *In* "Proceedings of the Fourth International Conference of Environmental Mutagens" (C. Ramel, B. Lambert, and J. Magnusson, eds.), pp. 376–381. Alan R. Liss, New York.

Paweletz, N., Mazia, D., and Finze, E. -M. (1984). The centrosome cycle in the mitotic cycle of sea urchin eggs. *Exp. Cell Res.* **152,** 47–65.

Paweletz, N., Mazia, D., and Finze, E. -M. (1987a). Fine structural studies of the bipolarization of the mitotic apparatus in the fertilized sea urchin egg. I. The structure and behavior of centrosomes before fusion of the pronuclei. *Eur. J. Cell Biol.* **44,** 195–204.

Paweletz, N., Mazia, D., and Finze, E. -M. (1987b). Fine structural studies of the bipolarization of the mitotic apparatus in the fertilized sea urchin egg. II. Bipolarization before the first mitosis. *Eur. J. Cell Biol.* **44,** 205–213.

Peterson, S. P., and Berns, M. W. (1980). The centriolar complex. *Int. Rev. Cytol.* **64,** 81–106.

Pickett-Heaps, J. D., and Tippit, D. H. (1978). The diatom spindle in perspective. *Cell* **14,** 455–467.

Rattner, J. B., and Berns, M. W. (1976a). Centriole behavior in early mitosis of rat kangaroo cells (PtK$_2$). *Chromosoma* **54,** 387–395.

Rattner, J. B., and Berns, M. W. (1976b). Distribution of microtubules during centriole separation in rat kangaroo (Potourus) cells. *Cytobios* **15,** 37–43.

Schatten, G. (1981a). The movements of the nuclei during fertilization. *In* "Mitosis/Cytokinesis" (A. M. Zimmerman and A. Forer, eds.), pp. 59–82. Academic Press, New York.

Schatten, G. (1981b). The movements and fusion of the pronuclei at fertilization of the sea urchin *Lytechinus variegatus:* Time-lapse video microscopy. *J. Morphol.* **167,** 231–247.

Schatten, G. (1981c). Sperm incorporation, the pronuclear migrations, and their relation to the establishment of the first embryonic axis: Time-lapse video microscopy of the movements during fertilization of the sea urchin *Lytechinus variegatus. Dev. Biol.* **86,** 426–437.

Schatten, G., and Schatten, H. (1981). Effects of motility inhibitors during sea urchin fertilization. *Exp. Cell Res.* **135,** 311–330.

Schatten, H., Schatten, G., Mazia, D., Balczon, R., and Simerly, C. (1986). Behavior of centrosomes during fertilization and cell division in mouse oocytes and in sea urchin eggs. *Proc. Natl. Acad. Sci. U.S.A.* **83,** 105–109.

Sluder, G. (1978). The reproduction of mitotic centers: New information on an old experiment.

In "Cell Reproduction. In Honor of Daniel Mazia" (E. R. Dirksen, D. M. Prescott, and C. F. Fox, eds.), pp. 563–569. Academic Press, New York.

Wassermann, F. (1929). "Die lebendige Masse. II. Wachstum und Vermehrung der lebendigen Masse." Julius Springer, Berlin.

Wheatley, D. N. (1982). "The Centriole: A Central Enigma of Cell Biology." Elsevier, Amsterdam.

Wick, S. M. (1985). Immunofluorescence microscopy of tubulin and microtubule arrays in plant cells. III. Transition between mitotic/cytokinesis and interphase microtubule arrays. *Cell Biol. Int. Rep.* **9,** 357–371.

Wilson, E. B. (1925). "The Cell in Development and Heredity." Macmillan, New York.

Wilson, E. B., and Leaming, E. (1895). "An Atlas of the Fertilization, and Karyokinesis of the Ovum." Macmillan, New York.

8

Intermediate Filaments during Fertilization and Early Embryogenesis

HARALD BIESSMANN AND MARIKA F. WALTER

Developmental Biology Center
University of California, Irvine
Irvine, California 92717

I. INTRODUCTION

The cytoskeleton is a complex, three-dimensional array of possibly inter-linked components, consisting of microtubules, actin microfilaments, and intermediate-sized filaments (IF). Specific proteins are associated with each

189

component, adding another level of complexity to the system. The cytoskeleton is involved in numerous intracellular movements that occur during and immediately after fertilization and in the early cleavage stages of embryogenesis. For example, cytoskeletal elements may be required in distributing "morphogenetic determinants" or mRNA asymmetrically in the egg cytoplasm, and an involvement of cytoskeletal components in this "ooplasmic segregation" process has been proposed in a variety of species, such as ascidians (Reverberi, 1975; Sawada and Osanai, 1981; Jeffery and Meier, 1983), mollusks (Raff, 1972; Guerrier et al., 1978; Schmidt et al., 1980), annelids (Eckberg, 1981; Shimizu, 1982; Peaucellier et al., 1974; Swalla et al., 1985), and nematodes (Strome and Wood, 1983). Moreover, rotation experiments with Xenopus eggs and treatment with pressure, low temperature, and UV light suggest that the cytoskeletal framework may play a functional role in determining the dorsoventral axis of the developing vertebrate embryo (Ubbels et al., 1983; Scharf and Gerhart, 1983).

While considerable work has been done to elucidate the function of microtubules and microfilaments in early embryogenesis, little is known about the contribution of the IF cytoskeleton. This is probably due to the fact that no obvious function could yet be assigned to the IF cytoskeleton (see Section II,B). The apparent absence of IF polypeptides in a few cultured cell lines (Venetianer et al., 1983; Traub et al., 1983) suggests that the presence of IF may not be required in all cell types at all times, yet their evolutionary conservation and their widespread occurrence in vertebrate and invertebrate animals (see Section V), as well as in plants (Dawson et al., 1985), argues for their involvement in basic cellular functions. In this chapter, we review the current state of knowledge on IF in early embryonic development and present some results of our own work on the characterization of IF polypeptides in early Drosophila embryogenesis.

II. STRUCTURE AND FUNCTION OF INTERMEDIATE FILAMENTS

A. Classification, Primary Sequence, and Structure of Vertebrate Intermediate Filaments

Intermediate filaments exhibit a diameter of 7–11 nm and, together with the 6-nm actin microfilaments and the 25-nm microtubules, comprise the cytoskeletal architecture of higher eukaryotic cells. Immunological and biochemical work has established the existence of five subclasses of IF proteins that are expressed in a tissue-specific manner in almost all vertebrate cells: some 30 different keratins (40–70 kDa) in epithelia, a triplet of neurofilament proteins (135,105 and 65 kDa) in neuronal cells, glial fibrillar acidic protein (50 kDa) in astroglial cells, vimentin (53 kDa) in cells of mesenchymal origin, and desmin (52 kDa) in muscle cells. These proteins, although different in their molecular weights, share more or less amino acid homology, but more importantly, all IF subunits are built according to a common plan: each has a central α-helical rod domain of 311–314 amino acids that has a highly conserved secondary structure, which is flanked by end domains of variable size and chemical character. Several genes coding for different IF polypeptides have recently been cloned from various vertebrate organisms, and nucleotide sequence analyses have confirmed earlier immunological, biochemical, and structural findings. IF proteins in vertebrates share considerable sequence homology with the nuclear lamins (McKeon et al., 1986), which gives them similar physical–chemical properties, resulting in coisolation of keratin IF proteins and lamins from BHK-21 cells (Goldman et al., 1986). Detailed reviews on IF polypeptides and the IF cytoskeleton have been published recently (Steinert et al., 1985; Lazarides, 1980, 1982; Traub, 1985; Wang et al., 1985).

B. Possible Function of Intermediate Filaments

Electron microscopic studies have revealed that IF, at least the neurofilaments, are cross-linked to other cytoskeletal components (e.g., Ellisman and Porter, 1980; Hodge and Adelman, 1980; Schliwa and van Blerkom, 1981; Hirokawa, 1982; Schnapp and Reese, 1982). This observation is corroborated by biochemical (Aamodt and Williams, 1984), as well as by immunocytochemical, evidence (Bloom and Vallee, 1983), particularly suggesting the involvement of microtubule-associated proteins as cross-linkers between IF and microtubules. IF may also be connected to the nuclear and the cellular membranes (Blose and Chacko, 1976; Lehto et al., 1978; Jones et al., 1982, 1985; Green and Goldman, 1986). These observations suggest that it may be the main function of the IF cytoskeleton to provide some stable and solid support to the cytoplasmic architecture, and the presence of neurofilaments in axons and of keratins in epithelial cells is consistent with this notion.

The functions of IF in structuring the cytoplasm may include anchoring the nucleus within the cell (Lehto et al., 1978; Small and Celis, 1978; Wang et al., 1979; Henderson and Weber, 1980), as well as maintaining the proper intracellular structure during mitosis (Ishikawa et al., 1968; Blose, 1979; Zieve et al., 1980). However, while vimentin filaments remain intact during mitosis and form a dense cage around the mitotic spindle apparatus (Hynes and Destree, 1978; Gordon et al., 1978; Blose, 1979; Aubin et al., 1980; Zieve et al., 1980; Blose and Bushnell, 1982), cytokeratin filaments are transiently disaggregated during cell division in several epithelial cells, and the subunits aggregate into many small cytoplasmic spheres (Horwitz *et al.*, 1981; Lane *et al.*, 1982; Franke *et al.*, 1982b).

Experimental disruption of the IF cytoskeleton in cultured cells by various means has been used to elucidate its function. It has been known for some time that the IF cytoskeleton collapses at the nucleus after treatment with Colcemid (Goldman and Knipe, 1972; Blose and Chacko, 1976; Starger and Goldman, 1977; Hynes and Destree, 1978). Other chemicals and treatments affecting the IF cytoskeleton include vanadate (Wang and Choppin, 1981), uncouplers of oxidative phosphorylation (Maro and Bornens, 1982), acrylamide (Eckert, 1985), toxins (Sharpe *et al.*, 1980), virus infections (Murti and Goorha 1983; Ben-Ze've, 1984), and heat shock (Falkner *et al.*, 1981; Thomas *et al.*, 1982; Welch and Suhan, 1985). Specific disaggregation of the IF cytoskeleton and aggregation of the IF material in perinuclear caps can also be achieved by microinjection of antibodies directed to specific IF polypeptides or to IF-associated proteins (Klymkowsky, 1981, 1982; Gawlitta *et al.*, 1981; Lin and Feramisco, 1981; Lane and Klymkowsky, 1982; Klymkowsky *et al.*, 1983; Tölle *et al.*, 1986). These experiments showed that collapse of the IF cyto-skeleton at the nucleus after injection of specific antibodies apparently did not affect cell morphology, organelle or granule transport, cell locomotion, membrane ruffling, or mitosis. However, treatment with acrylamide (Eckert, 1985) or vanadate (Wang and Choppin, 1981) resulted in a concentration of cytoplasmic organelles at the center of the cell near the nucleus, suggesting interaction of IF with mitochondria (Mose-Larsen *et al.*, 1983). The effects of heat shock, which also destroys the IF cytoskeleton (Falkner *et al.*, 1981; Thomas *et al.*, 1982; Biessmann *et al.*, 1982; Walter and Biessmann, 1984a), are more difficult to assess, because heat shock also causes a number of additional changes in the cytoplasm (Welch and Suhan, 1985).

The exact function of the IF cytoskeleton is not known, but it seems most likely to provide a structural component to the cytoplasm as proposed by Lazarides (1980). Since the effects of disruption of the IF cytoskeleton on the developing embryo have not yet been studied, published work on the expression of IF polypeptides and the distribution of the IF cytoskeleton in oogenesis and embryogenesis that will be reviewed in the following sections, remains descriptive.

III. INTERMEDIATE FILAMENT PROTEINS IN SPERMATOZOA

Earlier work by Franke *et al.*, (1979) using tissue sections of guinea pig testis failed to detect vimentin in spermatozoa, but showed an elaborate vimentin cytoskeleton in Sertoli cells. Only recently have IF been discovered in spermatozoa (Virtanen *et al.*, 1984). Indirect immunofluorescence staining of methanol-fixed human sperm demonstrated a unique localization of vimentin in the sperm head: vimentin is found in a narrow band encircling the sperm head in the equatorial segment region that is located at the more posterior region of the acrosome. This specialized surface region persists after completion of the acrosomal reaction and appears to initiate the fusion of the sperm with the surface membrane of the egg (Barros and Franklin, 1968; Bedford, 1971). Moreover, binding of rhodamine-conjugated *Helix pomatia* lectin showed similarly compartmentalized distribution of some *N*-acetylgalactosamine glycoconjugates at the surface of the equatorial segment region. Both staining patterns are unaffected by Triton X-100 treatment, suggesting some cytoskeletal–cell surface interaction that stabilizes this configuration of the equatorial segment. Other cytoskeletal elements (actin, tubulin, and α-spectrin) are also quite distinctly distributed in the spermatozoan cell (Virtanen *et al.*, 1984). These studies suggest that specific interactions between the cytoskeleton and specialized cell surface domains may be involved in the regulation of cell surface events associated with the acrosomal reaction and fertilization.

IV. EXPRESSION OF INTERMEDIATE FILAMENT PROTEINS IN EARLY EMBRYONIC DEVELOPMENT OF VERTEBRATES

A. Mouse

The temporal and spatial expression of IF proteins has been studied extensively in mouse embryogenesis. The first IF proteins that can be detected in the mouse embryo are members of the cytokeratin family. Slightly different results have been obtained, possibly due to different antibodies and techniques. Lehtonen and collaborators (Lehtonen and Badley, 1980; Lehtonen *et al.*, 1983b), using a polyclonal serum to smooth muscle 10-nm filament protein or to human cytokeratin polypeptides, found cross-reacting polypeptides in cytoskeletal preparations of oocytes and all cleavage stages, as well as evidence for 10-nm-thick filaments in detergent-extracted 2- or 4-cell-stage embryos, that become more abundant in later stages, especially in blastocysts (Fig. 1 a and e). The cytokeratins are probably initially present in a nonfilamentous configuration in cleavage-stage embryos, but later organize into well-defined filaments. This notion is supported by indirect immunofluorescence staining of mouse oocytes and early cleavage-stage embryos with a monoclonal antibody (OCS-1), directed against cytokeratin proteins from mouse oocytes (Lehtonen, 1985). This antibody gave punctuate staining of oocytes and morulae (Fig. 1 b and c), suggesting association of the antigen with paracrystalline

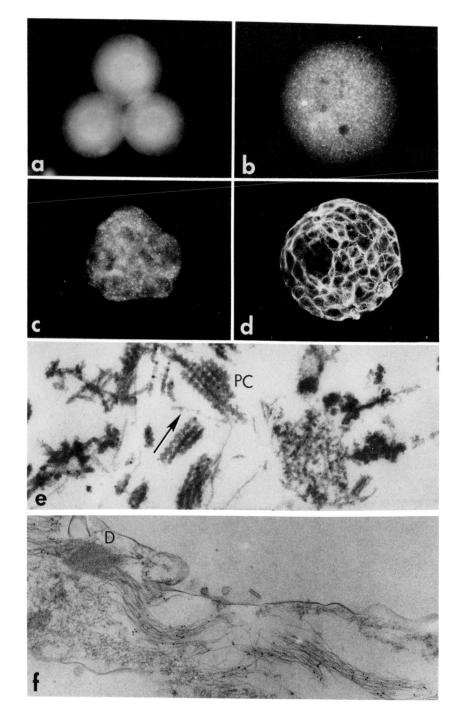

arrays, found abundantly in the cytoplasm of mouse oocytes. A distinct fibrillar organization was revealed in the trophectoderm cells of expanded blastocysts (Fig. 1 d and f). Oshima *et al.* (1983) found that members of the cytokeratin family (endo A and B) are expressed as early as the 4- to 8-cell stage, and other laboratories have also reported good evidence for the presence of cytokeratin polypeptides and 10-nm fibers in blastula and morula, but failed to detect keratins in early cleavage-stage embryos (Jackson *et al.*, 1980). It seems likely that failure to detect cytokeratins in early cleavage-stage mouse embryos may be due to lack of sensitivity in the methods employed and that cytokeratins are indeed expressed at these stages. This notion is supported by quantitative S1 mapping experiments with a DNA fragment of the cloned endo A cytokeratin gene, which revealed the presence of endo A mRNA in 8-cell morulae (Duprey *et al.*, 1985).

The transition from the 8-cell morula to the blastocyst is an important morphological event in mammalian embryogenesis, when the embryo undergoes compaction, and junctions between blastomeres are formed (Ducibella and Anderson, 1975). The involvement of cortical cytoskeletal elements in this process has been investigated by Ducibella *et al.* (1977). After one or two more cleavages, cavitation begins, and by day 3–4, the forming blastocyst is composed of an outer layer of differentiated cells (the trophectoderm) and an inner cell mass composed of a few multipotent cells. Cytokeratin-type IF proteins are abundantly synthesized at blastocyst stage (van Blerkom *et al.*, 1976; Brulet *et al.*, 1980; Jackson *et al.*, 1980, 1981; Oshima *et al.*, 1983) and form typical IF bundles in the cells of the trophectoderm, as visualized by indirect immunofluorescence or electron microscopy (Paulin *et al.*, 1980; Kemler *et al.*, 1981; Jackson *et al.*, 1980, 1981; Lehtonen *et al.*, 1983b; Oshima *et al.*, 1983; Lehtonen, 1985). Expression of keratin-type IF in the trophoblastic cells is concomitant with the formation of desmosome junctions (Ducibella *et al.*, 1975; Jackson *et al.*, 1980, 1981) and demonstrates that these cells represent

---

Fig. 1. Cytokeratin proteins in mouse oocytes and cleavage-stage embryos as revealed by indirect immunofluorescence with anti-human cytokeratin antibodies (a) and the monoclonal antibody OCS-1, directed against mouse oocyte cytoskeletal proteins (b, c, and d). Three blastomeres of a 4-cell-stage embryo (a, ×385), an oocyte (b, ×333), a morula (c, ×296), and a blastocyst (d, ×259). Electron micrograph of a detergent-extracted 4-cell-stage embryo showing 10-nm-thick filaments (arrow) that are often associated with the paracrystalline arrays (PC) present in mouse oocytes and early embryos (e, ×44,400). Immunoelectron micrograph of a detergent-extracted mouse blastocyst stained with the OCS-1 antibody and gold-conjugated secondary antibody. The label is associated with 10-nm-thick filaments, apparently representing tonofilaments attached to a desmosome (D) (f, ×32,560). [a and e are reprinted, with permission from *Dev. Biol.*, Lehtonen *et al.* (1983b); b, c, d, and f are reprinted, with permission from *J. Embryol. Exp. Morphol.*, Lehtonen (1985).]

the earliest embryonic differentiation event of epithelial character. By contrast, no cytokeratin mRNA or proteins are detectable in the inner cell mass (Paulin *et al.*, 1980; Brulet *et al.*, 1980; Jackson *et al.*, 1980, 1981; Kemler *et al.*, 1981; Oshima *et al.*, 1983; Duprey *et al.*, 1985).

Antibodies against vimentin and desmin failed to detect any expression of these types of IF proteins at the blastocyst stage (Jackson *et al.*, 1980; Paulin *et al.*, 1980). Vimentin is first expressed in the mesoderm cells, which form between the two epithelial cell layers at days 8–9 (Jackson *et al.*, 1981), and vimentin protein can then be detected by polyacrylamide gel electrophoresis (Franke *et al.*, 1982a, 1983), indicating early differentiation of mesoderm for which vimentin is a marker protein. Since these cells are mostly derived from the embryonic ectoderm and have expressed cytokeratins, they must terminate the synthesis of cytokeratins and then express another type of IF protein (vimentin). This switch from cytokeratin to vimentin expression occurs within a few hours (Franke *et al.*, 1982a, 1983), and coexpression of both types of IF proteins in parietal endoderm cells has been reported (Lane *et al.*, 1983; Lehtonen *et al.*, 1983a). Later in development, vimentin is replaced by glial fibrillar acidic protein in the prospective astroglial cells, and neurofilament proteins are expressed in maturing neurons.

B. *Xenopus*

In contrast to mouse, *Xenopus* oocytes and eggs contain a large pool of maternally stored IF protein. There is no conclusive explanation for this maternal supply of IF proteins, and one can only speculate about its function: it may provide a framework for microfilaments and microtubules to allow extensive movements of the cortex relative to the cytoplasm that occur after fertilization (Gerhart *et al.*, 1981). Alternatively, it might be required for maintaining proper cytoplasmic organization during oocyte maturation and the exceptionally fast cleavage of amphibian embryos.

In previtellogenic oocytes, two classes of IF proteins seem to be present, vimentin and keratin. By two-dimensional gel electrophoresis of cytoskeletal preparations, one basic (type II) and two acidic (type I) cytokeratin polypeptides have been identified in stage VI vitellogenic oocytes (Franz *et al.*, 1983). These cytokeratins are not specific to the oocyte, but appear again later in development in intestinal mucosa cells of the adult, yet they are different from the cytokeratins expressed from the zygotic nuclei of the epithelial ectoderm during the gastrula stage (Jonas *et al.*, 1985; Miyatani *et al.*, 1986; Jamrich, *et al.*, 1987). With the basic cytokeratin cDNA, cloned from an ovary cDNA library, the presence of its mRNA was demonstrated in vitellogenic oocytes (Franz and Franke, 1986), confirming earlier *in vivo* labeling translation experiments (Franz *et al.*, 1983).

Whereas unfixed, aldehyde-, or ethanol-fixed oocytes failed to stain with antivimentin antibodies (Franz *et al.*, 1983; Godsave *et al.*, 1984b), TCA-fixed material showed positive reaction with three different antivimentin antibodies, recognizing a slightly smaller protein than mammalian vimentin, and a possibly IF-associated 66-kDa protein (Godsave *et al.*, 1984b). Moreover, an anti-bovine hoof prekeratin serum (Gall *et al.*, 1983), as well as three different monoclonal antikeratin antibodies, recognized antigens in *Xenopus* oocytes (Godsave *et al.*, 1984a). The distribution of vimentin staining differs markedly from that of keratin staining (Wylie *et al.*, 1985): in the previtellogenic oocyte, the majority of vimentin is associated with the mitochondrial cloud surrounding the nucleus, and some staining was observed throughout the cytoplasm as a network of filaments. Most of the keratin is localized in the subcortical cytoplasm, but some keratin staining was found encircling the perinuclear mitochondrial cloud. In postvitellogenic stages, vimentin appears to be codistributed with the expanding mitochondrial cloud and shows asymmetric distribution in the animal and vegetal hemispheres, probably due to differences in yolk distribution. Keratin staining becomes more intense in the perinuclear region during vitellogenic stages, and filaments are stained in the animal hemisphere, which project radially from the nucleus to the keratin-rich cortex. In the mature oocyte a dense sheet of cytokeratin filaments is located underneath the cortex with some fibrils extending from the peripheral cytoplasm into the inner part of the oocyte, whereas others project toward the plasma membrane (Fig. 2). This IF cytoskeleton is interwoven with the cortical actin filaments (Gall *et al.*, 1983). In whole-mount preparations, quite regular cytokeratin arrays can be seen in the vegetal hemisphere, whereas cytokeratins in the animal hemisphere appear to be less regularly organized (Klymkowsky *et al.*, 1987). Upon maturation of the oocyte, accompanied by breakdown of the germinal vesicle, both types of IF proteins undergo rearrangements: in the mature egg, vimentin becomes almost evenly distributed, and a fine network of vimentin staining is observed throughout the cytoplasm that is inherited by all of the blastomeres (Godsave *et al.*, 1984b). The cortical cytokeratin filaments of the mature oocyte disappear, and cytokeratin staining is found in amorphous aggregates and some isolated filament fragments (Godsave *et al.*, 1984a; Klymkowsky *et al.*, 1987), reminiscent of keratin fiber breakdown during mitosis (Horwitz *et al.*, 1981; Franke *et al.*, 1982b; Lane *et al.*, 1982).

With fertilization or upon prick activation, a gradual transition from a punctate to fibrous cytokeratin staining pattern is observed, beginning in the vegetal hemisphere (Klymkowsky *et al.*, 1987). By the time of first cleavage, large bundles of cytokeratin filaments are apparent in the vegetal hemisphere, giving rise to a quite regular meshwork. The reorganization of keratins in the animal hemisphere is similar, but less dramatic, and apparently does not result in an interconnected filament system. Since keratin fibers reside almost exclusively in the cortical layer, the most superficial cells of the blastula will inherit the

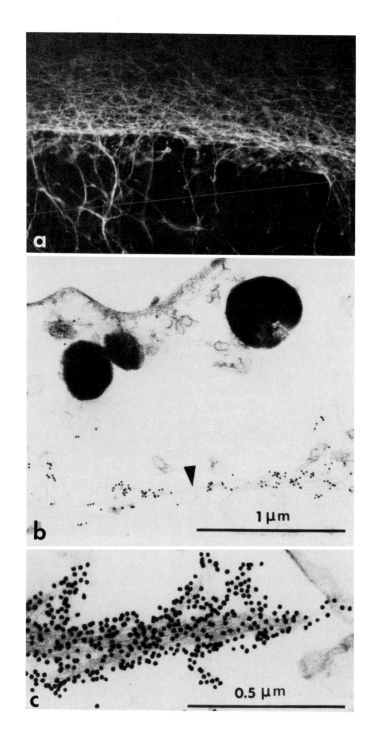

bulk of cytokeratin as cleavage proceeds. The fate of these early cytokeratins is unclear, and during gastrulation, a new set of cytokeratins is expressed in the epithelial ectoderm (Jonas *et al.*, 1985; Miyantani *et al.*, 1986; Jamrich, *et al.*, 1987).

V. INTERMEDIATE FILAMENTS IN INVERTEBRATES

A. Immunological and Ultrastructural Evidence

Until recently, IF proteins have been found only in vertebrate cells, but there is now good evidence for several types of IF in invertebrates. Even though they form filaments of about 10 nm in diameter, the monomeric polypeptides vary considerably in size from species to species. But as pointed out earlier, the different classes of IF proteins in vertebrates also exhibit quite different molecular weights, yet are built according to the same general structural principle. To date, nothing is known about amino acid sequence and secondary structure of any invertebrate IF protein, but it would not be surprising to find common structural elements and regions that are conserved in evolution, since two different monoclonal antibodies have been generated that recognize IF polypeptides in vertebrates, as well as in invertebrates (Pruss *et al.*, 1981; Falkner *et al.*, 1981). These two antibodies have been used to demonstrate the presence of IF in invertebrates by indirect immunofluorescence, immunoelectron microscopy, and protein blotting. Moreover, IF in invertebrates exhibit characteristics similar to their vertebrate homologs, such as insolubility in buffers containing 1% Triton X-100, sensitivity to Colcemid, to heat shock, and other stress, and the capability to reassemble *in vitro*. Using a combination of these techniques, IF have been found in axons of mollusks, such as squid (Gilbert *et al.*, 1975; Gilbert, 1976; Huneeus and Davison, 1970; Lasek *et al.*, 1979; Roslansky *et al.*, 1980; Zackroff and Goldman, 1980), *Aplysia* (Lasek *et al.*, 1979; Pruss *et al.*, 1981; Lasek *et al.*, 1983), and *Loligo* (Lasek *et al.*, 1979; Philips *et al.*, 1983), and in axons of the annelid

←_____

Fig. 2. Cytokeratin filaments in *Xenopus* oocytes and activated eggs. Indirect immunofluorescence staining of mature oocytes with antikeratin antibody of isolated pieces of peripheral cytoplasm including cortex (a, ×450). Electron micrograph of activated eggs showing immunogold staining of isolated cortices (b, ×40,000); higher magnification of a bundle of intermediate filaments labeled with antikeratin antibody and gold-conjugated secondary antibody (c, ×84,000). [Reprinted, with permission from *Biologie Cellulaire*, Gall *et al.* (1983).]

Myxicola (Gilbert *et al.*, 1975; Lasek and Hoffman, 1976; Lasek *et al.*, 1979; Eagles *et al.*, 1981; Philips *et al.*, 1983). Electron microscopic evidence for neurofilaments in arthropods has been published (Foelix and Hauser, 1979; Keil and Steinbrecht, 1984), and intermediate filaments in nonneuronal tissues have been detected by immunological methods and electron microscopy in *Drosophila* (Falkner *et al.*, 1981; Walter and Biessmann, 1984a; Walter and Alberts, 1984). These proteins are possibly related to vimentin and desmin. Keratin-type IF proteins have been detected in the molusk, *Helix pomatia* (Bartnik *et al.*, 1985), and in muscle and epithelial cells of mematodes (Bartnik *et al.*, 1986). IF have been described in the sea urchin tube foot (Harris and Shaw, 1984). Taken together, these data provide strong evidence for evolutionary conservation of IF cytoskeletons.

B. The Monoclonal Anti-*Drosophila* Intermediate Filament Antibody Ah6

Our work in the past several years has been concerned with the characterization of the IF cytoskeleton in *Drosophila* tissue-culture cells (Falkner *et al.*, 1981; Biessmann *et al.*, 1982; Walter and Biessmann, 1984a). We have prepared several monoclonal antibodies (e.g., Ah6 and Ah3) against a 46-kDa major cytoplasmic protein from *Drosophila* tissue-culture cells (Falkner *et al.*, 1981) and demonstrated that the antigen is related to the vertebrate IF proteins vimentin (Walter and Biessmann, 1984a). These findings provided evidence for the presence of IF-related proteins in *Drosophila* and also opened a new search for similar polypeptides in other invertebrate phyla, since the Ah6 antibody recognizes a highly conserved epitope (Walter and Biessmann, 1984b). The Ah6 epitope was localized to the first 116 amino acids at the amino-terminus of chicken desmin (K. Weber, personal communication), which positions it outside the central α-helical rod domain of IF polypeptides that is recognized by the highly cross-reacting anti-human glial fibrillar acidic protein antibody of Pruss *et al.* (1981; Geisler *et al.*, 1983). We also studied the effects of heat shock, Colcemid, and ionophores on the *Drosophila* IF cytoskeleton (Falkner *et al.*, 1981; Biessmann *et al.*, 1982; Walter and Biessmann, 1984a) and found that these treatments caused the collapse of the IF cytoskeleton at the cell nucleus.

The 46 kDa IF polypeptide that was originally used as antigen represents only one member of a group of six related *Drosophila* polypeptides that all carry the Ah6 epitope. The nature of these other proteins is not clear, but we have shown (Walter and Biessmann, 1984a) that at least three of them (40, 46, and 68 kDa) also carry the IF-specific epitope located in the α-helical rod domain recognized by the antiglial fibrillar acidic protein antibody (Pruss *et al.*, 1981) and that the 40-kDa protein reacts with a monoclonal antibody directed against vertebrate desmin (M. Walter, unpublished observation). Moreover, our Ah6 antibody clearly recognizes vimentin and desmin in vertebrate cells (Falkner *et al.*, 1981; Walter and Biessmann, 1984a), and we concluded that *Drosophila* tissue culture cells (Kc line) express a family of related IF polypeptides. In early *Drosophila* embryos, only five (80, 68, 53, 46, and 40 kDa) of the six IF polypeptides are expressed, as demonstrated by immunoblotting (Walter and Alberts, 1984), and the 110-kDa polypeptide expressed in Kc cells is apparently absent in embryos.

It is not clear to which class of vertebrate IF the five embryonic *Drosophila* polypeptides are related. Since molecular weights of invertebrate IF polypeptides are very different from those of their vertebrate counterparts (see Refs. in Section V,A), molecular weights cannot be used for classification. Tissue-specific expression (e.g., epidermal, neuronal, or mesenchymal) may be used as an additional criterion as soon as monoclonal antibodies against a subset or individual *Drosophila* IF polypeptides become available, but definite classification of these invertebrate IF proteins will require amino acid sequence comparison with their vertebrate homologs. In our case, we have obtained preliminary evidence (as described above) that the 46-kDa *Drosophila* IF is related to vertebrate vimentin and that the 40-kDa *Drosophila* protein is related to vertebrate desmin; the other members of the IF family that are recognized by the Ah6 antibody are as yet unclassified. It seems therefore very likely that IF proteins, other than cytokeratins, are expressed very early in *Drosophila* embryos.

VI. THE CYTOSKELETON IN *DROSOPHILA* AND OTHER DIPTERAN EMBRYOS

A. Cytoplasmic Events in Early Insect Embryogenesis

In insect embryonic development, the early nuclear divisions occur very rapidly (every 8–10 min in *Drosophila*) within a syncytium (Fig. 3). In *Drosophila*, all of the nuclei are located in the interior of the embryo until nuclear cycle 8 and migrate outward to the periphery during nuclear cycles 9 and 10. A few pole cell nuclei, giving rise to the germ cells, reach the posterior end of the embryo during nuclear cycle 9, while the rest (except for the yolk cell nuclei) arrive in the subcortical yolk-free plasma layer in interphase of cycle 10. Cell membranes are formed later at nuclear cycle 14. During the first 10 nuclear division cycles that occur synchroneously in *Drosophila*, no significant nuclear RNA synthesis is detectable, some genes are transcribed in cycles

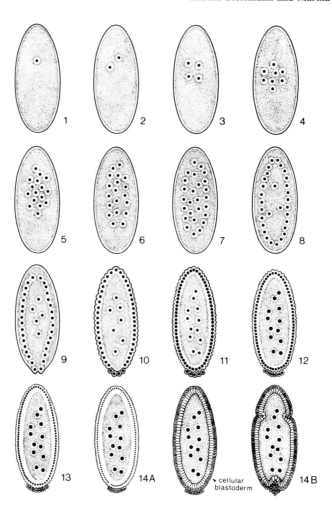

Fig. 3. Schematic drawing of early embryonic stages of *D. melanogaster* before gastrulation. This figure is modified from Zalokar and Erk (1976) to show the appearance of polar and somatic buds. The number beside each embryo corresponds to the total number of nuclear division cycles. Embryos are shown in longitudinal section and with their anterior ends at the top. They are depicted without vitelline membranes to emphasize the changes in the surface morphology. Solid black circles represent nuclei, stippled regions denote yolk, and nontextured regions denote the yolk-free regions of the cytoplasm. The average time required for stages 1–9 is 8 min at 25°C; stages 10, 11, 12, 13, and 14 occupy about 9, 10, 12, 21, and more than 65 min, respectively. [Reprinted, with permission from *J. Cell Sci.*, Foe and Alberts (1983.)]

11–13, but the bulk of zygotic genes is becoming actively transcribed during the cellular blastoderm stage in nuclear cycle 14 (Zalokar, 1976).

After the mature egg is fertilized, several processes occur in the cytoplasm that can be attributed to the action of cytoskeletal elements: migration of female and male pronuclei, formation of a cytoplasmic yolk-free island around every zygotic nucleus (energid), migration of these energids to the cortex of the embryo in nuclear cyles 9 and 10, yolk contractions, cytoplasmic streaming, saltatory particle movements, and the formation of polar caps and cleavage furrows. In many insects, some of these cytoplasmic events can occur in the absence of cleavage nuclei, suggesting the action of a nucleus-independent cytoplasmic program in the early cleavage stages of embryogenesis (reviewed by Krause and Sander, 1962; Counce, 1973). This phenomenon, which was later termed "pseudocleavage," was first described in insects by Seiler (1924). Subsequently, contractions of the cytoplasm of unfertilized eggs have been observed by Imaizumi (1954) and by Schnetter (1965), and cytoplasmic islands that form and divide in the absence of cleavage nuclei have been observed in a small percentage of unfertilized eggs in nonparthenogenetic insects (Mahr, 1960a,b; Scriba, 1964; Schnetter, 1965). Pseudocleavage also occurs in certain female sterile mutants that give rise to haploid blastoderms (Zalokar et al., 1975), and structures are formed in the peripheral cytoplasm that resemble wild-type structures in the absence of the normal number of zygotic nuclei (Freeman et al., 1986). Moreover, in experimentally constricted embryos, where one-half does not contain any nuclei, cytoplasmic aggregates within the reticular cytoplasm are formed (Mahr, 1960a). Wolf reported (1980, 1985) that migration cytasters, consisting of large, radially oriented microtubules, can appear and even migrate in the absence of nuclei. Finally, irradiation of early embryos that causes arrest of nuclear divisions does not interfere with the formation and cleavage of cytoplasmic islands; however, no migration of these "empty" energids occurs (Schneider-Minder, 1966).

Except for the formation of mitotic spindles by microtubules, no cytoskeletal element has yet unambiguously been associated with the above-mentioned cytoplasmic events, and no function at all has yet been assigned to the IF cytoskeleton. Some attempts in this direction, similar to those done in other systems, have been made by injection of drugs that affect microtubules (colchicine) or microfilaments (cytochalasin) into Drosophila (Zalokar and Erk, 1976; Foe and Alberts, 1983) and other dipteran embryos (Wolf, 1978, 1985; Kaiser et al., 1982). These experiments indicate that nuclear migration to the periphery, nuclear elongation at blastoderm stage, saltatory movements of yolk particles, and yolk contractions are inhibited by colchicine and that nuclear pulsation, cytoplasmic streaming, yolk contractions, the longitudinal migration of energids, and the spacing of nuclei at blastoderm are affected by cytochalasin.

B. Distribution of Intermediate Filament Proteins, Tubulin, and Actin in *Drosophila* Embryos

In this chapter, we describe the distribution of cytoskeletal components in early *Drosophila* embryogenesis. As we have summarized above, nuclear cleavages and nuclear migration occur within a syncytium, which allows one to study the distribution and rearrangement of all cytoskeletons simultaneously within this giant cell. Moreover, experimental manipulations, as well as microinjection techniques, are greatly facilitated in the developing insect embryo, and together with recombinant DNA techniques and genetic analyses, this system will certainly be very powerful in elucidating the role of the different cytoskeletal components in early embryogenesis.

1. Ultrastructure

Unfortunately, very few ultrastructural studies addressing the distribution of cytoskeletal elements in insect eggs and early cleavage-stage embryos are available, which is mainly due to technical difficulties caused by the presence of yolk and the impermeable vitelline membrane. Early electron microscopic studies failed to detect any cytoskeletal elements in the *Drosophila* egg and embryo (Okada and Waddington, 1959; Mahowald, 1963a), possibly due to poor conservation of these structures. Later, Zissler and Sander (1973, 1977, 1982), investigating eggs and embryos of the chironomid midge *Smittia* that contains less yolk than *Drosophila*, reported the occurrence of isolated microtubules in the periplasm, while microtubules associated with centrioles and cytasters were observed more frequently. A well-conserved cytaster in the anterior part of the fertilized gall midge *(Wachtliella)* egg has been observed by Wolf (1978, 1980). Its microtubules seem to extend into the subcortical cytoplasm and migration of this cytaster to the periphery occurs even in the absence of an attached nucleus (Wolf, 1980).

Isolated microfilament bundles have been found in the newly laid egg of *Chironomus anthracinus* (Zissler and Sander, 1982) that may be derived from the ring canal connecting the oocyte with the nurse cells. In the oocyte of the paedogenetic gall midge *Heteropeza pigmaea*, a meshwork of microfilaments is associated with the nucleus, suggesting its involvement in the observed nuclear pulsations (Went *et al.*, 1978).

With respect to the distribution of cytoskeletal elements, developmental stages shortly before and during cellularization of the blastoderm have been analyzed more extensively. At these stages, all nuclei, except the vitellophages, have migrated to the periphery and densely populate the subcortical region of the embryo. The initially spherical nuclei then elongate dramatically, and cleavage furrows begin to grow inward from the egg surface between the elongated nuclei. At this time, bundles of microtubules appear that are oriented parallel to the direction of nuclear elongation (Fullilove and Jacobson, 1971). In the paedogenetically developing embryo of the gall midge *Heteropeza pyg-maea,* which develop within the hemocoel of the mother larva and lack the impermeable vitelline membrane and chorion, numerous microtubules have been observed that are mainly associated with cleavage furrows, but also occur in the peripheral layer of the egg and around the cleavage nuclei (Junquera, 1985).

Fibrillar material that supposedly consists of actin microfilaments has been found underneath the inward growing cleavage furrows in *Drosophila* embryos (Mahowald, 1963b; Fullilove and Jacobson, 1971), and a band of microfilaments has been observed adjacent to the plasma membrane in cytoplasmic connections between the embryonic cells and the yolk sac during gastrulation (Rickoll, 1976).

IF have not yet been described in any stage of *Drosophila* embryogenesis. Our attempts to localize IF have been successful in heptane-permeabilized embryos (see below), and although this technique does not preserve the ultrastructure very well because it extracts many cytoplasmic components, it leaves the insoluble cytoskeleton behind. Using the IF-specific monoclonal antibody Ah6 (see Section V,B) and gold-conjugated secondary antibody in immunoelectron microscopy, 10-nm filaments, as well as amorphous material, were labeled in the periphery of the embryo (Fig. 4).

It is evident that the ultrastructural analysis of the cytoskeletal architecture

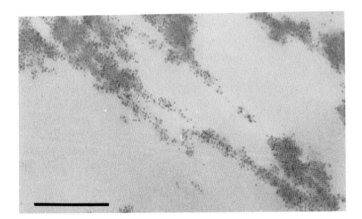

Fig. 4.   Electron micrograph of a section through a *Drosophila* embryo before nuclear migration. Intermediate filaments are labeled in these heptane-permeabilized embryos by the anti-IF antibody Ah 6 and gold-conjugated secondary antibody. Bar, 0.5 μm.

in insect eggs and embryos remains quite fragmentary, mainly due to the above-mentioned technical problems with fixation and sectioning of this material. A different approach became feasible with the discovery of Zalokar and co-workers that heptane permeabilizes the vitelline membrane and makes the interior of the egg accessible to fixatives (Limbourg and Zalokar, 1973; Zalokar and Erk, 1976, 1977). This technique, as modified by Mitchison and Sedat (1983), has proved very useful in localizing antigenic determinants in whole-mount *Drosophila* embryos. Previously, embryos had to be punctured with a needle to permit the fixative to enter and then devitellinized manually, which only allowed analysis of a much smaller number of specimens.

2. Immunofluorescence

Reaction with specific antibodies or labeled phalloidin revealed arrays of tubulin (Warn and Warn, 1986; Karr and Alberts, 1986), actin (Warn and Magrath, 1983; Warn *et al.*, 1984, 1985), as well as IF proteins (Walter and Alberts, 1984; 1988; Miller *et al.*, 1985) in the subcortical layer of the early *Drosophila* embryo. In this section, we want to summarize these observations, attempting to reconstruct a three-dimensional picture of the rearrangements of these three cytoskeletal elements during the early stages of embryogenesis. This description will provide the basis for our understanding of how the cytoskeletal elements may interact with each other and what their function might be during the phase of extremely rapid nuclear divisions. Specific emphasis will be given to the distribution of the IF proteins.

While monoclonal antibodies raised against vertebrate or yeast tubulin and actin cross-react with their *Drosophila* homologs, anti-vertebrate IF antibodies showed no cross-reactivity, with the notable exception of the anti-human glial fibrillar acidic protein antibody (Pruss *et al.*, 1981; see also Section V,A). Although this monoclonal antibody has been shown to recognize IF poly-peptides in protein blots from *Drosophila* tissue-culture cells (Walter and Biessmann, 1984a), it fails to convincingly stain IF by indirect immunofluorescence in embryos (our own observation) and in frozen tissue sections of *Drosophila* (Bartnik *et al.*, 1985). This may be due to a particular sensitivity of the epitope in the *Drosophila* IF cytoskeleton to standard fixation procedures. We have therefore used our own monoclonal antibody Ah6 that was made against one of the *Drosophila* IF polypeptides (Falkner *et al.*, 1981).

a. The Cytoskeleton before Nuclear Migration. During the early nuclear cleavage stages (nuclear cycles 1–8), tubulin, actin, and IF proteins are localized in the subcortical yolk-free cytoplasm (Fig. 5). Microtubules seem to form a filamentous network, actin shows a punctate distribution, and IF proteins do not appear to be organized into particular structures, but evenly occupy the subcortical cytoplasm. The unstained regions might be due to the

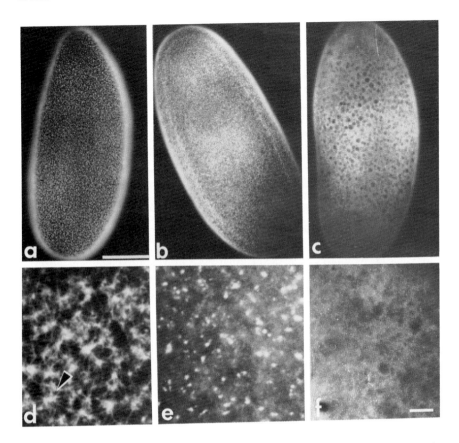

Fig. 5. Indirect immunofluorescence staining of whole-mount *Drosophila* embryos in nuclear cycle 4, prior to nuclear migration. Staining with antitubulin (a and d), antiactin (b and e) and antiintermediate filament protein antibody Ah 6 (c and f) reveals organization of these three cytoskeletal elements in the cortical cytoplasm. Bar in a, 100 μm in a, b, and c; bar in f, 5 μm in d, e, and f. [a, b, d, and e are reprinted, with permission from *J. Cell Biol.*, Karr and Alberts (1986).]

presence of multivesicular bodies that are also found at that stage in the cortex (Mahowald *et al.*, 1981). From the positive immunofluorescence with the Ah6 antibody in nuclear cycle 4 embryos, as well as in unfertilized eggs, we conclude that, similar to *Xenopus*, the *Drosophila* egg is maternally supplied with a large pool of IF proteins.

b. Nuclear Migration and Pole Cell Formation. During telophase of nuclear cycle 8 and interphase of nuclear cycle 9, the nuclei begin their outward migration. Migrating nuclei appear to be associated with a microtubular array

(Karr and Alberts, 1986) reminiscent of the migration cytaster described in *Wachtliella* (Wolf, 1978). The subcortical arrays of IF and actin remain unchanged, and no association of IF with migrating nuclei could be observed at the resolution of indirect immunofluorescence.

A group of seven nuclei on average reach the posterior end of the embryo during interphase of nuclear cycle 9, one nuclear division before the rest of the nuclei populate the subcortical cytoplasm. These "pole cell nuclei" are the progenitors of the germ-cell line. Where these nuclei reach the surface, highly folded protrusions ("polar buds") form above each nucleus. These buds grow rapidly in size, and a distinct F-actin-rich layer becomes associated with the plasma membrane over the area of the bulge, as visualized by rhodamine-coupled phalloidin. Polar nuclei, as well as their corresponding "polar buds," divide twice before they are cleaved off from the embryo to form the pole cells. Distribution of actin during the shape changes and cleavages of polar buds have been described by Warn *et al.* (1985). The IF proteins gather into a denser area at the base of the polar buds in which the nuclei insert into the caps and form a ring around each pole cell nucleus (Fig. 6 a and b). This rearrangement of IF occurs prior to the reorganization in the rest of the embryo at cycle 10.

   c. *Cytoskeletal Rearrangements during Syncytial Blastoderm Stages.* During interphase of nuclear cycle 10, the somatic nuclei reach the subcortical region, and "somatic buds" form above each nucleus. These somatic buds are much smaller than the polar buds described above. As the migrating energids enter the periplasm, dramatic rearrangements of the three cytoskeletal elements in the cortex occur. This rearrangement, the relative spatial distribution of the cytoskeletons, and their relationship to each other can best be described by double-label immunofluorescence.

Nuclear cycle 10 represents a transition period during which cytoskeletal reorganization occurs throughout the embryo. One of the first noticeable events is the reorganization of actin in the somatic buds, when the nuclei reach the periplasm. The actin distribution at nuclear cycle 10 differs from that at later mitotic cycles (Karr and Alberts, 1986): at interphase, bright areas of actin staining appear above each nucleus. Actin forms a cap over each nucleus (Warn *et al.*, 1984), while the area between energids seems to be depleted of actin stain. During the following prophase, actin is distributed in a bipolar fashion above each centrosome. At metaphase, actin forms a diffuse cap above each spindle, and by the end of the mitotic cycle, actin is again concentrated over each daughter nucleus.

The tubulin pattern during nuclear cycle 10 reveals an array of microtubules that closely surrounds each nucleus at interphase, while the cytoplasm between energids is depleted of microtubules except for a few filaments connecting

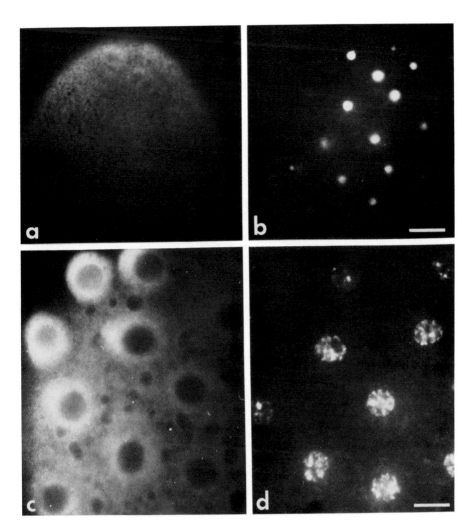

Fig. 6. Indirect immunofluorescence of whole-mount *Drosophila* embryos stained with anti-IF antibody Ah 6, demonstrating rearrangements of IF proteins as migrating nuclei reach the subcortical IF layer (a and c). Nuclei are visualized in the same preparation by staining with the fluorescent Hoechst dye 33258 (b and d). Pole cell nuclei just beneath the subcortical cytoplasm at the posterior end of the embryo (a and b) and concentration of intermediate filament proteins in pole cells and around somatic nuclei after they have reached the subcortical cytoplasm (c and d) are indicated. Bar in b, 7 μm in a and b; bar in d, 15 μm in c and d.

each energid to its neighbor. When the nuclei enter prophase, the centrosomes move to the opposite sides of each nucleus, and the array of microtubules splits into two halves as the microtubules form the mitotic spindles. At metaphase, most microtubules are organized in the spindles, and by the end of telophase of cycle 10, microtubules again radiate from the newly formed nuclei.

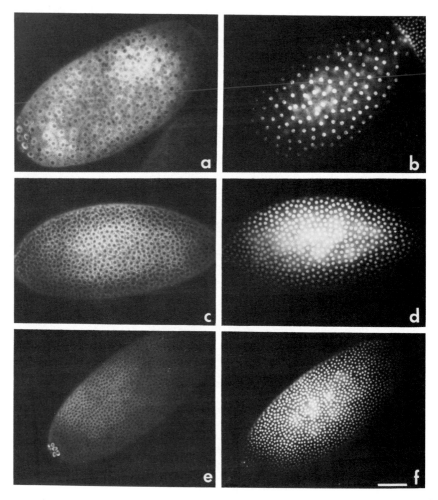

Fig. 7. Whole-mount *Drosophila* embryos at nuclear cycles 10 (a and b), 12 (c and d), and 14 (e and f) stained with the anti-IF antibody Ah 6 (a, c, and e) and Hoechst dye (b, d, and f). Although no cell membranes are present at these syncytial stages, intermediate filaments appear highly concentrated in cytoplasmic areas around each nucleus. Bar, 100 μm.

The IF rearrange at nuclear cycle 10 into a dense array around each nucleus (Fig. 6 c and d). At interphase and prophase, each nucleus becomes enclosed by IF proteins, and the staining between the energids decreases. At metaphase, IF proteins occupy the area around each spindle, and at telophase, IF proteins are again located in a cagelike structure around each nucleus, leaving a dark, unstained area above each nucleus.

During the following nuclear divisions in the syncytium, IF proteins localize preferentially in the cytoplasmic island around each nucleus. Since the number

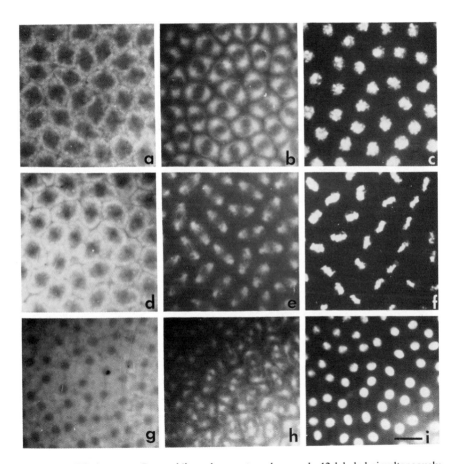

Fig. 8. Whole-mount *Drosophila* embryos at nuclear cycle 12 labeled simultaneously with anti-IF antibody Ah 6 (a, d, and g), visualized with rhodamine-conjugated secondary antibody, antitubulin antibody (b, e, and h), visualized with fluorescein-conjugated secondary antibody, and Hoechst dye (c, f, and i). Different stages of mitosis are shown: prophase (a, b, and c), metaphase (d, e, and f), and telophase (g, h, and i). Bar, 12.5 μm.

of nuclei increases, the space between them decreases with each nuclear division, resulting in a highly structured IF staining pattern that resembles cell borders, even though no cell membranes are present yet (Fig. 7). IF proteins do not rearrange dramatically during mitoses, as do actin and tubulin. The distribution of tubulin is essentially the same as described for nuclear cycle 10, while IF occupy the space around each spindle. Figure 8 shows pro-, meta-, and telophase in a triple-stained embryo. In telophase, the borders between the energids are not as obvious, but they reform as soon as the nuclei reach the next interphase. During nuclear cycles 11–14, the actin distribution is different from that in cycle 10: it does not show a bipolar distribution that follows the centrosomes. Instead, after the cap over each nucleus in interphase is formed (Fig. 9), where IF show the unstained nucleus as a black hole, the actin breaks down in prophase, and a fine line of bright actin staining delineates the border between the closely spaced domains in metaphase. This creates a

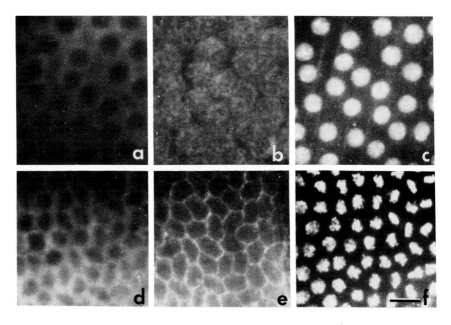

Fig. 9. Whole-mount *Drosophila* embryos at nuclear cycle 12 (a, b, and c) and 13 (d, e, and f) labeled simultaneously with anti-IF antibody Ah 6 (a and d), visualized with rhodamine-conjugated secondary antibody, antiactin antibody (b and e), visualized with fluorescein-conjugated secondary antibody, and Hoechst dye (c and f). To demonstrate the redistribution of cytoskeletal elements during mitosis, interphase (a, b, and c) and metaphase (d, e, and f) are shown. Bar, 12.5 μm.

distinct honeycomb pattern that extends over the whole embryo surface. The IF proteins remain within the honeycomb pattern, as can be seen in Fig. 9. By the end of mitosis, actin again reorganizes into cap structures over each nucleus.

To investigate how far the tubulin, IF, or actin meshworks reach down into the cytoplasm from the plasma membrane, embryos that had previously been stained by indirect immunofluorescence were embedded into Epon, and 5-μm sections were cut and observed (Fig. 10). While actin is closely associated with the plasma membrane of the embryos and reaches down into the region of adjacent spindles at metaphase, microtubules are located in the spindles (Karr and Alberts, 1986). IF are mostly localized near the plasma membrane, evenly staining the whole embryonic surface, and organize close to the nuclei after they have reached the periplasm at cycle 10. A stronger staining of IF proteins on each side of the nucleus can be seen at a later stage, which confirms

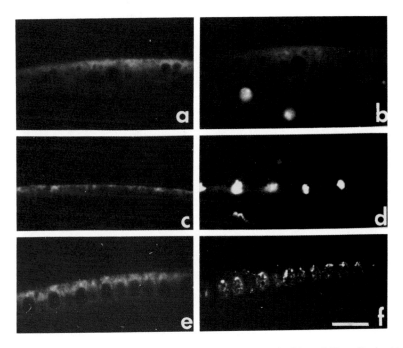

Fig. 10. Sections of fixed *Drosophila* embryos prestained with anti-IF antibody Ah 6 (a, c, and e) and Hoechst dye (b, d, and f). In premigration stages (a and b), staining of intermediate filament proteins is found evenly distributed in the subcortical cytoplasm (a); at cycle 10 (c and d), after the somatic nuclei have reached the subcortical cytoplasm, IF proteins begin to concentrate on both sides of each nucleus (c), and this IF distribution becomes more pronounced at cycle 14 (e and f). Bar, 7 μm.

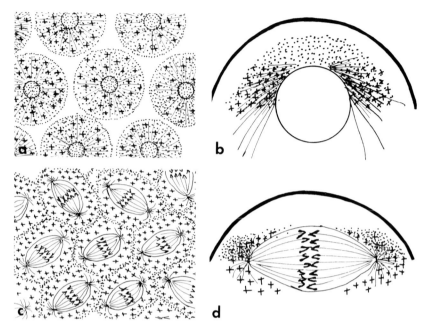

Fig. 11.   Schematic representation of the distribution of tubulin (lines), actin (stipples), and intermediate filaments (crosses) in the subcortical cytoplasm of a *Drosophila* embryo at synctial blastoderm stage as derived from whole-mount stained embryos (a and c) and sections (b and d). The distribution of the three cytoskeletal elements is shown in interphase (a and b) and in metaphase (c and d). At interphase (a and b), the nuclear membrane is intact, and the nuclei are surrounded by IF, sparing the area above each nucleus. Actin forms a cap over each nucleus that reaches to the borders of the cytoplasmic islands. Microtubules radiate outward from the nucleus and down toward the center of the embryo. At metaphase (c and d), the chromosomes are arranged in the metaphase plate between the two halves of the spindle formed by microtubules. Actin has moved to the periphery of the cytoplasmic islands, whereas IF lie between the actin array and the spindle.

the observation on whole-mount embryos. Figure 11 is a schematic representation of the reorganization of all three cytoskeletal elements during mitosis in the syncytial blastoderm.

VII. CONCLUSION

The syncytial *Drosophila* embryo provides an ideal system to study the distribution and rearrangements of the three major cytoskeletal components (IF, tubulin, and actin) by indirect immunofluorescence. These studies showed that the three cytoskeletal elements occupy distinct regions in the subcortical cytoplasm and that they are redistributed quite differently during mitosis. A group of IF polypeptides is deposited maternally into the egg and remains apparently unaltered until the nuclei have migrated into the subcortical cytoplasm. Then a dramatic rearrangement of IF material occurs, and IF proteins accumulate around each nucleus. It remains to be shown whether this spatial reorganization is accompanied by assembly of filaments from IF monomers. Very little is known about the dynamic aspects of the IF cytoskeleton in general.

These observations during *Drosophila* embryogenesis are consistent with the notion that IF serve some structural function to the cytoplasm, which may be achieved by anchoring the nuclei in the syncytium, by providing a solid attachment site for microtubules or microfilaments, or by organizing cytoplasmic domains around each nucleus well before cellularization. These presumed functions are not mutually exclusive; in fact, they are rather closely related. It is not unlikely that keratins found in early mouse embryos and in *Xenopus* eggs serve the same functions. Possible quantitative differences of IF amounts among these species may reflect different needs for IF structural functions, since in *Drosophila* and *Xenopus* embryos nuclear divisions and cleavage occur at a much faster rate than in mouse embryos.

IF are usually regarded as tissue-specific marker proteins. For example, expression of neurofilaments occurs at a characteristic time in differentiating neurons during rat (Raju *et al.*, 1981) and chicken (Tapscott *et al.*, 1981a,b) development, and the keratin composition is a characteristic marker for various epithelia. However, keratins expressed in early frog embryonic development are not embryo specific, but are simply a subset of keratins that are expressed again at a much later stage in certain simple epithelia (Franz *et al.*, 1983). We do not yet know whether all or a subset of the embryonic *Drosophila* IF are expressed in any adult tissue (we did, however, find them in larval salivary glands; Falkner *et al.*, 1981), but it seems reasonable to assume that, in analogy to the situation in *Xenopus*, the same IF polypeptides that we find in embryos are expressed again later in development. Thus, presence of IF in undifferentiated embryos may be reconciled by assuming a very general function for IF in embryogenesis.

Since eggs and embryos of a limited number of species have so far been tested for the presence of IF and since Franz *et al.* (1983) reported the absence of certain keratins in *Triturus* and *Pleurodeles* embryos, it is impossible to generalize from our observations on *Drosophila* to other organisms. However, evidence is accumulating that IF play an important role in structuring the cytoplasm and may have important influence on early developmental processes in vertebrate and invertebrate embryos.

ACKNOWLEDGMENTS

The indirect immunofluorescence localization of IF in *Drosophila* embryos represents unpublished work by Dr. Marika Walter, performed in Dr. Bruce Alberts' laboratory in the Department of Biochemistry and Biophysics, University of California, San Francisco. We wish to thank Dr. Alberts for his continuous interest and support. We enjoyed stimulating discussions with Drs. Tim Karr and Richard Warn, and we thank Hans Walter for artwork. Marika Walter was partly supported by a postdoctoral fellowship from the Deutsche Forschungsgemeinschaft, Bonn, Federal Republic of Germany, and by a NIH-Fogarty Foundation fellowship.

REFERENCES

Aamodt, E. J., and Williams, R. C., Jr. (1984). Microtubule-associated proteins connect microtubules and neurofilaments in vitro. *Biochemistry* **23**, 6023–6031.

Aubin, J. E., Osborn, M., Franke, W. W., and Weber, K. (1980). Intermediate filaments of the vimentin-type and the keratin-type are distributed differently during mitosis. *Exp. Cell Res.* **129**, 149–165.

Barros, C., and Franklin, I. E. (1968). Behavior of the gamete membranes during sperm entry into the mammalian egg. *J. Cell Biol.* **37**, C13–C18.

Bartnik, E., Osborn, M., and Weber, K. (1985). Intermediate filaments in non-neuronal cells of invertebrates: Isolation and biochemical characterization of intermediate filaments from the esophageal epithelium of the mollusc *Helix pomatia*. *J. Cell Biol.* **101**, 427–440.

Bartnik, E., Osborn, M., and Weber, K. (1986). Intermediate filaments in muscle and epithelial cells of nematodes. *J. Cell Biol.* **102**, 2033–2041.

Bedford, J. M. (1971). An electron microscopic study of sperm penetration into the rabbit egg after natural mating. *Am. J. Anat.* **133**, 213–254.

Ben-Ze'ev, A. (1984). Inhibition of vimentin synthesis and disruption of intermediate filaments in simian virus 40-infected monkey kidney cells. *Mol. Cell. Biol.* **4**, 1880–1889.

Biessmann, H., Falkner, F. G., Saumweber, H., and Walter, M. F. (1982). Disruption of the vimentin cytoskeleton may play a role in heat shock response. *In* "Heat Shock: From Bacteria to Man" (M. J. Schlesinger, M. Ashburner, and A. Tissieres, eds.), pp. 275–281. Cold Spring Harbor Lab., Cold Spring Harbor, New York.

Bloom, G. S., and Vallee, R. B. (1983). Association of microtubule-associated protein 2 (MAP 2) with microtubules and intermediate filaments in cultured brain cells. *J. Cell Biol.* **96**, 1523–1531.

Blose, S. H. (1979). Ten nanometer filaments and mitosis: Maintenance of structural continuity in dividing endothelial cells. *Proc. Natl. Acad. Sci. U.S.A.* **76**, 3372–3376.

Blose, S. H., and Bushnell, A. (1982). Observation of the vimentin-10-nm filaments during mitosis in BHK21 cells. *Exp. Cell Res.* **142**, 57–62.

Blose, S. H., and Chacko, S. (1976). Rings of intermediate (100 Å) filament bundles in the perinuclear region of vascular endothelial cells. Their motilization by colcemid and mitosis. *J. Cell Biol.* **70**, 459–466.

Brulet, P., Babinet, C., Kemler, R., and Jacob, F. (1980). Monoclonal antibodies against trophectoderm-specific markers during mouse blastocyst formation. *Proc. Natl. Acad. Sci. U.S.A.* **77**, 4113–4117.

Counce, S. J. (1973). The causal analysis of insect embryogenesis. *In* "Development Systems: Insects" (S. J. Counce and C. H. Waddington, eds.), Vol. 2, pp. 1–156. Academic Press, New York.

Dawson, P. J., Hulme, J. S., and Lloyd, C. W. (1985). Monoclonal antibody to intermediate filament antigen cross-reacts with higher plant cells. *J. Cell Biol.* **100,** 1793–1798.

Ducibella, T., and Anderson, E. (1975). Cell shape and membrane changes in the eight-cell stage mouse embryo: Prerequisites for morphogenesis in the blastocyst. *Dev. Biol.* **47,** 47–58.

Ducibella, T., Albertini, D. F., Anderson, E., and Biggers, J. D. (1975). The preimplantation mammalian embryo: Characterization of intercellular junctions and their appearance during development. *Dev. Biol.* **45,** 231–250.

Ducibella, T., Ukena, T., Karnovsky, M., and Anderson, E. (1977). Changes in cell surface and cortical cytoplasmic organization during early embryogenesis in the preimplantation mouse embryo. *J. Cell Biol.* **74,** 153–167.

Duprey, P., Morello, D., Vasseur, M., Babinet, C., Condamine, H., Brulet, P., and Jacob, F. (1985). Expression of the cytokeratin endo A gene during early mouse embryogenesis. *Proc. Natl. Acad. Sci. U.S.A.* **82,** 8535–8539.

Eagles, P. A., M., Gilbert, D. S., and Maggs, A. (1981). The polypeptide composition of axoplasm and neurofilaments from the marine worm *Myxicola infundibulum. Biochem. J.* **199,** 89–100.

Eckberg, W. R. (1981). The effects of cytoskeleton inhibitors on cytoplasmic localization in *Chaetopterus pergamentaceus. Differentiation* **19,** 55–58.

Eckert, B. S. (1985). Alteration of intermediate filament distribution in PtK1 cells by acrylamide. *Eur. J. Cell Biol.* **37,** 169–174.

Ellisman, M. H., and Porter, K. R. (1980). Microtrabecular structure of the axoplasmic matrix: Visualization of cross-linking structures and their distribution. *J. Cell Biol.* **87,** 464–479.

Falkner, F. G., Saumweber, H., and Biessmann, H. (1981). Two *Drosophila melanogaster* proteins related to intermediate filament proteins of vertebrate cells. *J. Cell Biol.* **91,** 175–183.

Foe, V. E., and Alberts, B. M. (1983). Studies of nuclear and cytoplasmic behaviour during the five mitotic cycles that precede gastrulation in *Drosophila* embryogenesis. *J. Cell Sci.* **61,** 31–70.

Foelix, R. F., and Hauser, M. (1979). Helically twisted filaments in giant neurons of a whip spider. *Eur. J. Cell Biol.* **19,** 303–306.

Franke, W. W., Grund, C., and Schmid, E. (1979). Intermediate-sized filaments present in Sertoli cells are of vimentin type. *Eur. J. Cell Biol.* **19,** 269–275.

Franke, W. W., Grund, C., Kuhn, C., Jackson, B. W., and Ilmensee, K. (1982a). Formation of cytoskeletal elements during mouse embryogenesis. III. Primary mesenchymal cells and the first appearance of vimentin filaments. *Differentiation* **23,** 43–59.

Franke, W. W., Schmid, E., Grund, C., and Geiger, B. (1982b). Intermediate filament proteins in nonfilamentous structures: Transient disintegration and inclusion of subunit proteins in granular aggregates. *Cell (Cambridge, Mass.)* **30,** 103–113.

Franke, W. W., Grund, C., Jackson, B. W., and Ilmensee, K (1983). Formation of cytoskeletal elements during mouse embryogenesis. IV. Ultrastructure of primary mesenchymal cells and their cell–cell interactions. *Differentiation* **25,** 121–141.

Franz, J. K., and Franke, W. W. (1986). Cloning of cDNA and amino acid sequence of a cytokeratin expressed in oocytes of *Xenopus laevis. Proc. Natl. Acad. Sci. U.S.A.* **83,** 6475–6479.

Franz, J. K., Gall, L., Williams, M. A., Picheral, B., and Franke, W. W. (1983). Intermediate-sized filaments in a germ cell: Expression of cytokeratins in oocytes and eggs of the frog *Xenopus. Proc. Natl. Acad. Sci. U.S.A.* **80**, 6254–6258.

Freeman, M., Nüsslein-Volhard, C., and Glover, D. M (1986). The dissociation of nuclear and centrosomal division in *gnu*, a mutation causing giant nuclei in *Drosophila. Cell (Cambridge, Mass.)* **46**, 457–468.

Fullilove, S. L., and Jacobson, A. G. (1971). Nuclear elongation and cytokinesis in *Drosophila montana. Dev. Biol.* **26**, 560–577.

Gall, L., Picheral, B., and Gounon, P. (1983). Cytochemical evidence for the presence of intermediate filaments and microfilaments in the egg of *Xenopus laevis. Biol. Cell.* **47**, 331–342.

Gawlitta, W., Osborn, M., and Weber, K. (1981). Coiling of intermediate filaments induced by microinjection of vimentin specific antibody does not interfere with locomotion and mitosis. *Eur. J. Cell Biol.* **26**, 83–90.

Geisler, N., Kaufmann, E., Fischer, S., Plessmann, U., and Weber, K. (1983). Neurofilament architecture combines structural principles of intermediate filaments with carboxy-terminal extensions increasing in size between triplet proteins. *EMBO J.* **2**, 1295–1302.

Gerhart, J., Ubbels, G., Black, S., Hara, K., and Kirschner, M. (1981). A reinvestigation of the role of the grey crescent in axis formation in *Xenopus laevis. Nature (London)* **191**, 511–516.

Gilbert, D. S. (1976). Neurofilament rings from giant axons. *J. Physiol. (London)* **266**, 81P–83P.

Gilbert, D. S., Newby, B. J., and Anderton, B. H. (1975). Neurofilament disguise, destruction, and discipline. *Nature (London)* **256**, 586–589.

Godsave, S. F., Wylie, C. C., Lane, E. B., and Anderton, B. H. (1984a). Intermediate filaments in the *Xenopus* oocyte: The appearance and distribution of cytokeratin-containing filaments. *J. Embryol. Exp. Morphol.* **83**, 157–167.

Godsave, S. F., Anderton, B. H., Heasman, J., and Wylie, C. C. (1984b). Oocytes and early embryos of *Xenopus laevis* contain intermediate filaments which react with anti-mammalian vimentin antibodies. *J. Embryol. Exp. Morphol.* **83**, 169–187.

Goldman, A. E., Maul, G., Steinert, P., Yang, H. Y., and Goldman, R. D. (1986). Keratin-like proteins that coisolate with intermediate filaments of BHK-21 cells are nuclear lamins. *Proc. Natl. Acad. Sci. U.S.A.* **83**, 3839–3843.

Goldman, R. D., and Knipe, D. M. (1972). Functions of cytoplasmic fibers in non-muscle cell motility. *Cold Spring Harbor Symp. Quant. Biol.* **37**, 523–534.

Gordon, W. E., III, Bushnell, A., and Burridge, K. (1978). Characterization of the intermediate (10 nm) filaments of cultured cells using an autoimmune rabbit antiserum. *Cell (Cambridge, Mass.)* **13**, 249–261.

Green, K. J., and Goldman, R. D. (1986). Evidence for the interaction between the cell surface and intermediate filaments in cultured fibroblasts. *Cell Motil. Cytoskel.* **6**, 389–405.

Guerrier, P., van den Biggelaar, J. A. M., van Dongen, C. A. M., and Verdonk, N. H. (1978). Significance of the polar lobe for the determination of dorsoventral polarity in *Dentalium vulgare* (da Costa). *Dev. Biol.* **63**, 233–242.

Harris, P., and Shaw, G. (1984). Intermediate filaments, microtubules and microfilaments in epidermis of sea urchin tube foot. *Cell Tissue Res.* **236**, 27–33.

Henderson, D., and Weber, K. (1980). Immunoelectron microscopic studies of intermediate filaments in cultured cells. *Exp. Cell Res.* **129**, 441–453.

Hirokawa, N. (1982). Cross-linker system between neurofilaments, microtubules, and membraneous organelles in frog axons revealed by the quick-freeze, deep-etching method. *J. Cell Biol.* **94**, 129–142.

Hodge, A. J., and Adelman, W. J. (1980). The neuroplasmic network in *Loligo* and *Hermissenda nerusons*. *J. Ultrastruct. Res.* **70**, 220–241.

Horwitz, J., Kupfer, H., Eshar, Z., and Geiger, B. (1981). Reorganization of arrays of prekeratin filaments during mitosis. *Exp. Cell Res.* **134**, 281–290.

Huneeus, F. C., and Davison, P. F. (1970). Fibrillar proteins from squid axons. I. Neurofilament proteins. *J. Mol. Biol.* **52**, 415–428.

Hynes, R. O., and Destree, A. T. (1978). 10 nm filaments in normal and transformed cells. *Cell (Cambridge, Mass.)* **13**, 151–163.

Imaizumi, T. (1954). Recherches sur l'expression des facteurs létaux héréditaires chez l'embryon de la drosophile. I. La variation du volume de l'embryon pendant la premiere période du développement. *Protoplasma* **44**, 1–10.

Ishikawa, H., Bischoff, R., and Holtzer, H. (1968). Mitosis and intermediate filaments in developing skeletal muscle. *J. Cell Biol.* **38**, 538–555.

Jackson, B. W., Grund, C., Schmid, E., Bürki, K., Franke, W. W., and Ilmensee, K. (1980). Formation of cytoskeletal elements during mouse embryogenesis. I. Intermediate filaments of the cytokeratin type and desmosomes in preimplantation embryos. *Differentiation* **17**, 161–179.

Jackson, B. W., Grund, C., Winter, S., Franke, W. W., and Ilmensee, K. (1981). Formation of cytoskeletal elements during mouse embryogenesis. II. Epithelial differentiation and intermediate-size filaments in early postimplantation embryos. *Differentiation* **20**, 203–216.

Jamrich, M., Sargent, T. D., and Dawid, I. B. (1987). Cell-type-specific expression of epidermal cytokeratin genes during gastrulation of *Xenopus laevis*. *Genes Dev.* **1**, 124–132.

Jeffery, W. R., and Meier, S. (1983). A yellow crescent cytoskeletal domain in ascidian eggs and its role in early development. *Dev. Biol.* **96**, 125–143.

Jonas, E., Sargent, T. D., and Dawid, I. B. (1985). Epidermal keratin gene expressed in embryos of *Xenopus laevis*. *Proc. Natl. Acad. Sci. U.S.A.* **82**, 5413–5417.

Jones, J. C. R., Goldman, A. E., Yuspa, S., Steinert, P., and Goldman, R. D. (1982). Dynamic aspects of the supramolecular organization of intermediate filament networks in cultured epidermal cells. *Cell Motil.* **2**, 197–213.

Jones, J. C. R., Goldman, A. E., Yang, H., and Goldman, R. D. (1985). The organizational fate of intermediate filament networks in two epidermal cell types during mitosis. *J. Cell Biol.* **100**, 93–102.

Junquera, P. (1985). Cleavage and blastoderm formation in normal and experimentally deformed naked eggs of a dipteran insect. *Wilhelm Roux's Arch. Dev. Biol.* **194**, 155–165.

Kaiser, J., Lang, A. B., and Went, D. F. (1982). Pulsation of nuclei in insect oocytes is reversibly inhibited by cytochalasin B. *Exp. Cell Res.* **139**, 460–463.

Karr, T. L., and Alberts, B. M. (1986). Organization of the cytoskeleton in early *Drosophila* embryos. *J. Cell Biol.* **102**, 1494–1509.

Keil, T. A., and Steinbrecht, R. A. (1984). Mechanosensitive and olfactory sensilla of insects. *In* "Insect Ultrastructure" (R. C. King and H. Akai, eds.), Vol. 2, pp. 477–516. Plenum, New York.

Kemler, R., Brulet, P., Schnebelen, M. T., Gaillard, J., and Jacob, F. (1981). Reactivity of monoclonal antibodies against intermediate filament proteins during embryonic development. *J. Embryol. Exp. Morphol.* **64**, 45–60.

Klymkowsky, M. W. (1981). Intermediate filaments in 3T3 cells collapse after intracellular injection of monoclonal anti-intermediate filament antibody. *Nature (London)* **291**, 249–251.

Klymkowsky, M. W. (1982). Vimentin and keratin intermediate filament systems in cultured PtK1 epithelial cells are interrelated. *EMBO J.* **1**, 161–165.

Klymkowsky, M. W., Miller, R. H., and Lane, E. B. (1983). Morphology, behavior and interaction of cultured epithelial cells after the antibody induced disruption of keratin filament organization. *J. Cell Biol.* **96**, 494–509.

Klymkowsky, M. W., Maynell, L. A., and Polson, A. G. (1987). Polar asymmetry in the organization of the cortical cytokeratin system of *Xenopus laevis* oocytes and embryos. *Development* **100**, 543–557.

Krause, G., and Sander, K. (1962). Ooplasmic reaction systems in insect embryogenesis. *Adv. Morpho.* **2**, 259–303.

Lane, E. B., and Klymkowsky, M. W. (1982). Epithelial tonofilaments: investigating their form and function using monoclonal antibodies. *Cold Spring Harbor. Symp. Quant. Biol.* **46**, 387–402.

Lane, E. B., Goodman, S. L., and Trejdosiewicz, L. K. (1982). Disruption of the keratin filament network during epithelial cell division. *EMBO J.* **1**, 1365–1372.

Lane, E. B., Hogan, B. L. M., Kurkinen, M., and Garrels, J. L. (1983). Co-expression of vimentin and cytokeratins in parietal endoderm cells of early mouse embryos. *Nature (London)* **303**, 701–704.

Lasek, R. J., and Hoffmann, P. N. (1976). The neuronal cytoskeleton, axonal transport, and axonal growth. *In* "Cell Motility" (R. Goldman, T. Pollard, and J. Rosenbaum, eds.), pp. 1021–1050. Cold Spring Harbor Lab., Cold Spring Harbor, New York.

Lasek, R. J., Krishnan, N., and Kaiserman-Abramof, I. R. (1979). Identification of the subunit proteins of 10 nm neurofilaments isolated from axoplasm of squid and *Myxicola* giant axons. *J. Cell Biol.* **82**, 336–346.

Lasek, R. J., Oblinger, M. M., and Drake, P. F. (1983). Molecular biology of neuronal geometry: Expression of neurofilament genes influences axonal diameter. *Quant. Biol.* **48**, 731–744.

Lazarides, E. (1980). Intermediate filaments as mechanical integrators of cellular space. *Nature (London)* **283**, 249–256.

Lazarides, E. (1982). Intermediate filaments: a chemically heterogeneous, developmentally regulated class of proteins. *Annu. Rev. Biochem.* **51**, 219–250.

Lehto, V. P., Virtanen, I., and Kurki, P. (1978). Intermediate filaments anchor the nuclei in nuclear monolayers of cultured human fibroblasts. *Nature (London)* **272**, 175–177.

Lehtonen, E. (1985). A monoclonal antibody against mouse oocyte cytoskeleton recognizing cytokeratin-type filaments. *J. Embryol. Exp. Morphol.* **90**, 197–209.

Lehtonen, E., and Badley, R. A. (1980). Localization of cytoskeletal proteins in preimplantation mouse embryos. *J. Embryol. Exp. Morphol.* **55**, 211–225.

Lehtonen, E., Lehto, V. P., Paasivuo, R., and Virtanen, I. (1983a). Pariatal and visceral endoderm differ in their expression of intermediate filaments. *EMBO J.* **2**, 1023–1028.

Lehtonen, E., Lehto, V. P., Vartio, T., Bradley, R. A., and Virtanen, I. (1983b). Expression of cytokeratin polypeptides in mouse oocytes and preimplantation embryos. *Dev. Biol.* **100**, 158–165.

Limbourg, B., and Zalokar, M. (1973). Permeabilization of *Drosophila* eggs. *Dev. Biol.* **35**, 382–387.

Lin, J. J., and Feramisco, J. R. (1981). Disruption of the in vivo distribution of the intermediate filaments in fibroblasts through the microinjection of a specific monoclonal antibody. *Cell (Cambridge, Mass.)* **24**, 185–193.

McKeon, F. D., Kirschner, M. W., and Caput, B. (1986). Homologies in both primary and secondary structure between nuclear envelope and intermediate filament proteins. *Nature (London)* **319**, 463–468.

Mahowald, A. P. (1963a). Ultrastructural differentiations during formation of the blastoderm in the *Drosophila melanogaster* embryo. *Dev. Biol.* **8**, 186–204.

Mahowald, A. P. (1963b). Electron microscopy of the formation of the cellular blastoderm in *Drosophila melanogaster*. *Exp. Cell Res.* **32**, 457–468.

Mahowald, A. P., Allis, C. D., and Caulton, J. H. (1981). Rapid appearance of multivesicular bodies in the cortex of *Drosophila* eggs at ovulation. *Dev. Biol.* **86**, 505–509.

Mahr, E. (1960a). Normale Entwicklung, Pseudofurchung und die Bedeutung des Furchungszentrums im Ei des Heimchens *(Gryllus domesticus.)* Z. *Morphol. Oekol. Tiere* **49**, 263–311.

Mahr, E. (1960b). Struktur und Entwicklungsfunktion des Dotter-Entoplasmasystems in Ei des Heimchens *(Gryllus domesticus). Wilhelm Roux' Arch. Entwicklungs mech. Org.* **152**, 263–302.

Maro, B., and Bornens, M. (1982). Reorganization of HeLa cell cytoskeleton induced by an uncoupler of oxidative phosphorylation. *Nature (London)* **295**, 334–336.

Miller, K. G., Karr, T. L., Kellogg, D. R., Mohr, I. J., Walter, M. F., and Alberts, B. M. (1985). Studies on the cytoplasmic organization of early *Drosophila* embryos. *Cold Spring Harbor Symp. Quant. Biol.* **50**, 79–90.

Mitchison, T. J., and Sedat, J. (1983). Localization of antigenic determinants in whole *Drosophila* embryos. *Dev. Biol.* **99**, 261–264.

Miyatani S., Winkles, J. A., Sargent, T. D., and Dawid, I. B. (1986). Stage-specific keratins in *Xenopus laevis* embryos and tadpoles: the XK81 gene family. *J. Cell Biol.* **103**, 1957–1965.

Mose-Larsen, P., Bravo, R., Fey, S. J., Small, J. V., and Celis, J. E. (1983). Putative association of mitochondria with a subpopulation of intermediate-sized filaments in cultured human skin fibroblasts. *Cell (Cambridge, Mass.)* **31**, 681–692.

Murti, K. G., and Goorha, R. (1983). Interactions of frog virus-3 with the cytoskeleton. I. Altered organization of microtubules, intermediate filaments, and microfilaments. *J. Cell Biol.* **96**, 1248–1257.

Okada, E., and Waddington, C. H. (1959). The submicroscopic structure of the *Drosophila* egg. *J. Embryol. Exp. Morphol.* **7**, 583–597.

Oshima, R. G., Howe, W. E., Klier, G., Adamson, E. D., and Shevinsky, L. H. (1983). Intermediate filament protein synthesis in preimplantation murine embryos. *Dev. Biol.* **99**, 447–455.

Paulin, D., Babinet, C., Weber, K., and Osborn, M. (1980). Antibodies as probes of cellular differentiation and cytoskeletal organization in the mouse blastocyst. *Exp. Cell Res.* **130**, 297–304.

Peaucellier, G., Guerrier, P., and Bergerard, J. (1974). Effects of cytochalasin B on meiosis and development of fertilized and activated eggs of *Sabellaria alveolata* L. (Polychaete annelid). *J. Embryol. Exp. Morphol.* **31**, 61–74.

Philips, L. L., Antilio-Gambetti, L., and Lasek, R. J. (1983). Bodian's silver method reveals molecular variation in the evolution of neurofilament proteins. *Brain Res.* **278**, 219–223.

Pruss, R. M., Mirsky, R., Raff, M. C., Thorpe, R., Dowding, A. J., and Anderton, B. H. (1981). All classes of intermediate filaments share a common antigenic determinant defined by a monoclonal antibody. *Cell (Cambridge, Mass.)* **27**, 419–428.

Raff, R. A. (1972). Polar lobe formation by embryos of *Ilyanassa obsoleta*: Effects of inhibitors of microtubule and microfilament function. *Exp. Cell Res.* **71**, 455–459.

Raju, T. R., Bignami, A., and Dahl, D. (1981). In vivo and in vitro differentiation of neurons and astrocytes in the rat embryo. Immunofluorescence study with neurofilament and glial filament antisera. *Dev. Biol.* **85**, 344–357.

Reverberi, G. (1975). On some effects of cytochalasin B on the eggs and tadpoles of the ascidians. *Acta Embryol. Exp.* **2**, 137–158.

Rickoll, W. L. (1976). Cytoplasmic continuity between embryonic cells and the primitive yolk sac during early gastrulation in *Drosphila melanogaster*. *Dev. Biol.* **49**, 304–310.

Roslansky, P. F., Cornell-Bell, A., Rice, R. V., and Adelman, W. J. (1980). Polypeptide composition of squid neurofilaments. *Proc. Natl. Acad. Sci. U.S.A.* **77**, 404–408.

Sawada, T., and Osanei, K. (1981). The cortical contractions related to the ooplasmic segregation in *Ciona intestinalis* eggs. *Wilhelm Roux's Arch. Dev. Biol.* **190**, 208–214.

Scharf, S. R., and Gerhart, J. C. (1983). Axis determination in eggs of *Xenopus laevis:* A critical period before cleavage identified by the common effects of cold, pressure, and ultraviolet irradiation. *Dev. Biol.* **99**, 75–87.

Schliwa M., and van Blerkom, J. (1981). Structural interaction of cytoskeletal components. *J. Cell Biol.* **90**, 222–235.

Schmidt, B. A., Kelley, P. T., May, M. C., Davis, S. E., and Conrade, G. W. (1980). Characterization of actin from fertilized eggs of *Ilyanassa obsoleta* during polar lobe formation and cytokinesis. *Dev. Biol.* **76**, 126–140.

Schnapp, B. J., and Reese, T. S. (1982). Cytoplasmic structure in rapid-frozen axons. *J. Cell Biol.* **94**, 667–679.

Schneider-Minder, A. (1966). Cytologische Untersuchung der Embryonalentwicklung von *Drosophila melanogaster* nach Röntgenbestrahlung in frühen Entwicklungsstadien. *Int. J. Radiat. Biol.* **11**, 1–20.

Schnetter, W. (1965). Experimente zur Analyse der morphogenetischen Funktion der Ooplasmabestandteile in der Embryonalentwicklung des Kartoffelkäfers (*Leptinotarsa decemlineata* Say). *Wilhelm Roux' Arch. Entwicklungs mech. Org.* **155**, 637–692.

Scriba, M. E. L. (1964). Beeinflussung der frühen Embryonalentwicklung von *Drosophila melanogaster* durch Chromosomenaberration. *Zool. Jahrb., Abt. Anat. Ontog. Tiere* **81**, 435–490.

Seiler, J. (1924). Furchung des Schmetterlingeies ohne Beteiligung des Kerns. *Biol. Zentralbl.* **44**, 69–71.

Sharpe, A. L., Chen, B., Murphy, J. R., and Fields, B. N. (1980). Specific disruption of vimetin filament organization in monkey kidney CV-1 cells by diphteria toxin, exotoxin A, and cycloheximide. *Proc. Natl. Acad. Sci. U.S.A.* **77**, 7267–7271.

Shimizu, T. (1982). Ooplasmic segregation in the *Tubifex* egg: Mode of pole plasm accumulation and possible involvement of microfilaments. *Wilhelm Roux's Arch. Dev. Biol.* **191**, 246–256.

Small, J. V., and Celis, J. E. (1978). Direct visualization of the 10 nm (100 Å) filament network in whole and enucleated cultured cells *J. Cell. Sci.* **31**, 393–409.

Starger, J. M., and Goldman, R. D. (1977). Isolation and preliminary characterization of 10 nm filaments from baby hamster kidney (BHK-21) cells. *Proc. Natl. Acad. Sci. U.S.A.* **74**, 2422–2426.

Steinert, P. M., Steven, A. C., and Roop, D. R. (1985). The molecular biology of intermediate filaments. *Cell (Cambridge Mass.)* **42**, 411–419.

Strome, S., and Wood, W. B. (1983). Generation of asymmetry and segregation of germline specific granules in *C. elegans* embryos. *Cell (Cambridge, Mass.)* **35**, 15–25.

Swalla, B. J., Moon, R. T., and Jeffery, W. R. (1985). Developmental significance of a cortical cytosktal domain in *Chaetopterus* eggs. *Dev. Biol.* **111**, 434–450.

Tapscott, S. J., Bennett, G. S., and Holtzer, H. (1981a). Neuronal precursor cells in the chick neural tube express neurofilament proteins. *Nature (London)* **292**, 836–838.

Tapscott, S. J., Bennett, G. S., Toyama, Y., Kleinbart, F., and Holtzer, H. (1981b). Intermediate filament proteins in the developing chick spinal cord. *Dev. Biol.* **86**, 40–54.

Thomas, G. P., Welch, W. J., Mathews, M. B., and Feramisco, J. R. (1982). Molecular and cellular effects of heat shock and related treatments of mammalian tissue culture cells. *Cold Spring Harbor Symp. Quant. Biol.* **46**, 985–996.

Tölle, H. G., Weber, K., and Osborn, M. (1986). Microinjection of monoclonal antibodies to vimentin, desmin and GFA in cells which contain more than one IF type. *Exp. Cell Res.* **162**, 462–472.

Traub, P. (1985). "Intermediate Filaments: A Review." Springer-Verlag, Berlin and New York.

Traub, U. E., Nelson, W. J., and Traub, P. (1983). Polyacrylamide gel electrophoretic screening of mammalian cells cultured in vitro for the presence of the intermediate filament protein vimentin. *J. Cell Sci.* **62**, 129–147.

Ubbels, G. A., Hara, K., Koster, C. H., and Kirschner, M. W. (1983). Evidence for a functional role of the cytoskeleton in determination of the dorsoventral axis in *Xenopus laevis* eggs. *J. Embryol. Exp. Morphol.* **77**, 15–37.

Van Blerkom, J., Barton, S. C., and Johnson, M. H. (1976). Molecular differentiation in the preimplantation mouse embryo. *Nature (London)* **259**, 319–321.

Venetianer, A., Schiller, D. L., Magin, T., and Franke, W. W. (1983). Cessation of cytokeratin expression in a rat hepatome cell line lacking differentiated functions. *Nature (London)* **305**, 730–733.

Virtanen, I., Badley, R. A., Paasivuo, R., and Lehto, V. P. (1984). Distinct cytoskeletal domains revealed in sperm cells. *J. Cell Biol.* **99**, 1083–1091.

Walter, M. F., and Alberts, B. M. (1984). Intermediate filaments in tissue culture cells and early embryos of *Drosophila melanogaster*. *UCLA Symp. Mol Cell. Biol.* Molecular Biology of Development" **19**, pp, 263–271.

Walter, M. F., and Alberts, B. M. (1989). In preparation.

Walter, M. F., and Biessmann, H. (1984a). Intermediate-sized filaments in *Drosophila* tissue culture cells. *J. Cell Biol.* **99**, 1468–1477.

Walter, M. F., and Biessmann, H. (1984b). A monoclonal antibody that detects vimentin-related proteins in invertebrates. *Mol. Cell. Biochem.* **60**, 99–108.

Wang, E., and Choppin, P. W. (1981). Effect of vanadate on intracellular distribution and function of 10 nm filaments. *Proc. Natl. Acad. Sci. U.S.A.* **78**, 2363–2367.

Wang, E., Gross, P. K., and Choppin, P. W. (1979). Involvement of microtubules and 10 nm filaments in the movement and positioning of nuclei in syncytia. *J. Cell Biol.* **83**, 320–337.

Wang, E., Fischman, D., Liem, R. K. H., and Sun, T. T. (eds.) (1985). Intermediate filaments. *Ann. N.Y. Acad. Sci.* **455**.

Warn, R. M., and Magrath, R. (1983). F-actin distribution during the cellularization of the *Drosophila* embryo visualized with FL-phalloidin. *Exp. Cell Res.* **143**, 103–114.

Warn, R. M., and Warn, A. (1986). Microtubule arrays present during the sycytial and cellular blastoderm stages of the early *Drosophila* embryo. *Exp. Cell Res.* **163**, 201–210.

Warn, R. M., Magrath, R., and Webb, S. (1984). Distribution of F-actin during cleavage of the *Drosophila* syncytial blastoderm. *J. Cell Biol.* **98**, 156–162.

Warn, R. M., Smith, L., and Warn, A. (1985). Three distinct distributions of F-actin occur during the divisions of polar surface caps to produce pole cells in *Drosphila* embryos. *J. Cell Biol.* **100**, 1010–1015.

Welch, W. J., and Suhan, J. P. (1985). Morphological studies of the mammalian stress response: Characterisation of changes in cytoplasmic organelles, cytoskeleton, and nucleoli, and appearance of intranuclear actin filaments in rat fibroblasts after heat-shock treatment. *J. Cell Biol.* **101**, 1198–1211.

Went, D. F., Fux, T., and Camenzind, R. (1978). Movement pattern and ultrastructure of pulsating oocyte nuclei of the paedogenetic gall midge, *Heteropeza pygmaea* Winnerz (Diptera: Cecidomyiidae). *Int. J. Insect Morphol. Embryol.* **7**, 301–314.

Wolf, R. (1978). The cytaster, a colchicine-sensitive migration organelle of cleavage nuclei in an insect egg. *Dev. Biol.* **62**, 464–472.

Wolf, R. (1980). Migration and division of cleavage nuclei in the gall midge *Wachtliella persicariae. Wilhelm Roux's Arch. Dev. Biol.* **188,** 65–73.

Wolf, R. (1985). Migration and division of cleavage nuclei in the gall midge *Wachtliella persicariae.* III. Pattern of anaphase-triggering waves altered by temperature gradients and local gas exchange. *Wilhelm Roux's Arch. Dev. Biol.* **194,** 257–270.

Wylie, C. C., Brown, D., Godsave, S. F., Quarmby, J., and Heasman, J. (1985). The cytoskeleton of *Xenopus* oocytes and its role in development. *J. Embryol. Exp. Morphol.* **89,** Supplement 1–15.

Zackroff, R. V., and Goldman, R. D. (1980). In vitro reassembly of squid brain intermediate filaments (neurofilaments): Purification by assembly-disassembly. *Science* **208,** 1152–1155.

Zalokar, M. (1976). Autoradiographic study of protein and RNA formation during early development of *Drosophila* eggs. *Dev. Biol.* **49,** 425–437.

Zalokar, M., and Erk, I. (1976). Division and migration of nuclei during early embryogenesis of *Drosophila melanogaster. J. Microsc. Biol. Cell* **25,** 97–106.

Zalokar, M., and Erk, I. (1977). Phase-partition fixation and staining of *Drosophila* eggs. *Stain Technol.* **52,** 89–95.

Zalokar, M., Audit, C., and Erk, I. (1975). Developmental defects of female sterile mutants of *Drosophila melanogaster. Dev. Biol.* **47,** 419–432.

Zieve, G. W., Heidemann, S. R., and McIntosh, J. R. (1980). Isolation and partial characterization of a cage of filaments that surrounds the mammalian mitotic spindle. *J. Cell Biol.* **87,** 160–169.

Zissler, D., and Sander, K. (1973). The cytoplasmic architecture of the egg cell of *Smittia* spec. (Diptera, Chironomidae). I Anterior and posterior pole regions. *Wilhelm Roux's Arch. Dev. Biol.* **172,** 175–186.

Zissler, D., and Sander, K. (1977). The cytoplasmic architecture of the egg cell of *Smittia* spec. (Diptera, Chironomidae). II. Periplasm and yolk-endoplasm. *Wilhelm Roux's Arch. Dev. Biol.* **183,** 233–248.

Zissler, D., and Sander, K. (1982). The cytoplasmic architecture of the insect egg cell. *In* "Insect Ultrastructure" (R. C. King and H. Akai, eds.), Vol. 1, pp. 189–221. Plenum, New York.

9

Nuclear Architectural Changes during Fertilization and Development

STEPHEN STRICKER, RANDALL PRATHER, CALVIN SIMERLY,
HEIDE SCHATTEN, AND GERALD SCHATTEN

Integrated Microscopy Resource for Biomedical Research
The University of Wisconsin–Madison
Madison, Wisconsin 53706

I. INTRODUCTION: THE KARYOSKELETON AND ITS DYNAMIC NATURE

Fertilization and development exact a series of demands on the intracellular machinery that regulates nuclear architecture. During fertilization, the highly condensed sperm nucleus that is incorporated into the cytoplasm of the egg decondenses and subsequently becomes surrounded by an annulated nuclear envelope of predominately maternal origin. Since most eggs are fertilized as oocytes, changes in sperm nuclear architecture during fertilization are accompanied by dramatic alterations in the karyoskeleton of the oocyte nucleus. Such alterations include the decondensation of the maternal meiotic chro-

225

THE CELL BIOLOGY OF
FERTILIZATION

mosomes and the subsequent formation of the female pronucleus and polar body nucleus. Fertilization is typically completed prior to the first mitotic division as the male and female pronuclei fuse. During the complicated process of pronuclear fusion, the two nuclear membranes that constitute the nuclear envelope of the female pronucleus must coalesce with corresponding membranes in the male pronucleus. As these events are being completed, a series of related changes takes place in the organization of the cytoskeleton. Various alterations in nuclear architecture also occur during subsequent development, as the cytoplasm of the zygote becomes partitioned into cellular units with varying developmental fates.

Descriptions of the molecular biology of the karyoskeleton are rapidly advancing, and it is now clear that several classes of proteins constitute discrete parts of the karyoskeleton. The nucleus is separated from the cytoplasm by a double-membraned nuclear envelope that is studded with octagonal nuclear pores. Recently, a new class of glycoproteins in the nuclear pore complex, called nucleoporins, has been investigated (Davis and Blobel, 1986; Park *et al.*, 1987; Snow *et al.*, 1987). At least some of these glycoproteins are unusual in that they contain O-linked *N*-acetylglucosamine exposed on the cytoplasmic and nuclear faces of the pores, rather than on the luminal surfaces (Holt *et al.*, 1987; Park *et al.*, 1987).

The nuclear lamina, discovered by Fawcett (1966), resides subjacent to the inner nuclear membrane and comprises polypeptides that range from 60 to 80 kDa (for reviews, see Gerace, 1986; Krohne and Benavente, 1986; Franke, 1987). Lamins share considerable homology in primary and secondary structure with intermediate filament proteins (Fisher *et al.*, 1986; McKeon *et al.*, 1986; Krohne *et al.*, 1987; Wolin *et al.*, 1987), and the supramolecular organization of these proteins has been shown to be an elaborate quiltwork of 10-nm filaments that resemble intermediate filaments in ultrastructure (Aebi *et al.*, 1986). In mammals, the three major types of lamins are designated A, B, and C (Gerace and Blobel, 1982). At the onset of mitosis, the nuclear lamina becomes hyperphosphorylated (Ottaviano and Gerace, 1985). Concomitant with this hyperphosphorylation, lamins A and C dissociate from the nuclear envelope and disperse into the cytoplasm. Lamin B, however, retains its association with fragments of the inner nuclear membrane.

Between the lamins and the chromatin resides another class of proteins referred to as peripheral nuclear antigens, Pl (Chaly *et al.*, 1984), or perichromin (McKeon *et al.*, 1983). These proteins are 27–33 kDa and characteristically undergo a change at mitosis. Unlike lamins A and C, however, peripheral nuclear antigens condense around each chromosome at mitosis, effectively ensheathing the chromatids.

Knowledge regarding the nuclear matrix is relatively incomplete, but advances are being made. For example, a primary component of the nuclear matrix, topoisomerase II, provides a remarkable example of a protein that combines structure and function. Topoisomerase II is an enzyme that occurs in the chromosome scaffold and is capable of cutting both strands of DNA. The cut strands are then swiveled into tighter or looser coils before being ligated at their correct ends. As a major constituent of the chromosome scaffold, topoisomerase II is well situated for conducting its ligating activities during coiling and uncoiling of the chromosomes (Earnshaw and Heck, 1985; Earnshaw et al., 1985).

Another active area of research in nuclear architecture involves centromeres and kinetochores. Although these terms have been used interchangeably, a centromere is the actual portion of chromatin that corresponds to the primary constriction of a condensed chromosome. A kinetochore, on the other hand, is often closely associated with the centromere and consists of a proteinaceous plaque to which the spindle fibers attach. When cloned centromeric DNA is inserted into plasmid DNA, the exogenous DNA is carried from generation to generation with high fidelity and thus provides a bioassay for the accuracy of the cloned sequence (Bloom and Carbon, 1982). An understanding of the molecules constituting the kinetochore was initially advanced by the use of autoimmune sera (Moroi et al., 1980; Tan et al., 1980), and it now appears that kinetochores contain 18-, 80-, and 140-kDa subunits, referred to as CENP-A, CENP-B, and CENP-C, respectively (Earnshaw and Rothfield, 1985; Valdivia and Brinkley, 1985; Balczon and Brinkley, 1987; Palmer et al., 1987). Cloning of kinetochore genes has begun (Earnshaw et al., 1987).

The scope of this chapter is to review knowledge about the organization of the karyoskeleton and consider its changing nature during fertilization and development.

II. NUCLEAR LAMINS

A. Comparative Biochemistry of Lamins

Lamins isolated from several types of cells have been characterized biochemically in vertebrates of the Class Mammalia (humans, mice, and rats), Class Aves (chicken), and Class Amphibia (*Xenopus*). Biochemical data are also available on the lamins of two invertebrates, *Drosophila* (Phylum Arthropoda, Class Insecta) and *Spisula* (Phylum Mollusca, Class Bivalvia). Most of the biochemical studies conducted on lamins have utilized nuclei isolated from rat livers. To provide a foundation for understanding the dynamic nature of the nuclear lamina during fertilization and development, this section summarizes the salient biochemical properties of rat lamins and compares them with the data available on lamins of other organisms.

The three major lamins of the rat, designated A, B, and C, can be distinguished by differences in their molecular weights (M_r) and isoelectric points (pI) (Gerace and Blobel, 1982; Krohne and Benavente, 1986). Lamin A has a molecular weight of approximately 70,000, whereas lamins B and C have M_rs of about 67,000 and 60,000, respectively. As noted by Stick and Hausen (1985), however, the apparent molecular weights ascertained for each type of lamin can be greatly affected by the gel system used in the analysis of these proteins.

The isoelectric points of lamins A and C are nearly neutral (Krohne and Benavente, 1986). In addition to a correspondence in pIs, lamins A and C exhibit similar peptide digestion maps (Lam and Kaspar, 1979; Gerace and Blobel, 1982) and substantial regions of identity in primary and secondary structure based on sequencing of human cDNA clones (Fisher *et al.*, 1986; McKeon *et al.*, 1986). It is not yet possible to discriminate between lamins A and C based on immunological affinities (Lehner *et al.*, 1986). Hence, these two lamins are often referred to as a single A/C complex (Schatten *et al.*, 1985).

Unlike lamins A/C, lamin B has an acidic pI of <5.5 (Krohne and Benavente, 1986). Moreover, the peptide map of lamin B differs markedly from those of lamins A and C (Gerace and Blobel, 1982; Kaufmann *et al.*, 1983). Although some cross-reactivity exists among all three types of mammalian lamins (Burke *et al.*, 1983), lamin B can be routinely distinguished from lamins A/C using various monoclonal antibodies (McKeon *et al.*, 1983) or autoimmune sera (Maul *et al.*, 1987). Recently, a quantitatively minor type of non-B lamin with an acidic pI has been identified in mammalian nuclei (Lehner *et al.*, 1986). It has been suggested that this lamin be referred to as lamin D (Lehner *et al.*, 1987).

As in mammals, three distinct lamins have been characterized in studies of chicken fibroblasts (Lehner *et al.*, 1986, 1987). Lamin A of chickens (M_r 69,000) has a neutral pI and appears to correspond to the lamin A of mammals. The other two avian lamins, referred to as B_1 and B_2, migrate to acidic positions, when subjected to two-dimensional gel electrophoresis and have M_rs of 68,000 and 66,000, respectively. Lamin B_1 is a quantitatively minor component of the nuclear lamina that apparently corresponds in its biochemical properties to mammalian lamin B. Lamin B_2, on the other hand, forms the bulk of the acidic lamins in chickens and seems to be related in structure to the minor "lamin D" of mammalian nuclei. No counterpart of mammalian lamin C has been identified in chickens.

In the frog *Xenopus*, five types of lamins, designated L_I, L_{II}, L_{III}, L_{IV}, and A, have been characterized (Krohne *et al.*, 1981, 1987; Benavente and Krohne, 1985). Lamin L_I (M_r 72,000) and lamin L_{II} (M_r 68,000) are found in various adult somatic cells, as well as in the blastomeres of postgastrula developmental stages. Lamin L_I apparently corresponds to lamin B of mammals, whereas lamin L_{II} is probably the counterpart of mammalian lamin C (Krohne *et al.*, 1987; Wolin *et al.*, 1987). Lamin L_{III} resembles lamin L_{II} in that it is like the A/C lamins of mammals and has a M_r of 68,000. Lamin L_{III}, however, is found predominantly in oocytes or in prelarval developmental stages and occurs in only a few adult somatic cells (e.g., muscles, neurons, and Sertoli cells). Lamin L_{IV} has a M_r of 75,000 and is found exclusively in spermatids or spermatozoa (Benavente and Krohne, 1985). This lamin exhibits a specific cross-reactivity with lamin L_{III}. Thus, it may be more like lamins A/C than lamin B, but the exact relationship of lamin L_{IV} to mammalian lamins A, B, or C remains to be determined. Lamin A has a predicted M_r of 75,000 and occurs in all adult somatic cells examined, except for erythrocytes (Krohne *et al.*, 1987). Amino acid sequences obtained from cDNA clones indicate that lamin A of *Xenopus* corresponds to mammalian lamin A (Krohne *et al.*, 1987).

According to Smith *et al.* (1987), the fruitfly *Drosophila melanogaster* has two major types of lamins that are called Dm_1 and Dm_2 in reference to the organism's scientific name. Dm_1 has a M_r of 74,000, whereas Dm_2 has a M_r of 76,000. The isoelectric points of these two lamins appear to be intermediate between those of lamins A/C and B (Smith *et al.*, 1987), and the overall relationship of the two insect lamins to the three lamins of mammals remains unclear.

Lamin-like polypeptides with M_rs that differ from those of Dm_1 and Dm_2 have been reported in other studies of *Drosophila*. According to Risau *et al.* (1981), the nuclear envelopes of Kc cells in *Drosophila* contain an 80,000 M_r protein that may represent a lamin. Lamin-like proteins with M_rs of 67,000 and 65,000 (Fuchs *et al.*, 1983) or 70,000 and 68,000 (McKeon *et al.*, 1983) have also been isolated. The 68,000 M_r protein cross-reacts with a human autoimmune serum that recognizes lamins A/C (McKeon *et al.*, 1983). The relationships of the other types of putative lamins in *Drosophila* to mammalian lamins has not been determined. A single lamin, referred to as lamin G (Maul and Schatten, 1986) or clamin (Maul *et al.*, 1987), has been isolated from the nuclear envelopes of primary oocytes produced by the clam *Spisula*. Lamin G has a molecular weight of 67,000 and an isoelectric point which is intermediate between those of mammalian lamins A/C and B (Maul *et al.*, 1984). Immunological studies indicate that lamin G resembles mammalian lamins A/C more closely than it does lamin B (Maul *et al.*, 1984).

B. Lamins during Fertilization

Changes in the composition of lamins during gametogenesis have been recently reviewed (Krohne and Benavente, 1986; Maul and Schatten, 1986; Moss *et al.*, 1987; Maul, 1988). In this section, the nature of the lamins in fully formed gametes of mammals, chicken, *Xenopus*, and sea urchins is summarized. In addition, the redistribution of lamins during fertilization is considered in the case of the mouse, *Xenopus*, sea urchin, and clam.

Based on ultrastructural observations (Fawcett, 1966) and immunofluorescence studies (Stick and Schwarz, 1982), a nuclear lamina appears to be lacking in the spermatozoa of mice. Maul *et al.* (1986), on the other hand, report that lamins can be identified by immunofluorescence microscopy in patchy locations along the nuclear envelope of mouse sperm. They propose that differences in fixation protocols can account for the apparent absence of lamins reported by previous workers. Similarly, Moss *et al.* (1987) report that lamin B polypeptides are present at all stages of mouse spermatogenesis. It should be noted, however, that detailed analyses of bovine sperm by immunocytochemical methods and immunoblotting failed to detect any lamins in the fully mature sperm of this mammal (Longo *et al.*, 1987). Instead of lamins, the major karyoskeletal elements in bovine sperm consist of two distinct sets of basic proteins, referred to as calicin (60 kDa) and multiple-band polypeptides (56–74 kDa) (Longo *et al.*, 1987).

Lamins appear to be lacking in the sperm of chicken, based on immunofluorescence microscopy (Stick and Schwarz, 1982). The absence of lamins in the fully formed sperm of chickens has recently been confirmed by immunoblotting methods (Lehner *et al.*, 1987).

In sperm of the frog *Xenopus*, no lamins are detectable by immunofluorescence microscopy (Stick and Schwarz, 1982). Using biochemical methods, however, the sperm-specific lamin L_{IV} can be isolated from purified preparations of sperm nuclear envelopes (Benavente and Krohne, 1985). A patchy distribution of lamin L_{IV} along the inner surface of the sperm nuclear envelope has been ascertained in immunolabeled specimens examined by electron microscopy (Benavente and Krohne, 1985).

In the sea urchin *Lytechinus variegatus*, lamins have been detected in sperm by immunofluorescence microscopy utilizing human autoimmune sera that recognize lamins A/C or B (Schatten *et al.*, 1985). In the fully formed sperm, lamins appear to occur only at the acrosomal and centriolar fossae.

The composition of lamins in fully developed female gametes varies greatly, depending on the stage of meiosis at which fertilization occurs. The mouse oocyte, for example, is arrested at metaphase of the second meiotic division directly prior to fertilization and thus lacks a nuclear envelope and polymerized lamina (Schatten *et al.*, 1985). Similarly, the nuclear lamina in the germinal vesicle (GV) of the fully formed primary oocyte in the frog *Xenopus* depolymerizes during GV breakdown and is presumably in a soluble state as the metaphase-arrested oocyte is fertilized. The GV of the clam *Spisula*, on the other, remains intact prior to fertilization and possesses (1) a polymerized lamina that is composed of lamin G and (2) a store of soluble B-like lamins dispersed throughout the nucleoplasm of the GV (Maul *et al.*, 1987). Unlike most other animals, sea urchins complete meiosis prior to fertilization, and the pronucleus of each unfertilized egg contains both lamins A/C and lamin B (Schatten *et al.*, 1985).

During fertilization in mice, lamins of the A/C and B types become associated with the nuclear envelope of the nascent female pronucleus (Fig. 1A–C; Schatten *et al.*, 1985). The male pronucleus that develops from the incorporated sperm head also acquires lamins A/C and B during fertilization (Fig. 1D–F). At the completion of syngamy during the first mitotic division in mice (Fig. 1G–I), lamins associated with the male and female pronuclei coalesce and form the lamina of the nuclei in the 2-cell embryo (Fig. 1J–L).

Lamins surrounding the male pronucleus in the fertilized egg of mice are presumably derived from maternal sources, although direct confirmation of this supposition is lacking. In *Xenopus*, a maternal origin of male pronuclear lamins seems likely, based on the results of immunoblotting studies in which demembranated sperm heads are injected into fully formed primary oocytes (Stick and Hausen, 1985). In such investigations, multiple male pronuclei form in the cytoplasm of each injected oocyte, but only the oocyte-specific form lamin L_{III} is detected.

Recent studies of fertilization indicate the acquisition of lamins by the nascent pronuclei in mice can be blocked by various agents that depolymerize microtubules (Figs. 2 and 3; H. Schatten *et al.*, 1988). In sea urchins, however, the acquisition of the lamins around the zygote nucleus is not inhibited by microtubule-depolymerizing agents (H. Schatten *et al.*, 1988). One possible explanation for these results is that lamin acquisition during the meiotic divisions in mice requires polymerized microtubules in the egg cytoplasm, whereas no such need for cytoplasmic microtubules exists in the case of sea urchins, which complete meiosis prior to fertilization.

In the clam *Spisula*, the GV breaks down 7–10 min after sperm penetration (Maul and Schatten, 1986). No polymerized lamina is visible until the pronuclei are formed, since the nuclear envelope does not reform during the second meiotic division (Maul and Schatten, 1986).

C. Changes in Lamin Composition during Development

In all cases examined, the composition of the lamins associated with the nuclear envelope undergoes some sort of a change during ontogeny. Such changes can involve the simple addition or deletion of a particular kind of lamin. Alternatively, a specific type of lamin may be lost at one stage in development and reacquired at a subsequent stage. Reviewed below are the data on ontogenetic changes in lamin composition in six systems—mouse, chicken, *Xenopus*, *Drosophila*, clam, and sea urchin. In addition to a consideration of

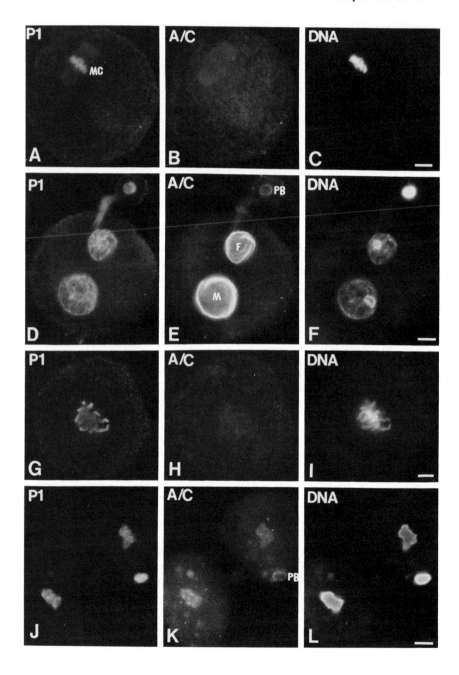

the strength and variability of the reported evidence, some overall trends in the ontogenetic variations in lamin composition are discussed at the end of the section.

1. Xenopus

Lamin L_{III} is the only lamin in the nuclei of blastomeres up to the tenth or eleventh cleavage (Stick and Hausen, 1985; Krohne and Benavente, 1986). Since transcription is halted after fertilization and is not reinitiated until midway through the blastula stage (Newport and Kirschner, 1982), it is clear that the lamins incorporated into the nuclei of the early embryo must arise from a stockpile of lamin L_{III} accumulated in the primary oocyte prior to fertilization. Immunoblotting studies of enucleated oocytes suggest that the surplus lamin L_{III} is stored as unactivated maternal transcripts within the GV rather than as cytoplasmic pools of masked mRNA or translated lamins (Stick and Hausen, 1985). Forbes et al. (1983) have calculated that the GV of the fully grown primary oocyte contains enough nuclear constituents to provide for approximately 1000 nuclear envelopes without requiring additional transcription. This value corresponds fairly well to the 2000–4000 nuclei that are present in the blastula at the time of transcription reinitiation (Newport and Kirschner, 1982) and lends further support to the notion that lamins of the early Xenopus embryo are derived from stores of maternal L_{III} transcripts.

At the midblastula transition stage (MBT), the B-like lamin L_I is present in addition to lamin L_{III} (Stick and Hausen, 1985). The appearance of lamin L_I at this stage in development seems to be due to a de novo synthesis involving the activation of maternal transcripts. This conclusion is based on the fact that (1) no cytoplasmic pools of lamins are detectable prior to MBT and (2) the potent inhibitor of transcription, α-amanitin, does not block the appearance

←————————————————————————————————

Fig. 1. Nuclear lamins and periperal nuclear antigens during mouse fertilization and early development. (A–C) Unfertilized oocyte. (A) The P1 periperal antigens ensheathe the surface of each meiotic chromosome (MC). (B) Lamin staining is lost in the ovulated oocyte, which is arrested at the second meiotic metaphase (lamins A/C). (C) Hoechst DNA fluorescence. (D–F) Pronucleate egg. (D) The P1 peripheral antigens are associated with the rims of the male and female pronuclei and with the polar body nucleus. (E) The lamins A/C reassociate with the nuclear surface, and characteristically the polar body nucleus (PB) stains only weakly. F, female; M. male. (F) Hoechst DNA fluorescence. (G–I) Mitotic egg. (G) At prophase, the P1 antibody against the peripheral antigens is redistributed from the pronuclear surfaces to cover each chromosome. (H) The lamins A/C dissociate from the mitotic chromosomes. (I) Hoechst DNA fluorescence. (J–L) Cleavage. (J) As the daughter nuclei reform after first division, the peripheral antigens dissociate from the decondensing chromosomes and reassociate with the nuclear periphery (P1 antigen in J). (K) The lamins A/C associate with the reformed nuclear envelope. (L) Hoechst DNA fluorescence. Bars, 10 μm. [From Schatten et al. (1985).]

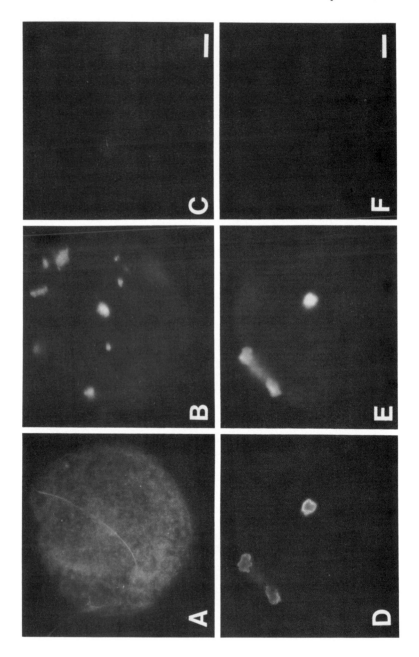

of lamin L_I at MBT. A similar *de novo* synthesis of lamin L_{III} from activated maternal transcripts occurs at MBT (Stick and Hausen, 1985).

At the gastrula stage of development, the C-like lamin L_{II} begins to be detected along with lamins L_{III} and L_I (Stick and Hausen, 1985). Whether the synthesis of lamin L_{II} at gastrulation involves the activation of maternal transcripts or a *de novo* transcription from embryonic genes remains to be determined.

During subsequent development, the levels of lamin L_{III} gradually diminish. By the time the larva hatches, large amounts of lamins L_I and L_{II} are found in the majority of somatic cell nuclei (Krohne *et al.*, 1981), and lamin L_{III} is present in only a few highly differentiated cells, such as muscles, neurons, and Sertoli cells (Benavente *et al.*, 1985). In addition to lamins L_I and L_{II}, lamin A occurs in most somatic cells of the adult, but it has not been found in preneurula developmental stages (Krohne *et al.*, 1987).

Such observations indicate the ontogenetic changes in lamin composition exhibited by *Xenopus* are more complex than those observed in other vertebrates. Unlike the situation in mice and chickens where only lamins A/C are acquired at later stages of development, both a B-like form of lamin and the embryonic-specific lamin L_{III} are synthesized during the blastula stage in *Xenopus*. At gastrulation, production of a third lamin, the C-like lamin L_{II}, appears to be switched on, and lamin A begins to be synthesized at some undetermined time following neurulation.

2. Chicken

Lamins B_1 and B_2 are present in substantial amounts during early development in chickens (Lehner *et al.*, 1987). Lamin A, on the other hand, is essentially lacking in early embryos (Lehner *et al.*, 1987). During embryogenesis, lamin B_1 decreases in concentration within many tissues, and lamin

←——

Fig. 2. Microtubule inhibitors prevent the acquisition of nuclear lamins during mouse fertilization. Oocytes fertilized *in vitro* in the presence of colcemid permit sperm incorporation. The cytoplasmic array of microtubules in these oocytes is depolymerized, and only the incorporated axoneme of the sperm is identified by antitubulin antibodies (A). Colcemid treatment causes the maternal chromatin to disperse throughout the ooplasm, as seen by Hoechst-labeled DNA fluorescence (B). No lamins are evident around the incorporated male pronucleus or the meiotic chromosomes in colcemid-treated specimens (C). By 12-hr postinsemination, meiotic chromosomes occur in a highly condensed state, as judged by DNA fluorescence (E). By contrast, PI antigens surround the incorporated male nucleus and the condensed meiotic chromosomes in colcemid-treated specimens at 12-hr postinsemination (D). No lamins are detected at this time (F). (Bar, 10 μm). [From G. Schatten *et al.* (1988).]

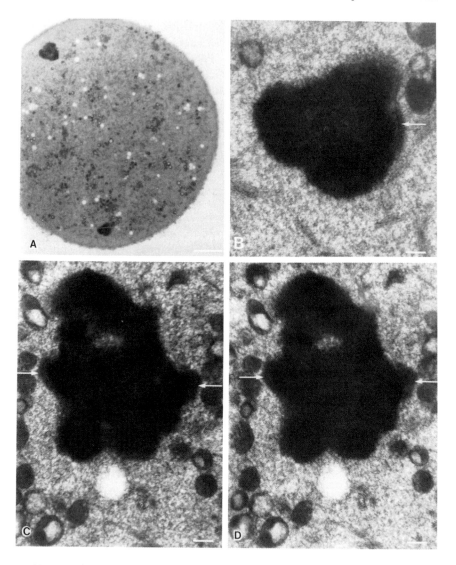

Fig. 3. High-voltage electron microscopy of colcemid-treated oocytes at 12-hr postin-semination. The sperm (top) and egg chromatin (bottom) remain as condensed masses at the oocyte cortex in the presence of colcemid (A). Control oocytes at this time display well-decondensed pronuclei at the egg center with annulated nuclear envelopes. (B) A wide plate is found in association with the sperm chromatin (arrow); it has the typical trilaminar appearance of a kinetochore. (C and D) Stereo pair of the region adjacent to that shown in B from a serial set of sections of condensed chromatin. A kinetochore-like structure is present on both sides (arrows) of this sperm nucleus. Another kinetochore could be observed at the apex of the sperm nucleus after tilting the specimen. Bar, 10 μm (A); bar, 1 μm (B–D) [From G. Schatten *et al.* (1988).]

A undergoes a progressive increase in expression. Lamin B_2, on the other hand, remains at a relatively constant level throughout development.

The sequence of variation in lamin composition during development of chickens suggests that lamin B_1 is characteristic of relatively undifferentiated cells, whereas lamin A tends to be found in differentiated cells. A similar distribution of lamin types occurs in the leukocytes of adult specimens, as precursors of lymphocytes and macrophages contain low amounts of lamin A, but increasingly greater amounts of this lamin are expressed during the differentiation of macrophages (Lehner *et al.*, 1987).

3. Mouse

Fertilized eggs of mice contain both lamins A/C and B, based on data obtained from immunofluorescence microscopy (Schatten *et al.*, 1985; Stewart and Burke, 1987; Houliston *et al.*, 1988). Correlative immunoblotting investigations by Stewart and Burke (1987) indicate that (1) lamins A/C are no longer present after the first 2–4 cleavages and (2) only lamin B is synthesized in later preimplantation stages. Immunofluorescence studies conducted by Schatten *et al.* (1985) also indicate that lamins A/C are lacking in morulae and blastocysts.

According to Schatten *et al.* (1985), lamins A/C are detectable in adult mice, and subsequent immunofluorescence investigations conducted by Stewart and Burke (1987) reveal that lamins A/C are detectable at 8-days postimplantation in a few cells of the trophoblast and at 9-days postimplantation in the embryo proper. Correlative immunoblotting analyses agree well with these findings, as A/C lamins begin to be detected in the embryo at about 10-days postimplantation (Stewart and Burke, 1987).

A recent report by Houliston *et al.* (1988) highlights the complexity of interpreting some of the reported results (discussed Section II,C,7). In support of the reports by Schatten *et al.* (1985) and Stewart and Burke (1987), Houliston *et al.* (1988), using immunoprecipitation, present evidence that virtually no lamins A/C are synthesized at either the 8-cell stage or in blastocysts. However, these workers are able to detect all lamins at all the stages they studied using immunofluorescence and immunoblotting protocols.

The distribution of lamins in developing mouse embryos indicates that lamin B is present throughout ontogeny. Lamins A/C, on the other hand, tend to be absent after the first few cleavages and only reappear much later in development. A similar acquisition of lamins A/C at relatively late stages of differentiation occurs in activated embryonal carcinoma cells of mice (Lebel *et al.*, 1987; Stewart at Burke, 1987). As in mouse embryogenesis, lamin B is expressed throughout differentiation, whereas lamins A/C are only evident in cell lines that are well differentiated.

4. Drosophila

In the fruitfly *Drosophila*, only lamin Dm_1 (M_r 74,000) is present in early embryos, according to the studies of Smith and Fisher (1984) and Smith *et al.* (1987). By hatching, lamin Dm_2 (M_r 76,000) is present in quantities that are equal to or slightly greater than the concentrations of lamin Dm_1. Fuchs *et al.* (1983) have identified lamin-like proteins in 0- to 2-hr-old embryos of *Drosophila* that localize at M_r 67,000 and 65,000 in immunoblots. Alternatively, lamin-like proteins with M_rs of 70,000 and 68,000 have been isolated from 16-hr-old embryos (McKeon *et al.*, 1983). The relationships of these lower molecular-weight nuclear envelope proteins to Dm_1 or Dm_2 remain unclear, although preliminary studies indicate a closer affinity to lamin B (Greunbaum *et al.*, 1988).

In addition to a switch in lamin composition that is observed during development in *Drosophila*, a rapid and quantitative conversion of Dm_2 to Dm_1 occurs in response to heat shock (Smith *et al.*, 1987). This conversion involves a decrease in phosphorylation and in turn results in a more rigid lamina (McConnell *et al.*, 1987). It is hypothesized that the change in lamin type and the concomitant increase in structural rigidity serve to compensate for an increased fluidity of the nuclear envelope brought about by the rise in temperature (Smith *et al.*, 1987).

5. Clam

In the fertilized egg of the clam *Spisula*, only the A/C-like lamin G is present in the lamina of the female pronucleus (Maul *et al.*, 1984, 1987; Maul and Schatten, 1986). During subsequent development, lamin G remains the exclusive type of lamin detectable from 2 hr of embryogenesis (Maul *et al.*, 1987) to the 48-hr-old larval stage (Maul and Schatten, 1986). Adult somatic cells, on the other hand, exhibit a B-like lamin reactivity, rather than lamin G reactivity (Maul *et al.*, 1987). The exact point in development, when the composition of the lamina switches from lamin G to lamin B, remains unknown.

6. Sea Urchin

Based on immunofluorescence studies using monoclonal antibodies and polyclonal antibodies that recognize lamins A/C or B (Schatten *et al.*, 1985), a complex sequence of ontogenetic changes in lamin composition occurs in sea urchins. The fertilized egg exhibits lamin B reactivity and an A/C reactivity that is recognized by a polyclonal antibody. By the blastula stage, no lamin B reactivity is observed, and the polyclonal antibody against lamins A/C fails to stain the lamina. Instead, a monoclonal antibody that is directed against lamins A/C gives positive results for the nuclear envelopes of blastula-stage embryos. In the absence of correlative immunoblotting, it is not yet clear whether this switch in staining pattern of the A/C lamins represents a secondary modification of immunoreactive epitopes or the complete replacement of one set of lamins by another type that is recognized only by the monoclonal antibody. In either case, it seems likely that the A/C lamins undergo some sort of a change by the blastula stage in development.

At the gastrula and pluteus larva stages, the staining pattern is identical to that observed in blastulae, as no lamin B is detected and the A/C lamins are recognized by the monoclonal antibody. At the adult stage, lamin B immunoreactivity reappears, and the A/C lamins are stained by the polyclonal antibody, but not by the monoclonal form. It is hypothesized that the reappearance of lamin B and the second switch in A/C immunoreactivity occurs at larval metamorphosis (Schatten *et al.*, 1985).

7. Overview

Before attempting to synthesize the results of studies on changes in lamin composition during development, several words of caution should be noted. First, much of the data on this subject consists solely of observations made by immunofluorescence microscopy. These observations can in turn be confounded by erroneous results that arise when analyzing antigens with epitopes of similar immunoreactivity. Thus, some of the so-called positive identifications of lamins A/C or B at certain stages of development may actually reflect the presence of another type of lamin or even a nonlamin macromolecule that cross-reacts with the sera used for analysis. The use of monoclonal antibodies tends to yield more specific results, but even monoclonal antibodies can exhibit cross-reactivity among the three major types of mammalian lamins (Burke *et al.*, 1983). Negative results obtained by immunofluorescence microscopy must also be interpreted conservatively, since posttranslational modifications or other alterations of epitope reactivity may mask the presence of lamins. Corroboration of immunofluorescence observations with data derived from immunoblotting methods helps to sort out some of the problems associated with immunofluorescence techniques. It should be noted, however, that immunoblotting may fail to detect forms of lamins that do not cross-react with the antibodies used for the blots. In sum, then, the reported identifications of a type of lamin being either present or absent at a particular stage in development may be misleading in some cases.

With these caveats in mind, a few basic trends can nevertheless be discerned in the data reviewed above. For example, lamins that resemble mammalian lamin B tend to be present during much of development, including the early stages. Embryos and embryonal carcinoma cells of mice possess lamin B throughout differentiation, and the frog *Xenopus* has a B-like lamin from the midblastula stage onward. Chickens lose their structural counterpart to mammalian lamin B during development, but retain relatively constant levels of another acidic lamin that may function analogously to lamin B (Lehner *et al.*, 1987). The two exceptions to the rule that B-like lamins are present throughout much of ontogeny are exhibited by (1) the clam *Spisula,* which apparently displays lamin B reactivity associated with the nuclear envelope only during postlarval stages, and (2) the sea urchin, which loses its B-like lamin during early embryonic stages and only reacquires lamin B reactivity in postlarval stages.

Another common feature of many developmental sequences is that the synthesis of A/C-like lamins is switched on at later stages of differentiation. Such a pattern occurs in chicken embryos and in embryonal carcinoma cells of mice. Slight variations on this theme occur during embryogenesis in mice, *Xenopus,* and sea urchins. In mouse embryos, lamins A/C reappear in postimplantation stages after having been lost in early cleavage stages. In *Xenopus,* an embryonic form of an A/C-like lamin is replaced by adult types of a C-like lamin and an A-like lamin at the gastrula stage and at postneurula stages, respectively. The sea urchin displays A/C-like reactivity throughout development, but may undergo a change in the exact type of A/C lamins that are present at later stages of embryogenesis. The clam *Spisula* represents an exception to the rule that A/C reactivity is correlated with later stages of differentiation, since the A/C-like form of lamin in *Spisula* is apparently present during all stages of development up to the larval stage, but then is lost in adult somatic tissues.

Such observations indicate that the developmental programs of animals typically include a distinct change in the types of lamins that are synthesized. In some cases, such as the midblastula stage in *Xenopus,* the change in lamin composition is clearly correlated with a significant turning point in the ontogeny of the organism. It is unclear, however, exactly what roles these changes in the types of lamins may play during development. For example, no unequivocal evidence is available to show that the expression of any of the different types of lamins during development is temporally or functionally correlated with cellular events that determine the developmental fate of a particular cell type. Even in the case of lamin L_{III}, which in adult somatic cells of *Xenopus,* appears to be highly specific for muscles, neurons, and Sertoli cells, the fates of these three types of cells are determined prior to MBT, the time at which expression of L_{III} is resumed (Krohne and Benavente, 1986).

In order to understand more fully the functional significance of ontogenetic changes in lamin composition, several major topics need to be addressed. First, corroborative immunoblotting investigations are required in cases such as sea urchin development in which the data available are based solely from immunofluorescence observations. Moreover, since lamin composition has been analyzed biochemically in only a few groups of animals, the additional characterizations of lamins and the concomitant production of antibodies against these lamins are required for various taxa. Such studies would provide valuable comparative data that may help to reveal trends that are not presently evident in the literature. In addition, the development of antibodies against novel forms of lamins in other groups of organisms may help to correct results of previous studies, since developmental stages that were once reported either to lack lamins or to contain only a particular type of lamin may actually reveal additional types of lamins based on immunoblotting analyses utilizing these new antilamin antibodies.

III. KINETOCHORES, THE NUCLEAR MATRIX, AND OTHER KARYOSKELETAL COMPONENTS

A. Kinetochores

Since kinetochores are essential for the attachment of chromosomes to spindle microtubules, their presence and behavior during the events of mammalian fertilization are of interest, since these events overlap with both meiosis and mitosis. Kinetochores are traced with immunofluorescence microscopy using autoimmune sera from patients with CREST scleroderma (calcinosis, Raynaud's phenomenon, esophageal dysmotility, sclerodactyly, telangiectasia). In mouse ovulated oocyte arrested at second meiotic metaphase, the kinetochores are initially detectable as paired structures aligned at the spindle equator (Fig. 4). At meiotic anaphase, the kinetochores separate and remain aligned at the distal sides of the chromosomes until telophase, when their alignment perpendicular to the spindle axis is lost. The female pronucleus and the second polar body nucleus each receive a detectable complement of kinetochores. Mature sperm have neither detectable centrosomes nor detectable kinetochores, and shortly after sperm incorporation, kinetochores become detectable in the decondensing male pronucleus. In pronuclei, the kinetochores are initially distributed randomly and later found in apposition with nucleoli. At mitosis, the kinetochores behave in a pattern similar to that observed at meiosis or mitosis in somatic cells: irregular distribution at prophase, alignment at metaphase, separation at anaphase, and redistribution at telophase. They are also detectable in later-stage embryos. Colcemid treatment disrupts the meiotic spindle and results in the dispersion of the meiotic chromosomes along the oocyte cortex; the chromosomes remain condensed with detectable kinetochores. Fertilization of colcemid-treatment oocytes results in the incor-

242

Stephen Stricker *et al.*

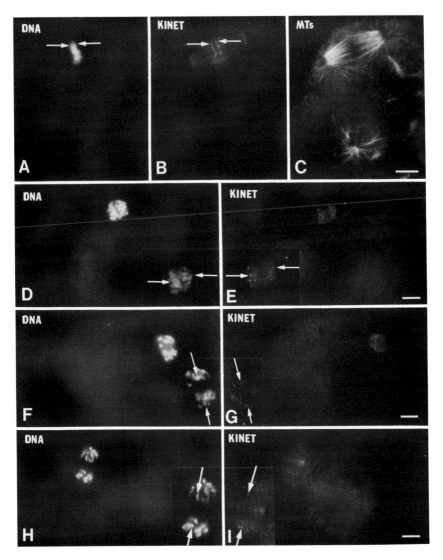

Fig. 4. Kinetochores (KINET) during the completion of the second meiotic division in mouse oocytes. Ovulated oocytes are arrested at second meiotic metaphase (A–C). The chromosomes, detected with Hoechst dye 33258 (DNA, A, D, F, and H), are aligned at the equator of the meiotic spindle. The kinetochores appear as paired structures associated with each chromosome (arrows in B). The kinetochores are juxtaposed between the spindle microtubules (MTs in C) and the chromosomes (A). Cytoplasmic asters (bottom of C) are not associated with kinetochores (compare B). At anaphase as the chromosomes separate (D), the kinetochores retain their alignment perpendicular to the spindle axis (KINET in E) and are found at the polar regions of the chromosomes (D). By telophase (F and G), the registration

poration of a sperm, which is unable to decondense into a male pronucleus. Remarkably kinetochores become detectable at 5-hr postinsemination, suggesting that the emergence of the paternal kinetochores is not strictly dependent on male pronuclear decondensation. These results suggest that the pathways leading to the exposure of the male kinetochores are regulated separately from those involved in pronuclear decondensation, nuclear lamin acquisition, and the onset of DNA synthesis.

B. The Nuclear Matrix and snRNP Localization

During fertilization and development, the nuclear matrix antigens undergo a dramatic redistribution (Schatten *et al.*, 1985; Prather *et al.*, 1988). In the mouse, the P1 antigens ensheathe the meiotic chromosomes of the oocyte and subsequently shift to a peripheral position lining the nuclear envelopes of the male pronucleus, female pronucleus, and polar body. Similarly, P1 reactivity is observed around the condensed chromosomes during the first mitosis, and a peripheral rim of staining is visible in the nuclei of the two daughter cells produced by this cell division.

Other nuclear matrix antigens—PI1, PI2, and I1—are present in the cytoplasm and nuclear matrix of mouse oocytes at the GV stage. These antigens are detectable in the cytoplasm of meiotic oocytes, but are not present in pronuclei (Fig. 5) or early 2-cell nuclei. They reappear during the mid 2-cell stage and are present in 8- to 16-cell nuclei (Fig. 6). The nuclear appearance of these antigens correlates with the transition from maternal control of development to zygotic control of development in the mouse at the 2-cell stage (Flach *et al.*, 1982). PI1 antigen colocalizes with anti-Sm which recognizes U1, U2, U4, U5, and U6 (Chaly *et al.*, 1987). The predominate form of U1 RNA in the oocyte and 2-cell-stage egg is U1a RNA (Lobo *et al.*, 1988). By the 8-cell stage, the total amount of U1 RNA has increased 2- to 3-fold, and the proportion of U1a has diminished from >85 to <60%. At the blastocyst stage, the amount of U1b RNA has risen to 60% of the total (Lobo *et al.*, 1988). The changes in the distribution of various antigens during fertilization and early development serve to illustrate the dramatic changes in RNA synthesis and RNA processing that the egg undergoes.

←―――

of the kinetochores (KINET in G) is weaker (arrows in the inset of G). As the second polar body forms (H and I), the kinetochores (KINET in I) become more randomly positioned over the chromosome masses (DNA, H). Bars, 10 μm. Insets: approximately 2× photographic enlargement over surrounding figure. Bars, 10 μm. [From G. Schatten *et al.* (1988).]

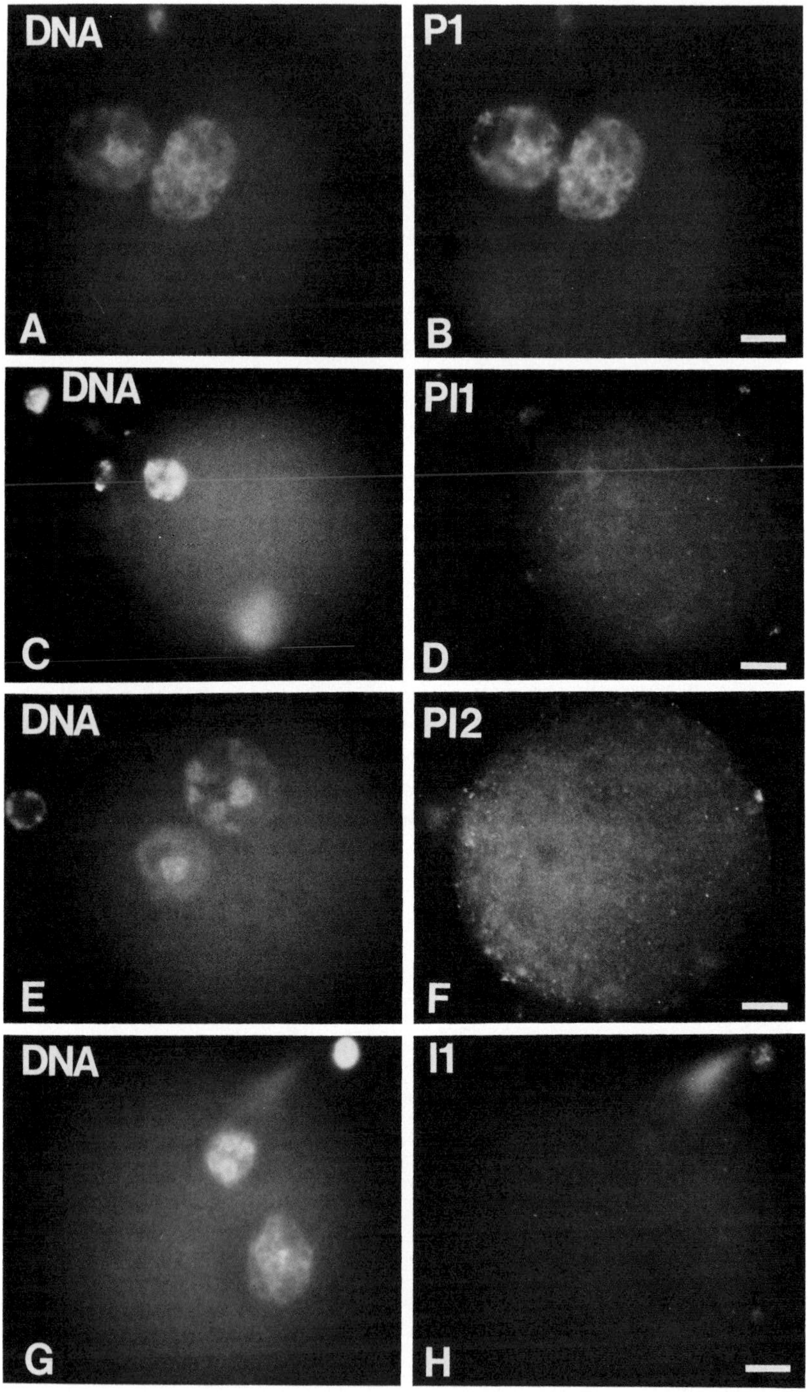

DNA
A
P1
B
DNA
C
PI1
D
DNA
E
PI2
F
DNA
G
I1
H

C. Other Karyoskeletal Constituents

It would be naive to assume that we understand the complexity of nuclear architecture. Further, the discovery of additional components is predicted as is the prognostication that their expression or insertion will be found to be regulated during fertilization and during development. Topoisomerase II (Earnshaw and Heck, 1985; Earnshaw et al., 1985) and the nucleoporins (Davis and Blobel, 1986; Holt et al., 1987; Snow et al., 1987) might be expected to be karyoskeletal proteins of developmental interest. Indeed as knowledge about the properties and components of the nuclear matrix become better identified, investigation of the behavior of these proteins at fertilization and during development will be of considerable interest.

IV. FUTURE RESEARCH

In attempting to conduct more direct analyses of the functions of karyo-skeletal proteins during fertilization and development, several types of manipulative experiments could be performed. For example, the microinjection of antilamin antibodies into mitotic cells has been shown in some cases to disrupt the normal production of daughter nuclei (Benavente and Krohne, 1986). Similarly, the microinjection of antibodies against lamins, kinetochores, or structural proteins of the nucleus into blastomeres of developing embryos may help to determine whether a particular kind of karyoskeletal component is necessary at a specific stage in differentiation in order for development to proceed normally.

A related type of experiment would be the isolation of lamins or other elements of the nuclear architecture from embryos at one stage in development and the introduction of fluorosceinated derivatives of these proteins into blastomeres at a different stage in development. Such heterologous proteins could be introduced by microinjection or by cell fusion to determine if stage-specific forms of the karyoskeletal components influence developmental patterns.

The molecular characterizations of lamins and kinetochores have recently been addressed. Supplemental cloning of karyoskeletal elements should provide valuable structural information and enable the synthesis of relatively large

←───

Fig. 5.   Immunofluorescence distribution of nuclear matrix antigens in pronucleate mouse eggs. (A, C, E, and G) Hoechst stained eggs. (B, D, F. and H) Corresponding nuclear matrix localizations. (B) Antibody to P1, (D) antibody to PI1, (F) antibody to PI2, and (H) antibody to I1. Bars, 10 μm. [From Prather et al. (1988).]

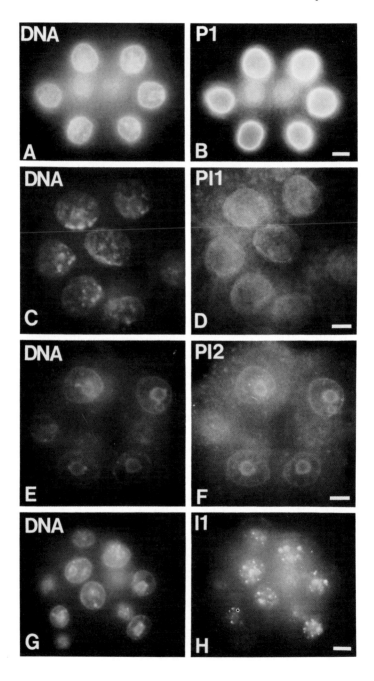

quantities of these proteins. The purified products could then be used as molecular probes to facilitate various studies on the roles of the karyoskeleton in fertilization and development.

REFERENCES

Aebi, U., Cohn, J., Buhle, L., and Gerace, L. (1986). The nuclear lamina is a meshwork of intermediate-type filaments. *Nature (London)* **323**, 560–564.

Balczon, R. D., and Brinkley, B. R. (1987). Tubulin interaction with kinetochore proteins: Analysis by *in vitro* assembly and chemical cross-linking. *J. Cell Biol.* **105**, 855–862.

Benavente, R., and Krohne, G. (1985). Change of karyoskeleton during spermatogenesis of *Xenopus:* Expression of lamin L_{IV}, a nuclear lamina protein specific for the male germ line. *Proc. Natl. Acad. Sci. U.S.A.* **82**, 6176–6180.

Benavente, R., and Krohne, G. (1986). Involvement of nuclear lamins in postmitotic reorganization of chromatin as demonstrated by microinjection of lamin antibodies. *J. Cell Biol.* **103**, 1847–1854.

Benavente, R., Krohne G., and Franke, W. W. (1985). Cell type-specific expression of nuclear lamina proteins during development of *Xenopus laevis. Cell (Cambridge, Mass.)* **41**, 177–190.

Bloom, K. S., and Carbon, J. (1982). Yeast centromere DNA is in a unique and highly ordered structure in chromosomes and small circular minichromosomes. *Cell (Cambridge, Mass.)* **29**, 305–317.

Burke, B., Tooze, J., and Warren, G. (1983). A monoclonal antibody which recognizes each of the nuclear polypeptides in mammalian cells. *EMBO J.* **2**, 361–367.

Chaly, N., Bladon, T., Setterfield, G., Little, J. E., Kaplan, J. G., and Brown, D. L. (1984). Changes in distribution of nuclear matrix antigens during the cell cycle. *J. Cell Biol.* **99**, 661–671.

Chaly, N., Bertin, J., and Venance, M. (1987). Restructuring of HeLa nuclear matrix antigens in response to adenovirus infection. *J. Cell Biol.* **105**, 178a.

Davis, L., and Blobel, G. (1986). Identification and characterization of a nuclear pore complex protein. *Cell (Cambridge, Mass.)* **45**, 699–709.

Earnshaw, W. C., and Heck, M. M. S. (1985). Localization of topoisomerase II in mitotic chromosomes. *J. Cell Biol.* **100**, 1716–1725.

Earnshaw, W. C., and Rothfield, N. F. (1985). Identification of a family of human centromere proteins using autoimmune sera from patients with scleroderma. *Chromosoma* **91**, 313–321.

Earnshaw, W. C., Halligan, B., Cooke, C. A., Heck, M. S. M., and Liu, L. F. (1985). Topoisomerase II is a structural component of mitotic chromosome scaffolds. *J. Cell Biol.* **100**, 1706–1715.

←———————————————————————————————————

Fig. 6.   Immunofluorescence distribution of nuclear matrix antigens in 8- to 16-cell mouse embryos. (A, C, E, and G) Hoechst stained embryos. (B, D, F, and H) Corresponding nuclear matrix localization. (B) Antibody to P1, (D) antibody to PI1, (F) antibody to PI2, and (H) antibody to I1. Bars, 10 µm. [From Prather *et al.* (1988).]

Earnshaw, W. C., Sullivan, K. F., Machlin, P. S., Cooke, C. A., Kaiser, D. A., Pollard, T. D., Rothfield, N. F., and Cleveland, D. W. (1987). Molecular cloning of cDNA for CENP-B, the major human centromere autoantigen. *J. Cell Biol.* **104**, 817–829.

Fawcett, D. W. (1966). On the occurrence of a fibrous lamina on the inner aspect of the nuclear envelope in certain cells of vertebrates. *Am. J. Anta.* **119**, 129–146.

Fisher, D. Z., Chaudhary, N., and Blobel, G. (1986). cDNA sequencing of nuclear lamins A and C reveals primary and secondary structural homology to intermediate filament proteins. *Proc. Natl. Acad. Sci. U.S.A.* **83**, 6450–6454.

Flach, G., Johnson, M. H., Braude, P. R., Taylor, R. A. S., and Bolton, V. N. (1982). The transition from maternal to embryonic control in the 2-cell mouse embryo. *EMBO J.* **1**, 681–686.

Forbes, D. J., Kirschner, M. W., and Newport, J. W. (1983). Spontaneous formation of nucleus-like structures around bacteriophage DNA microinjected into *Xenopus* eggs. *Cell (Cambridge, Mass.)* **34**, 13–23.

Franke, W. W. (1987). Nuclear lamins and cytoplasmic intermediate filament proteins: A growing multigene family. *Cell (Cambridge, Mass.)* **48**, 3–4.

Fuchs, J. -P., Giloh, H., Kuo, C. -H., Saumweber, H., and Sedat, J. (1983). Nuclear structure: Determination of the fate of the nuclear envelope in *Drosophila* during mitosis using monoclonal antibodies. *J. Cell Sci.* **64**, 331–349.

Gerace, L. (1986). Nuclear lamina and organization of nuclear architecture. *Trends Biol. Sci.* **11**, 443–446.

Gerace, L., and Blobel, G. (1982). Nuclear lamina and the structural organization of the nuclear envelope. *Cold Spring Harbor Symp. Quant. Biol.* **46**, 967–978.

Gruenbaum, Y., Landesman, Y., Drees, B. D., Bare, J. W., Saumweber, H., Paddy, M. R., Sedat, J. W., Smith, D. E., Benton, B. M., and Fisher, P. A. (1988). *Drosophila* nuclear lamin precursor Dm_a is translated from either of two developmentally regulated mRNA species apparently encoded by a single gene. *J. Cell Biol.* **106**, 585–596.

Holt, G. D., Snow, C. M., Senior, A., Haltiwanger, R. S., Gerace, L., and Hart, G. W. (1987). Nuclear pore complex glycoproteins contain cytoplasmically disposed O-linked N-acetylglucosamine. *J. Cell Boil.* **104**, 1157–1164.

Houliston, E., Guilly, M. -N., Courvalin, J. -C., and Maro, B. (1988). Expression of nuclear lamins during mouse preimplantation development. *Development* (in press).

Kaufmann, S. H., Gibson, W., and Shaper, J. H. (1983). Characterization of the major polypeptides of the rat nuclear envelope. *J. Biol. Chem.* **258**, 2710–2719.

Krohne, G., and Benavente, R. (1986). The nuclear lamins: A multigene family of proteins in evolution and differentiation. *Exp. Cell Res.* **162**, 1–10.

Krohne, G., Dabauvalle, M. -C., and Franke, W. W. (1981). Cell type-specific differences in protein composition of nuclear pore complex-lamin structures in oocytes and erythrocytes of *Xenopus laevis*. *J. Mol. Biol.* **151**, 121–141.

Krohne, G., Wolin, S. L., McKeon, F. D., Franke, W. W., and Kirschner, M. W. (1987). Nuclear lamin L_1 of *Xenopus laevis*: cDNA cloning, amino acid sequence and binding specificity of a member of the lamin B subfamily. *EMBO J.* **6**, 3801–3808.

Lam, K. S., and Kasper, C. B. (1979). Selective phosphorylation of a nuclear envelope polypeptide by an endogenous protein kinase. *Biochemistry* **18**, 307–311.

Lebel, S., Lampron, C., Royal, A., and Raymond, Y. (1987). Lamins A and C appear during retinoic acid-induced differentiation of mouse embryonal carcinoma cells. *J. Cell Biol.* **105**, 1099–1104.

Lehner, C. F., Kurer, V., Eppenberger, H. M., and Nigg, E. A. (1986). The nuclear lamin protein family in higher vertebrates. Identification of quantitatively minor lamin proteins by monoclonal antibodies. *J. Biol. Chem.* **261**, 13293–13301.

Lehner, C. F., Stick, R., Eppenberger, and Nigg, E. A. (1987). Differential expression of nuclear lamin proteins during chicken development. *J. Cell Biol.* **105**, 577–587.

Lobo, S. M., Marzluff, W. F., Seufert, A. C., Dean, W. L., Schultz, G. A., Simerly, C., and Schatten, G. (1988). Localization and expression of U1 RNA in early mouse embryo development. *Dev. Biol.* **127,** 349–361.

Longo, F. J., Krohne, G., and Franke, W. W. (1987). Basic proteins of the perinuclear theca of mammalian spermatozoa and spermatids: A novel class of cytoskeletal elements. *J. Cell. Biol.* **105,** 1105–1120.

McConnell, M., Whalen, A. M., Smith, D. E., and Fisher, P. A. (1987). Heat shock-induced changes in the structural stability of proteinaceous karyoskeletal elements *in vitro* and morphological effects *in situ. J. Cell Biol.* **105,** 1087–1098.

McKeon, F. D., Tuffanelli, D. L., Fukuyama, K., and Kirschner, M. W. (1983). Autoimmune response directed against conserved determinants of nuclear envelope proteins in a patient with linear scleroderma. *Proc. Natl. Acad. Sci. U.S.A.* **80,** 4374–4378.

McKeon, F. D., Kirschner, M. W., and Caput, D. (1986). Homologies in both primary and secondary structure between nuclear envelope and intermediate filament proteins. *Nature (London)* **319,** 463–468.

Maul, G. G. (1988). Redistribution on nuclear envelope, nucleolar and kinetochore antigens during mouse spermatogenesis and early development. In "The Molecular Biology of Fertilization" (H. Schatten and G. Schatten, eds.). Academic Press, New York (in press).

Maul, G. G., and Schatten, G. (1986). Nuclear lamins during gametogenesis, fertilization and early development. *In* "Nucleocytoplasmic Transport" (R. Peters and M. Trendelenburg, eds.), pp. 123–134. Springer-Verlag, Berlin and New York.

Maul, G. G., Baglia, F. A., Newmeyer, D. D., and Ohlsson-Wilhelm, B. M. (1984). The major 67 000 molecular weight protein of the clam oocyte nuclear envelope is lamin-like. *J. Cell Sci.* **67,** 69–85.

Maul, G. G., French, B. T., and Bechtol, K. B. (1986). Identification and redistribution of lamins during nuclear differentiation in mouse spermatogenesis. *Dev. Biol.* **115,** 68–77.

Maul, G. G., Schatten, G., Jimenex, S. A., and Carrera, A. E. (1987). Detection of nuclear lamin B epitopes in oocyte nuclei from mice, sea urchins, and clams using a human autoimmune serum. *Dev. Biol.* **121,** 368–375.

Moroi, Y., Peebles, C., Fritzler, M. J., Steigerwald, J., and Tan, E. M. (1980). Autoantibody to centromere (kinetochore) in scleroderma sera. *Proc. Natl. Acad. Sci. U.S.A.* **77,** 1627–1631.

Moss, S. B., Donovan, M. J., and Bellvé, A. R. (1987). The occurrence and distribution of lamin proteins during mammalian spermatogenesis and early embryonic development. *In* "Cell Biology of the Testis and Epididymis" (M. -C. Orgebin Crist and B. J. Danzo, eds.) Ann. N. Y. Acad. Sci, New York.

Newport, J., and Kirschner, M. (1982). A major developmental transition in early *Xenopus* embryos. I. Characterization and timing of cellular changes at the midblastula stage *Cell (Cambridge, Mass.)* **30,** 675–686.

Ottaviano, Y., and Gerace, L. (1985). Phosphorylation of the nuclear lamins during interphase and mitosis. *J. Biol. Chem.* **260,** 624–632.

Palmer, D. K., O'Day, K., Wener, M. H., Andrews, B. S., and Margolis, R. L. (1987). A 17-kD centromere protein (CENP-A) copurifies with nucleosome core particles and with histones. *J. Cell Biol.* **104,** 805–815.

Park, M. K., D'Onofrio, M., Willingham, M. C., and Hanover, J. A. (1987). A monoclonal antibody against a family of nuclear pore proteins (nucleoporins): O-linked N-acetylglucosamine is a part of the immunodeterminant. *Proc. Natl. Acad. Sci. U.S.A.* **84,** 6462–6466.

Prather, R., Chaly, N., Simerly, C., Schatten, H. and Schatten, G. (1988). Nuclear matrix appearance during preimplantation mouse embryogenesis. In preparation.

Risau, W., Saumweber, H., and Symmons, P. (1981). Monoclonal antibodies against a nuclear membrane protein of *Drosophila*. *Exp. Cell Res.* **133**, 47–54.

Schatten, G., and Schatten, H. (1987). Fertilization: Motility, the cytoskeleton and the nuclear architecture. *Oxford Rev. Reprod. Biol.* **9**, 322–378.

Schatten, G., Maul, G. G., Schatten, H., Chaly, N., Simerly, C., Balczon, R., and Brown, D. L. (1985). Nuclear lamins and peripheral nuclear antigens during fertilization and embryogenesis in mice and sea urchins. *Proc. Natl. Acad. Sci. U.S.A.* **82**, 4727–4731.

Schatten, G., Simerly, C., Palmer, D. K., Margolis, R. L., Maul, G., Andrews, B. S., and Schatten, H. (1988). Kinetochore antigen appearance during meiosis, fertilization and mitosis in mouse oocytes and zygotes. *Chromosoma* **96**, 341–352.

Schatten, H., Simerly, C., Maul, G., and Schatten, G. (1988). Assembled microtubules are essential for nuclear lamina acquisition, nuclear envelope formation, pronuclear decondensation and the onset of DNA synthesis during mouse, but not sea urchin, fertilization. Submitted.

Smith, D. E., and Fisher, P. A. (1984). Identification, developmental regulation, and response to heat shock of two antigenically related forms of a major nuclear envelope protein in *Drosophila* embryos: Application of an improved method for affinity purification of antibodies using polypeptides immobilized on nitrocellulose blots. *J. Cell Biol.* **99**, 20–28.

Smith, D. E., Gruenbaum, Y., Berrios, M., and Fisher, P. A. (1987). Biosynthesis and interconversion of *Drosophila* nuclear lamin isoforms during normal growth and in response to heat shock. *J. Cell Biol.* **105**, 771–790.

Snow, C. M., Senior, A., and Gerace, L. (1987). Monoclonal antibodies identify a group of nuclear pore complex glycoproteins. *J. Cell Biol* **104**, 1143–1156.

Stewart, C., and Burke, B. (1987). Teratocarcinoma stem cells and early mouse embryos contain only a single major lamin polypeptide closely resembling lamin B. *Cell (Cambridge, Mass.)* **51**, 383–392.

Stick, R., and Hausen, P. (1985). Changes in the nuclear lamina composition during early development of *Xenopus laevis*. *Cell (Cambridge, Mass.)* **41**, 191–200.

Stick, R., and Schwarz, H. (1982). The disappearance of the nuclear lamina during spermatogenesis: An electron microscopic and immunofluorescence study. *Cell Differ.* **11**, 235–243.

Tan, E. M., Rodnan, G. P., Garcia, I., Moroi, Y., Fritzler, M. J., and Peebles, C. (1980). Diversity of antinuclear antibodies in progressive systemic sclerosis: Anti-centromere antibody and its relationship to CREST syndrome. *Arthritis Rheum.* **23**, 617–625.

Valvidia, M. M., and Brinkley, B. R. (1985). Fractionation and initial characteristics of the kinetochore from mammalian metaphase chromosomes. *J. Cell Biol.* **101**, 1124–1134.

Wolin, S. L., Krohne, G., and Kirschner, M. W. (1987). A new lamin in *Xenopus* somatic tissues displays strong homology to human lamin A. *EMBO J.* **6**, 3809–3818.

10

Extracellular Remodeling during Fertilization

BENNETT M. SHAPIRO,[*] **CYNTHIA E. SOMERS,**[*] **AND PEGGY J. WEIDMAN**[†]

[*]Department of Biochemistry
University of Washington
Seattle, Washington 98195

[†]Department of Biochemistry
Princeton University
Princeton, New Jersey 08544

I. INTRODUCTION

Two major problems confront the egg at fertilization. One is to restore the diploid state and initiate the development of a new individual. The other is to overcome the inhospitable environment outside of the ovary. The newly ovulated egg needs to be vulnerable enough to fuse with a sperm, yet robust

251

THE CELL BIOLOGY OF
FERTILIZATION

enough to ensure the survival of the early embryo. In their delicate transitions from ovary to embryo, metazoan eggs protect themselves by the cortical reaction of fertilization.

The cortical reaction involves a modification of the egg surface in order to protect the early embryo from supernumerary sperm and other deleterious agents. The extracellular coat is converted to a modified shell from which the embryo later hatches, either to become independent (as with many invertebrates and some vertebrates) or to implant in the uterine wall (as with mammals). The extracellular remodeling attendant upon the cortical reaction was first described by Derbes over 130 years ago (1847) and has been recognized as one of the principal events of fertilization since the turn of the century. Jacques Loeb, a physical biochemist who began considering problems of development at that time, examined the relationship between the cortical changes of eggs and activation of the developmental sequence. Although these two events had appeared to be inextricably linked, Loeb provided evidence for dissociation of the two processes and speculated upon the biochemical changes that accompany formation of the fertilization membrane of invertebrate eggs. He suspected

. . . that the membrane formation is the result of a process of secretion of a liquid from the egg; and that this secretion or the throwing out of certain substances of the egg is the important feature, for the lifting up of the surface layer of the egg (the membrane formation proper) is only a mechanical consequence of this secretion but of no importance in itself. (Loeb, 1905, p. 154.)

Loeb went on to provide another hypothetical suggestion

. . . that the fertilization membrane is preformed in the unfertilized egg and is merely the peripheral film of protoplasm which is lifted up from the egg through the swelling and liquifaction of some protein lying underneath in the cortical layer of the egg. When lifted up from the egg the preformed membrane undergoes a modification; it becomes thicker and tougher. (Loeb, 1913, p. 218.)

These ideas emanating from Loeb's studies on fertilization and artificial parthenogenesis remain relevant to current molecular mechanisms for the extracellular remodeling that occurs with fertilization.

Although extracellular remodeling occurs in most cases where eggs leave the protection of the ovary to become fertilized, only certain systems have been characterized at a molecular level. The principal invertebrate system is the sea urchin egg, and among vertebrates the amphibian egg has provided the clearest picture. Less information exists about fish and mammalian fertilization. Several recent reviews deal with egg surface changes at fertilization and the block to polyspermy (Schmell *et al.*, 1983; Kay and Shapiro, 1985). In this chapter, we concentrate upon molecular alterations in the egg coats and try to identify certain common features in their modifications. As with many control systems in molecular and cell biology, strong similarities and certain interesting differences exist between species. We are just beginning to see the outlines of the regulatory events that allow an egg to be fertilized by only one sperm and then to be protected during early development. The shelter provided by the newly formed fertilization envelope allows the embryo to undergo critical, early determinative events in an isolated environment. The ubiquity of this cortical reaction indicates the importance of such a protective event.

II. SPECIFIC SYSTEMS OF EGG COAT MODIFICATION

A. The Sea Urchin Fertilization Envelope

This classic example of extracellular remodeling after fertilization involves a complex array of morphological changes that occur in more or less discrete stages (Veron et al., 1977; reviewed in Kay and Shapiro, 1985). The first stage is initiated upon mixing of the two cellular compartments that contain fertilization envelope precursors: the egg glycocalyx (vitelline layer) that is covalently attached to the plasma membrane (Kidd, 1978) and the contents of egg secretory vesicles, the cortical granules. Fertilization induces a wave of cortical granule exocytosis that causes the vitelline layer to elevate from the egg surface, starting at the site of sperm entry and propagating around the entire egg within a minute. The second stage involves a restructuring of the elevated vitelline layer and the deposition of paracrystalline arrays of secreted cortical vesicle proteins on the modified scaffold (Chandler and Heuser, 1980). The final stage of fertilization envelope morphogenesis is characterized by a change in its physical properties: it is converted from a soft, pliable structure to a hard, protective coat. The mature fertilization envelope is refractory to both chemical and mechanical disruption, but is degraded by a specific hatching enzyme secreted by the blastula larva (Ishida, 1936).

During the past two decades, significant advances have been made on biochemical mechanisms underlying this striking example of extracellular morphogenesis. Although the picture that has emerged is far from complete, the observed morphological transitions are associated with highly organized molecular events involving both spatial and temporal regulatory controls.

1. The Vitelline Layer

The vitelline layer plays several roles for the sea urchin egg. Species-specific sperm receptors of the vitelline layer mediate gamete association (e.g., see Rossignol et al., 1984). These sperm receptors are removed at fertilization by released cortical vesicle protease(s) that also hydrolyze attachments between the vitelline layer and the plasma membrane (Carroll and Epel, 1975; Carroll, 1976). The vitelline layer elevates due to osmotic effects of the cortical vesicle contents (e.g., Loeb, 1913). Transformation of the elevated vitelline layer into a scaffold for fertilization envelope assembly involves more extensive proteolysis. If eggs are isotopically labeled on their extracellular surfaces (the vitelline layer and plasma membrane) prior to fertilization, less than 5% of this surface material is found in the fertilization envelope (Shapiro, 1975; E. S. Kay, unpublished observations). In assembled, unhardened (soft) fertilization envelopes, most of the egg surface material is associated with very large insoluble components. One soluble vitelline layer component that ends up in the fertilization envelope is a $\sim305,000$ M_r peptide formed of two apparently identical $\sim170,000$ M_r, disulfide-linked subunits (E. S. Kay and P. J. Weidman, unpublished observations). These vitelline layer proteins are apparently critical to the assembly process, for eggs with disrupted vitelline layers do not form fertilization envelopes (Epel, 1970; Epel *et al.*, 1970). The extent to which proteolytic modification is necessary for fertilization envelope assembly per se is not clear. Although proteolysis is required to release the vitelline layer from the egg surface, it may not be needed for the secreted proteins to associate with the scaffold, for the unmodified vitelline layer of unfertilized eggs contains specific binding sites for a cortical vesicle-derived component (see below). The diameter of the fertilization envelope is nearly twice that of the egg, so that limited proteolytic cleavages or other glycocalyx modifications may allow the vitelline layer scaffolding to expand and/or may facilitate the penetration of released cortical granule proteins through this latticework. An egg surface transglutaminase activity facilitates this process (D. E. Battaglia and B. M. Shapiro, unpublished data).

2. Cortical Granule Vesicle[1] Exocytosis

The egg employs Ca^{2+} as a mediator of both exocytosis and fertilization envelope assembly. Fertilization induces a transient release of Ca^{2+} from intracellular stores; this propagates from the site of sperm fusion around the egg cortex, activating the exocytotic machinery in an explosive wave of secretion (reviewed in Whitaker and Steinhardt, 1985; Eisen and Reynolds, 1985; see also Turner and Jaffe, Chapter 12, this volume). The proteins secreted from the cortical vesicles fall into three categories: hyalin, an embryonic cell adhesive that forms a Ca^{2+}-dependent, amorphous gel on the surface of the

[1] A class of vesicles located beneath the eggs' surface that undergo exocytosis at fertilization was originally called cortical granules or cortical alveoli (in fish), on morphological grounds. We prefer the term cortical vesicle, but use them interchangeably.

nascent embryo (Stephens and Kane, 1970; Citkowitz, 1971); soluble proteins that are not incorporated into the fertilization envelope, including proteases that modify the vitelline layer (Carroll and Epel, 1975; Vacquier *et al.*, 1973) and a β-1,3-glucanohydrolase of unknown function (Epel *et al.*, 1969; Truschel *et al.*, 1986); and structural components of the fertilization envelope (Bryan, 1970a) that associate with the modified vitelline layer matrix under the influence of divalent cations in seawater (e.g., see Kay and Shapiro, 1985). The latter consist of approximately five principal polypeptides (Kay *et al.*, 1982; E. S. Kay, unpublished observations; Weidman and Kay, 1986) two of which, ovoperoxidase and proteoliaisin, have been purified and characterized.

3. Ovoperoxidase

Ovoperoxidase is the enzyme responsible for cross-linking, and thereby hardening, the fertilization envelope. It uses H_2O_2 formed in a CN^--insensitive respiratory burst at fertilization (Foerder *et al.*, 1978) to catalyze the formation of dityrosine cross-links between a subset of adjacent peptides in the assembled fertilization envelope (Foerder and Shapiro, 1977; Hall, 1978; Kay and Shapiro, 1987). Purified ovoperoxidase is a 70,000 M_r heme glycoprotein that is similar in catalytic and spectral properties to lactoperoxidase (Deits *et al.*, 1984). Like most peroxidases, it catalyzes the oxidation of halides and a variety of phenolic compounds, as well as tyrosine, and thus poses a potential hazard to the nascent embryo. The activity of ovoperoxidase appears to be regulated by a combination of mechanisms. The first is a timing delay that prevents the enzyme from achieving full activity until it is assembled into the fertilization envelope (Deits and Shapiro, 1985, 1986). The activity of ovoperoxidase is pH dependent, with pK_a of ~6.5. At low pH, i.e., conditions that exist in the secretion milieu of the egg (see Deits and Shapiro, 1985, for further discussion), the enzyme is inactive. When the inactive enzyme encounters a higher pH, as exists in the bulk sea water (pH 8.0), there is a lag in the onset of ovoperoxidase activity ($t_{1/2}$ = ~30 sec), thereby suppressing ovoperoxidase activity until the enzyme is well away from the surface of the nascent embryo (~90 sec postfertilization). In the assembled fertilization envelope, ovoperoxidase is fully active, without any slow transitions upon pH shift (Deits and Shapiro, 1986).

A second mechanism that regulates ovoperoxidase activity is the timing of H_2O_2 production. The burst of H_2O_2 synthesis begins ~2 min after fertilization and peaks at 7–8 min (Foeder *et al.*, 1978), well after assembly of the fertilization envelope is complete. Despite years of intensive investigation, the mechanism of H_2O_2 production at fertilization is just emerging. In the presence of NAD(P)H and an egg cytoplasmic 4-thiohistidine called ovothiol (Turner *et al.*, 1986), ovoperoxidase catalyzes Ca^2-dependent, CN^--resistant H_2O_2

production *in vitro* (Turner *et al.*, 1985). However, several observations made it clear that the ovoperoxidase–ovothiol system is not involved in H_2O_2 production *in vivo* (Turner *et al.*, 1987). Instead ovothiol, present at 5 mM in eggs, may protect the egg from reactive oxygen toxicity, since it reacts quickly with H_2O_2 (Turner *et al.*, 1986). An NADPH-specific oxidase leading to H_2O_2 production has recently been identified (J. W. Heinecke and B. M. Shapiro, unpublished data).

Ovoperoxidase also is regulated by spatial constraints. Ovoperoxidase is sequestered in the fertilization envelope, limiting its catalytic activities to substrates within this structure. The insertion of ovoperoxidase into the envelope is accomplished by the formation of a 1 : 1 complex between ovoperoxidase and a second cortical granule protein, proteoliaisin. This complex then binds to the vitelline scaffold, where proteoliaisin has specific attachment sites (Weidman and Shapiro, 1986).

4. Proteoliaisin

Proteoliaisin is a highly asymmetric protein of ~235,000 M_r (Weidman *et al.*, 1986). Unlike many extracellular proteins, it does not appear to be a glycoprotein, nor are any of its known interactions carbohydrate dependent. Its amino acid composition is distinctive: Gln/Glu, Asn/Asp, Gly, and Cys account for 50% of the residues; all of the Cys residues (>200) are in disulfides. Whereas proteoliaisin is a substrate for ovoperoxidase-catalyzed dityrosine formation *in vivo* (Kay *et al.*, 1982; Kay and Shapiro, 1987), it contains relatively few aromatic amino acids (~30 tyrosines/molecule). The amino acid composition and some other characteristics of proteoliaisin are strikingly similar to those of thrombospondin, a constituent of diverse mammalian extracellular matrices (Majack and Bornstein, 1986) and fibrin clots (Silverstein *et al.*, 1986).

The interaction of proteoliaisin with ovoperoxidase, both *in vitro* and *in vivo*, is dependent on, and specific for, Ca^{2+} ($K_{0.5}$ = 50 μM (Weidman *et al.*, 1985). Although its affinity for ovoperoxidase is not unusually high [K_d = ~1 μM (Weidman *et al.*, 1986)], the complex can be extracted with dilute buffer from uncross-linked fertilization envelopes and is stable to electrophoresis under nondenaturing conditions (Weidman *et al.*, 1985). Proteoliaisin binds to the vitelline layer of unfertilized eggs (and presumably the modified, elevated vitelline scaffold after fertilization) in a reversible, divalent cation-mediated interaction that is independent of its association with ovoperoxidase (Weidman and Shapiro, 1986). The major seawater divalent cations (Ca^{2+} and Mg^{2+}) have a synergistic effect on this binding interaction, which involves vitelline-layer binding sites of different affinities. The highest affinity sites (K_d = 0.2 μM) occur at the lowest frequency (8 × 10^7 sites/egg) and are found in the presence of Mg^{2+} ($K_{0.5}$ = 2.5 mM). Sites of lower affinity (K_d = 0.5 μM) are more than six times as abundant (5.5 $\times$ 10^8 sites/egg) and are specific for Ca^{2+} ($K_{0.5}$ = 200 μM). In the presence of both Ca^{2+} and Mg^{2+}, there are twice as many sites, but the affinity is lower (K = 1.4 μM). Although the mechanism of synergism for the divalent cation effects is unclear, one explanation is that when both cations are present binding occurs on both sides of the vitelline layer, whereas when only Ca^{2+} is present binding is restricted to the internal side of the vitelline layer, where the protein is provided at secretion. If this explanation is correct, then the small number of Mg^{2+}-specific sites might facilitate the transport of the ovoperoxidase–proteoliaisin complex across the vitelline scaffold during fertilization envelope assembly, thereby accounting for the 2-fold increase in site number.

5. Assembly and Cross-Linking

Ovoperoxidase associates with the vitelline layer of unfertilized eggs only when bound to proteoliaisin (Weidman and Shapiro, 1986). The physical characteristics of proteoliaisin (its asymmetry, independent binding domains, and the flexible structure suggested by its high glycine content) suggest that it acts as an extended tether that anchors ovoperoxidase to the vitelline layer. In the presence of H_2O_2, the bound ovoperoxidase catalyzes the irreversible association of proteoliaisin and the vitelline layer, presumably by forming dityrosine cross-links between these two natural substrates. This, coupled with the observation that ovoperoxidase catalyzes protein cross-linking most efficiently when assembled into the fertilization envelope (Kay and Shapiro, 1987), suggests that proteoliaisin juxtaposes ovoperoxidase and its substrates appropriately for the hardening reaction to occur. A limited subset of fertilization envelope components, in addition to proteoliaisin and the vitelline layer, are targets for cross-linking in vivo (Kay et al., 1982; Kay and Shapiro, 1987) and each molecule of ovoperoxidase catalyzes the formation of only 5–10 cross-links (Kay and Shapiro, 1987).

Aside from the ovoperoxidase–proteoliaisin complex, the other constituents of the fertilization envelope derived from the cortical granules are less well characterized. Two macromolecular aggregates present in uncross-linked fertilization envelopes, with distinct electrophoretic behavior, size, and composition, are substrates for cross-linking in vivo (E. S. Kay and P. J. Weidman, unpublished observations). The largest of these complexes is spherical, with a Stokes radius >100 Å. This complex comprises 5–10 glycoproteins that are substrates for ovoperoxidase-catalyzed dityrosine formation in vivo and cannot be extracted from cross-linked fertilization envelopes (Kay et al., 1982; Kay and Shapiro, 1987). The second macromolecular complex of soft fertilization membranes is spherical, with an apparent Stokes radius of 69 Å. It contains the remaining cortical granule-derived proteins of less than 110,000 Da (excluding ovoperoxidase) including two proteins of 108,000 and 56,000 Da that are held together by a linkage that is disrupted in 10 mM EGTA.

The relationship between the structural units seen by high-resolution electron microscopy on the surfaces of the fertilization envelope and the several macromolecular complexes described above is uncertain. In the presence of divalent cations, the cortical granule exudate proteins form a paracrystalline precipitate that has a surface ultrastructure similar to that of the fertilization envelope (Bryan, 1970b). This aggregation might be mediated in part by specific association between proteoliaisin and the vitelline scaffolding, but the mechanism by which the other protein complexes insert is still not clear, nor is the nature of the vitelline receptor for proteoliaisin.

B. The Amphibian Fertilization Envelope

Extracellular remodeling in amphibian fertilization has been studied most by the assembly of the fertilization envelope of *Xenopus laevis*. Although the detailed analysis of fertilization envelope assembly in this species began comparatively recently, a substantial amount of information is available concerning the mechanisms involved (see Schmell *et al.*, 1983, for review). The morphology of fertilization envelope assembly is similar, but not identical, to that of the sea urchin.

1. The Egg Coat

The *Xenopus* egg is surrounded by three layers of jelly coat that are deposited on the egg as it traverses the oviduct (Yurewicz *et al.*, 1975). The innermost of these, J1, is closely apposed to a thick vitelline envelope which, although attached to the plasma membrane, is separated from the egg surface by an ~1-μm perivitelline space (Wyrick *et al.*, 1974). Together these layers comprise the extracellular fertilization envelope precursor compartment (Grey *et al.*, 1974). The intracellular compartment is composed of two types of cortical granules "(vesicles)": 1.5-μm granules next to the plasma membrane in both hemispheres of the egg, and less closely associated 2.5-μm granules predominantly in the vegetal hemisphere (Grey *et al.*, 1974). At fertilization, a wave of exocytosis spreads around the egg from the site of sperm entry, and the vitelline envelope lifts away from the plasma membrane (Wolf, 1974a,b). Secreted proteins diffuse through the vitelline envelope to mix with the J1 layer of jelly and form a thin, electron-dense layer termed the F-layer between J1 and the vitelline envelope (Wyrick *et al.*, 1974). The resultant fertilization envelope is impenetrable to sperm (Grey *et al.*, 1976), more stable to disruption by aqueous solvents than the vitelline envelope (Wolf *et al.*, 1976), and resistant to tryptic digestion (Wolf, 1974b). Escape from this integument occurs when the larva releases a specific hatching enzyme that weakens the fertilization envelope and allows it to be ruptured (Urch and Hedrick, 1981a,b).

2. Cortical Vesicle (Granule) Exocytosis

As with the sea urchin egg, the cortical reaction in *Xenopus* eggs is inducible by the Ca^{2+} ionophore A23187 in the absence of external Ca^{2+}, suggesting that a release of intracellular Ca^{2+} is responsible for triggering exocytosis (Monk and Hedrick, 1986). An endoplasmic recticulum associated with the cortical granules in mature oocytes (Campanella and Andreuccetti, 1977) is suggested to be the source of intracellular Ca^{2+} involved in exocytosis (Chabonneau and Grey, 1984). Unlike the situation with sea urchins, the vitelline envelope does not appear to be extensively remodeled during the assembly process, for its morphology is largely conserved in the fertilization envelope (Grey *et al.*, 1974; see below). The conversion of the vitelline envelope to a fertilization envelope is thus morphologically defined by the formation of the F-layer. Both cortical vesicle exocytosis and an intact J1 layer are necessary to produce the F-layer. If the jelly coat is removed prior to fertilization, the F-layer is not formed, the secreted proteins diffuse away from the egg, and the solubility characteristics of the elevated vitelline envelope remain the same as those of the unfertilized egg (Wolf, 1974a,b; Wolf *et al.*, 1976).

3. The Cortical Granule Lectin

Approximately 70% of the protein released from cortical granules is composed of a Ca^{2+}-dependent, galactose-specific agglutinin (Grey *et al.*, 1974). This cortical granule lectin has an apparent monomer M_r of 43,000 (Gerton and Hedrick, 1986) and exists in soluble cortical granule exudate as multimeric aggregates, even in the presence of divalent cation chelators (Wolf, 1974b). The J1 layer consists of three minor, nonglycosylated proteins and one major sulfated 90 kDa (Wolf *et al.*, 1976) glycoprotein (Yurewicz *et al.*, 1975) that serves as the ligand for the cortical granule lectin (Birr and Hedrick, 1979). The interaction between the jelly component and the cortical granule lectin results in the formation of a Ca^{2+}-dependent precipitate that can be detected *in vitro* by the formation of a precipitin line in double diffusion studies (Wyrick *et al.*, 1974) and *in vivo* by the formation of the F-layer between the J1 layer and the vitelline envelope. The agglutinin reaction is not species specific, because the cortical granule lectin from *Xenopus* agglutinates egg jelly from other amphibians, although it does not affect sea urchin egg jelly (Wyrick *et al.*, 1974). The lectin does not interact with several other synthetic and natural carbohydrate polymers (Wyrick *et al.*, 1974).

4. Assembly Reactions

Although the formation of the F-layer is required for transformation of the vitelline envelope into a fertilization envelope, it does not constitute the sole block to sperm permeability nor does it appear solely responsible for the physicochemical properties of the fertilization envelope. In studies of the interaction of sperm with isolated egg envelopes (Wolf, 1974a,b), both sides of the vitelline envelope were receptive to sperm penetration, whereas in the fertilization envelope both the outer (F-layer) side and the inner (non F-layer) side were impenetrable, suggesting that the F-layer is not the only block to sperm entry. When the F-layer is extracted from the fertilization envelopes (using chelators and/or galactose) the residual envelope, termed a VE*, retains the modified physical and chemical properties that characterize the intact fertilization envelope (Wolf, 1974a,b). Only ~5% of the exocytosed cortical granule lectin is associated with the F-layer, and the remainder is found in soluble form within the perivitelline space (Wolf, 1974a,b; Urch and Hedrick, 1981b). Thus, the F-layer appears to act as a permeability barrier and may facilitate chemical conversion of the vitelline envelope by trapping other modifying agents or enzymes within the perivitelline space.

The molecular changes that accompany the alteration of vitelline envelope solubility and sperm penetrability are subtle and more difficult to characterize than F-layer formation. Although the fertilization envelope is generally more stable than the vitelline envelope, it can be solubilized by heat, detergents, and chaotropic agents (Wolf *et al.*, 1976; Gerton, and Hedrick, 1986) indicating that the formation of covalent bonds is probably not involved in its stabilization. Immunological analyses indicate that all vitelline envelope antigens are found in the fertilization envelope (Wolf *et al.*, 1976), but two minor vitelline envelope constituents are altered in the fertilization envelope (Gerton and Hedrick, 1986). The vitelline envelope consists of approximately seven peptides ranging from 37,000 to 120,000 M_r, with two components of 37,000 and 41,000 comprising more than two-thirds of the envelope protein (Wolf *et al.*, 1976; Gerton and Hedrick, 1986). Fertilization envelopes from which the F-layer has been extracted give an identical profile, with one exception: two minor vitelline envelope components of 69,000 and 64,000 M_r had been shifted to lower M_r (66,000 and 61,000, respectively) (Gerton and Hedrick, 1986). Peptide mapping reveals that the two vitelline envelope peptides are related and that the reduction in size is associated with a single glycopeptide fragment common to both proteins (Gerton and Hedrick, 1986). The cleavage site appears to be near the carboxy-terminus of these proteins, since both retain blocked amino-termini after vitelline envelope to fertilization envelope conversion (Gerton and Hedrick, 1986). Although this may be a limited proteolytic event, neither its role in altering fertilization envelope properties, nor the nature of the processing enzyme is known.

The agent responsible for affecting the hydrolysis of these proteins presumably is released from the cortical vesicles. The nonlectin portion of cortical granule exudate (~30% of the protein; Wolf, 1974b) contains five major and four minor glycoproteins ranging from 22,000 to 117,500 M_r (Grey *et al.*, 1976) that appear to reside in the perivitelline space after fertilization. When partially separated by gel filtration, all fractions could block fertilization, presumably by modifying the vitelline envelope of the unfertilized egg (Prody *et al.*, 1985). Despite intensive efforts, neither proteolytic (Wolf, 1974a,b; Gerton and Hedrick, 1986) nor peroxidase activities (Greve and Hedrick, 1978) have been found, although an ~40,000 M_r N-acetyl-β-D-glucosaminidase has been purified from cortical granule exudate (Prody *et al.*, 1985; Greve *et al.*, 1985). This enzyme and an anologous glucosaminidase from jack beans render eggs unfertilizable (Prody *et al.*, 1985) suggesting that cleavage of some carbohydrate component(s) of the vitelline envelope alters sperm–egg interactions.

The presence of multiple fertilization inhibitory factors in the cortical granule secretion (Prody *et al.*, 1985) suggests that still unidentified changes occur in the conversion of the vitelline envelope to the fertilization envelope. To test for conformational changes that would be undetected by compositional analyses, the relative accessibility of vitelline envelope and fertilization envelope components to iodination was assessed (Nishihara *et al.*, 1983). No qualitative differences were observed between the vitelline envelope and fertilization envelope. However, certain quantitative differences in labeling were seen, which may have been due to the F-layer blocking access to interior sites. In both vitelline envelope and fertilization envelopes, a principal protein of ~41,000 M_r was not isotopically labeled by peroxidase-catalysed iodination in intact or heat solubilized envelopes, but was readily iodinated after the envelopes were dissociated with guanidine-HCl, suggesting that this component either lies buried within the vitelline envelope or has buried tyrosyl residues. The relationship of such changes in structure to fertilization envelope assembly remains obscure.

C. The Fish Chorion

Most investigations of fish fertilization have been carried out on teleosts, a well studied representative of which is the fresh water fish *Oryzias latipes* (or more commonly the medaka). In contrast to the extensively explored fertilization envelope assembly of echinoderms and amphibians, relatively little research has examined the remodeling of the fish extracellular matrix, the chorion. Thus, the molecular mechanism by which fish chorions harden is still a matter of speculation.

The medaka egg is surrounded by a single, complex, 15-μm-thick protein-containing chorion that functions after fertilization in protecting the developing embryo from environmental hazards such as rapid changes in salinity and dessication. The chorion and its enclosed perivitelline fluid provide a sterile and stable environment for the developing embryo, as attested to by the fragility of dechorionated embryos (reviewed in Dumont and Brummett, 1985). Fish sperm have direct access to the egg plasma membrane only through a channel through the chorion, the micropyle, located at the animal pole. Medaka fertilization is normally monospermic, but removal of the chorion permits polyspermy (Sakai, 1961) by providing additional sites for sperm entry. Thus, the micropyle (usually only one sperm wide at its inner end), the chorion, and the cortical reaction act in concert as a block to polyspermy: only one target site exists, and supernumerary sperm are forced out of the micropyle which is closed after cortical granule exocytosis (Ginsburg, 1961, 1972).

1. Cortical Vesicle Exocytosis

Around the beginning of the century it was found that fish eggs could be activated by placing them in solutions of high calcium concentration, or by certain insults to the plasma membrane which were effective only if external calcium was present. It now appears that at fertilization a wave of elevated Ca^{2+} propagates through the cytoplasm (Ridgeway *et al.*, 1977) to cause the first visible change, a progressive breakdown of the cortical vesicles (or alveoli) that begins near the micropyle at the animal pole and ends at the vegetal pole (Yamamoto, 1961). The leading edge of the Ca^{2+} wave precedes cortical vesicle fusion, and no fusion occurs in regions that the Ca^{2+} wave does not enter (Gilkey *et al.*, 1978; Ridgeway *et al.*, 1977).

The cortical granules of the medaka range in size from 5–40 μm and are found in a closely packed layer beneath the plasma membrane, except for a small area adjacent to the animal pole (Yamamoto, 1961). Their contents vary depending on the species and include neutral polysaccharides, acid mucopolysaccarides, and glycoproteins (reviewed in Gilkey, 1981; Laale, 1980), as well as an uncharacterized "hardening enzyme" (Zotin, 1958; see below). Upon fusion, the cortical granules discharge their contents into the perivitelline space. The intact chorion is freely permeable to water, gases, salts, dyes, and amino acids, but larger molecules are retained in the perivitelline space (Yamamoto, 1961), so that material released by exocytosis forces the elastic chorion to move away from the plasma membrane by an osmotic mechanism (Yamamoto, 1962).

2. Chorion Modifications

Although there is no visual evidence that the contents of the granules are incorporated into the chorion, a reaction between the chorion and some released components must occur since the chorion loses its elasticity and hardens (reviewed in Yamamoto, 1961; Gilkey, 1981). This hardening process is completed in the medaka in about 30 min (Ohtsuka, 1957) and the hardened chorion remains around the developing embryo until hatching, when it is digested by a specific metalloprotease secreted by the embryo (Yamamoto and Yamagami, 1975; Yamagami, 1981).

The teleost chorion is composed of two main layers: a thick (15 μm) protein-containing inner layer in which a network of fibers is arranged in lamellae, and a thinner (0.3 μm) polysaccharide-rich outer layer with many short projections and attaching filaments. The inner layer changes most profoundly after fertilization. Following fertilization the inner layer, the contents of the cortical vesicles and Ca^{2+} interact to become a compact, hardened structure (Dumont and Brummett, 1985; Gilkey, 1981; Blaxter, 1969; Ginsburg, 1972; Yamamoto, 1961). At the time of hatching the soluble material released from the chorion contains high-molecular-weight glycoproteins (Yamamoto and Yamagami, 1975) that had been constituents of the inner layer (Yamagami, 1981). A biochemical analysis of the proteolytic products of the inner layer of the medaka chorion (Yamagami, 1981) suggested that six glycoproteins, from 86,000 to 210,000 M_r, were linked together to form the interconnected fibrillar network of the mature chorion. Upon ovulation a final jelly layer is deposited on the chorion surface. The jelly layer also undergoes physicochemical changes after egg activation; it develops an increased stickiness, to adhere eggs to environmental substrates or each other (Nakano, 1956, 1969).

Divalent cations play an important role in the remodeling of the chorion (reviewed in Yamamoto, 1961). If eggs are transferred to Ca^{2+}-free solution after insemination, hardening is inhibited. This Ca^{2+} requirement obtains for only a short fraction (5 min) of the total time required for complete chorion hardening (Zotin, 1958). The rate of chorion hardening is dependent on the Ca^{2+} concentration (Ohtsuka, 1957). Mg^{2+} is also effective.

A "hardening enzyme" (Zotin, 1958) was implicated in the hardening reaction, because perivitelline fluid from activated eggs causes hardening of other eggs. Hardening is inhibited by CN^- (Ohtsuka, 1957, 1960, 1964) suggesting that it may be an oxidative process, as with the sea urchin egg where a peroxidase is involved in cross-linking. In the trout egg, Glu-Lys isopeptides are present in the chorions of fertilized, but not unfertilized eggs (Lonning et al., 1984), indicating that the formation of covalent ε-(λ-glutamyl)lysine cross-links between adjacent polypeptides may also stabilize the envelope. Despite these hints, the biochemical nature of chorion hardening remains undefined.

Some investigators have speculated that hardening of medaka (Ohtsuka, 1960, 1964) and salmonid (Zotin, 1958) egg chorions may be caused by secretion of oxidizing and cross-linking agents by the egg independently of cortical vesicle fusion, but we feel that these ideas should be interpreted with care. In almost all such reports either (1) hardening appeared abnormal or was poorly characterized; (2) proper fertilization controls were not done; or (3) some cortical granule breakdown may have occurred, since some eggs always autoactivate (Yamamoto, 1955). Thus, it is likely that the extracellular coat modifications in fish eggs, as with other animals, are due to the cortical reaction.

D. The Mammalian Zona Pellucida

A characterization of the remodeling of the mammalian zona pellucida after fertilization has been hindered by the limited quantities of material available, but several recent approaches have been fruitful. Radiolabeling and mechanical isolation of mouse oocytes (Bleil and Wassarman, 1980a,b) and isolation of large amounts of material from ovarian oocytes by sieving techniques used for pig (Dunbar *et al.*, 1980, 1981), bovine (Gwatkin *et al.*, 1979, 1980), and rabbit (Dunbar *et al.*, 1981) oocytes have illuminated the macromolecular composition of the zona pellucida. Although oocytes isolated by sieving techniques have not been used for fertilization studies, mature mouse eggs have provided information about the biochemical changes in zona pellucida glycoproteins at fertilization (reviewed in Wassarman *et al.*, 1984, 1985a). So our discussion focuses on the mouse egg system.

The zona pellucida fulfills two important roles. First, the modified zona pellucida is a block against polyspermy (Gwatkin, 1977), which would lead to abnormal development and early embryonic death (Pikó, 1969). The zona reaction alters sperm receptors such that additional sperm can attach but will not bind. Furthermore, sperm that have partially penetrated the zona prior to fertilization are prevented from further entry (reviewed in Wasserman *et al.*, 1985b). Second, the modified zona pellucida serves a protective role. The hardened zona pellucida protects the early embryo before implantation and maintains a specific biochemical microenvironment in the perivitelline space (Kapur and Johnson, 1986; Okada *et al.*, 1986).

1. Cortical Vesicle Exocytosis

After penetrating the zona pellucida, the sperm that fuses with the egg triggers a complex series of events. Transient elevations in intracellular Ca^{2+} are seen in mouse (Cuthbertson *et al.*, 1981; Cutherbertson and Cobbold, 1985) and hamster eggs (Miyazaki *et al.*, 1986) after artificial activation and fertilization; they propagate as a wave from the point of sperm–egg fusion bidirectionally over the surface of the egg. The activation of hamster eggs by ionophore A23187, which is independent of external Ca^{2+} and thus due to Ca^{2+} release from internal stores, induces a zona reaction apparently identical to that of normal fertilization (Steinhardt *et al.*, 1974). Cortical granules fuse with the plasma membrane, leading to expulsion of their contents and zona hardening. In preparation for implantation in the uterus, the blastocyst rids itself of the zona through a specific hatching process (Perona and Wassarman, 1986).

Cortical granules are a general feature of the mammalian egg (reviewed in Gulyas, 1980; Guraya, 1983). They contain glycoproteins and two enzymes, a trypsin-like protease (Barros and Yanagimachi, 1971; Gwatkin et al., 1977) and a peroxidase (Gulyas and Schmell, 1980; Schmell and Gulyas, 1980), that have been implicated in modifying the zona pellucida (discussed below).

2. The Zona Pellucida

Although the zona pellucida varies between mammalian species, some general structural features are conserved. It is a porous structure composed of polysaccharides and glycoproteins in a three dimensional network that permits large molecules and small viruses to penetrate (see Gwatkin, 1977; Dunbar, 1983; Dunbar and Wolgemuth, 1984). Zona pellucidae of different species vary in their susceptibility to acid, base, heat, or enzymatic digestion (see Dunbar, 1983; Dunbar and Wolgemuth, 1984; Gwatkin, 1977). Heating disaggregates some zona pellucidae into high-molecular-weight complexes (Dunbar et al., 1980; Gwatkin et al., 1980; Ahuja and Bolwell, 1983) that can be isolated by gel filtration and further dissociated into their constituent glycoproteins by SDS and reducing agents (Bleil and Wassarman, 1980b). Zona pellucida glycoproteins from procine (Dunbar et al., 1981), hamster (Ahuja and Bolwell, 1983), mouse (Bleil and Wassarman, 1980b), and rabbit (Dunbar et al., 1981) eggs consist of discrete sets of protein families characterized by charge and molecular-weight heterogeneity. When porcine zona pellucidae are deglycosylated with endo F, the three major glycoprotein families decrease in molecular weight and heterogeneity (Dunbar and Wolgemuth, 1984) consistent with the idea that the zona pellucida is composed of a small set of polypeptide chains with discrete carbohydrate modifications.

Three unique mouse egg zona pellucida glycoproteins designated ZP1, ZP2, and ZP3 (200, 120, and 83 kDa, respectively) are well characterized. These are acidic sialoproteins that account for all of the protein in the mouse zona pellucida and about 10% of the protein of the mature mouse oocyte. The three glycoproteins are synthesized and secreted by the growing oocyte prior to ovulation (reviewed in Wassarman et al., 1984). Mature ZP1, ZP2, and ZP3 contain N-linked oligosaccharides (Greve et al., 1982; Roller and Wassarman, 1983) and ZP2 and ZP3 have O-linked carbohydrate residues (Florman and Wassarman, 1985; Wassarman et al., 1984) that might be important in the zona modification reaction (see below).

The three mouse zona pellucida proteins are organized into long, interconnected filaments with a recognizable structural repeat (Greve and Wassarman, 1985) that form a three-dimensional matrix. Exposure of the intact zona pellucida to chymotrypsin results in limited proteolysis of ZP1, whereas exposure to dithiothreitol (DTT) results in the reduction of intermolecular disulfide linkages of ZP1 (Greve and Wassarman, 1985). Both treatments give rise to unbranched zona filaments of similar lengths, suggesting that ZP1 acts as a cross-linking protein at filament branch points within the zona pellucida.

3. Modifications in the Zona at Fertilization

No marked changes in extracellular coat morphology are seen after mammalian fertilization in contrast to what is found in other animals (Phillips and Shalgi, 1980). However physicochemical changes occur, as evidenced by changes in resistance to various chemical (disulfide bond reducing agents and low pH), enzymatic (proteases), and physical (heat) treatments (see Dunbar, 1983; Dunbar and Wolgemuth, 1984; Gwatkin, 1977).

Ovoperoxidase-mediated dityrosine formation similar to that of the sea urchin egg has been implicated in zona hardening. Peroxidase activity is associated with intact cortical granules in unfertilized mouse eggs, and with the egg surface, zona pellucida, and particulate matter of the perivitelline space in artificially activated eggs (Gulyas and Schmell, 1980; Schmell and Gulyas, 1980). Moreover, peroxidase inhibitors and some tyrosine analogs inhibit zona hardening. However, no evidence yet exists for intermolecular cross-linking of zona pellucida glycoproteins in such a reaction.

Sperm receptor activity has been demonstrated using competition assays in solubilized preparations of zonae from mouse eggs (Bleil and Wassarman, 1980c). The ZP3 glycoprotein appears to be both the sperm receptor and the inducer of the acrosome reaction at fertilization (reviewed in Wassarman *et al.*, 1985a,b). The sperm receptor activity of ZP3 is dependent on its O-linked carbohydrate components only (Florman *et al.*, 1984; Wassarman *et al.*, 1985b), whereas the acrosome inducing activity of ZP3 is dependent upon both the carbohydrate and the polypeptide portion of the molecule (Florman *et al.*, 1984; Florman and Wassarman, 1985).

Zona pellucidae from 2-cell embryos no longer contain sperm-receptor activity, consistent with the idea that the sperm receptor becomes inactivated during fertilization (Bleil and Wassarman, 1980c; Inoue and Wolf, 1975). Purified ZP3 from unfertilized eggs competes for sperm binding to eggs, but ZP3 isolated from 2-cell embryos is ineffective in reducing sperm binding (Bleil and Wassarman, 1980c) and inducing the acrosome reaction. The mechanism for inactivation of ZP3 at fertilization is not known; its mobility on SDS–PAGE does not change (Bleil and Wassarman, 1980c; Bleil and Wassarman, 1983).

The major glycoprotein in the mouse zona pellucida, ZP2, is subject to limited proteolysis following fertilization or artificial activation of mouse eggs (Bleil *et al.*, 1981). Under reducing conditions, ZP2 from embryos, designated ZP2f, migrates on SDS–PAGE with an apparent molecular weight of 90,000 instead of 120,000 for ZP2 from unfertilized eggs. The cortical exudate alone is enough to trigger this conversion, suggesting that cortical granule proteases may be responsible for the modification. The role that this modification plays in the physicochemical changes of the zona reaction, or indeed how any of the other changes are reflected in altered zona properties, remain interesting subjects for study.

III. MECHANISM OF EXTRACELLULAR REMODELING DURING FERTILIZATION

Having considered the current knowledge of specific egg coat alterations at fertilization, we begin to glimpse several molecular mechanisms that are retained with remarkable constancy in the diverse systems explored. To provide a summary of these mechanisms is hazardous, in that not all have been found in all egg systems; nonetheless, it is useful in focusing attention on problems that require further biochemical exploration.

A. Increased Intracellular Ca^{2+}

Gamete membrane fusion initiates a propagating wave of secretion. The current model for this secretion involves inositol trisphosphate (IP_3) acting as an effector of Ca^{2+} release from the ooplasmic reticulum. The resultant Ca^{2+} wave induces exocytosis, thereby serving as the transmitter of the fertilization signal from the site of sperm entry throughout the spherical egg. Aside from being the agent of the cortical reaction, Ca^{2+} initiates a portion of the developmental sequence and thus serves as a branch point between the pathways of developmental initiation and embryo protection that were the subjects of Loeb's investigations.

The secretion of the cortical reaction is an example of a cellular control strategy that resembles laboratory chemistry. Two cellular compartments are physically separated as if in different reaction flasks; when their contents are combined, a chemical reaction occurs with new product formation. This allows the egg to remodel its surface only after entry of the fertilizing sperm, in order to end up with a special extracellular environment and a protective shell.

B. Limited Proteolysis

Limited proteolysis seems to be one of the early events after secretion, even though cortical vesicle proteases are not linked unambiguously to one or another function. Proteolysis has several potential roles: it may remove sperm receptors from the glycocalyx; it may allow the vitelline layer to move away from the plasma membrane; or it may modify the glycocalyx as one step in the assembly or hardening process. For example, such a proteolytic cleavage may be required where a hardened egg coat is formed without covalent cross-linking. Noncovalent assembly is initiated by proteolysis in other systems, such as the conversion of fibrinogen to polymerized fibrin.

C. Assembly

Some secreted proteins are assembled into the extracellular matrix, as best shown in the amphibian and sea urchin systems. The amphibian egg secretes a lectin that becomes part of the fertilization envelope, its ligand being a component of the jelly coat. Together these form the F-layer of the fertilization envelope. Proteoliaisin secreted by the sea urchin egg has binding sites on the unfertilized, unmodified egg vitelline layer. Since it also binds ovoperoxidase, proteoliaisin inserts the enzyme into a limited class of sites in the assembled fertilization envelope. Ovoperoxidase is a stoichometric component of the fertilization envelope, and it is also the enzyme that cross-links it. The assembly reaction seems to be an important coordinating step whereby the fertilization envelope achieves new properties. We still do not know the extent to which the several physicochemical alterations in the extracellular coat are dependent upon these non-convalent assembly interactions or the mechanisms by which they occur.

D. Cross-Linking

Although hardening of the egg coat is a frequent sequela of fertilization, the biochemical pathway of the cross-linking reaction has been demonstrated only for the sea urchin egg. The dityrosine cross-links introduced by ovoperoxidase form a large ($\sim 10^{16}$ Da) covalently stabilized matrix. Preliminary results from mammalian fertilization suggest that similar cross-linking may occur, but this has yet to be documented. In using H_2O_2 to effect cross-linking, the egg has employed a two-edged sword. Because it must accomplish an oxidation–reduction extracellularly, it uses this diffusable reactive oxygen intermediate; however if H_2O_2 diffusion were to occur internally, the potential is great for severe genetic and cytoplasmic damage. It is interesting in this regard that sea urchin eggs have high concentrations of a novel thiohistidine, ovothiol, that reacts with hydrogen peroxide. Ovothiol derivatives are found in other types of eggs, including those from two species of teleost fish, suggesting that fish may employ a similar peroxidative mechanism and use ovothiol as the protectant against cellular damage from reactive oxygen.

E. The Perivitelline Milieu

The result of the above processes is to form a fertilization envelope around the egg. This is one component of the late block to polyspermy, because it clearly is a barrier to sperm penetration. The fertilization envelope also protects the developing early embryo by acting as a shell. Yet another important role for this structure may be in providing a unique extracellular environment. The milieu between the plasma membrane of the egg and the extracellular coat (the perivitelline space) may have distinct properties. The fertilization coats of most eggs are permeant to only small molecules; thus, the perivitelline environment could be both sterile and of defined composition. The barrier to bacteria may be of special importance where eggs are living in the open ocean; in the sea urchin egg, elevation of the vitelline layer sweeps away microorganisms and ensures sterility within the perivitelline space, allowing the embryo to develop in a germ-free environment. Some of the secreted macromolecules trapped within the perivitelline space may play a role in determining the nature of this special extracellular environment. Since they are localized around the egg, they may modify other components that pass through the fertilization envelope. In the case of the sea urchin egg, for example, the peroxidase assembled in the fertilization envelope is locked into a high-activity conformation. It could potentially act as a source of peroxidative damage to bacteria or other organisms that encounter the egg, by using hydrogen peroxide either released from the egg or from a potential predator. Thus, the fertilization envelope might serve as a chemical shield, by bombarding pathogens or predators with highly reactive peroxidative products. It is noteworthy that early sea urchin embryos with intact fertilization envelopes can pass through the digestive tract of predator tunicates without damage (M. Levine and D. Epel, unpublished data), whereas unfertilized eggs or embryos without fertilization envelopes are destroyed. Although these ideas are speculative and the details of protective effects (if any) effected by the environment of vitelline envelope and perivitelline space are unknown, an exploration of the unique perivitelline microenvironment of eggs that employ the cortical reaction is warranted. It is a common observation that embryos lacking fertilization envelopes do not develop as well as those that kept their coats on. This might be due to a special environment in the perivitelline space as well as to the hardened protective fertilization envelope. Both may be critical extracellular modifications that allow early embryos to become viable animals.

Note Added in Proof

The peroxidase activity suggested to harden fish chorions in section IIC has now been demonstrated ultracytochemically in the vitelline and fertilization envelopes of eggs of the fish *Tribolodon hakonensis* (S. Kudo *et al.*, 1988).

ACKNOWLEDGMENTS

We would like to thank Dave Battaglia and Donner Babcock for useful comments and discussions, and Mary Patella and Robert Gollehon for typing the manuscript. Supported by NIH grant GM23910 and PHS NRSA 5 T32 GM07270 from NIGM.

REFERENCES

Ahuja, K. K., and Bolwell, G. P. (1983). Probable asymmetry in the organization of components of the hamster zona pellucida. *J. Reprod. Fertil.* **69**, 1–7.

Barros, C., and Yanagimachi, R (1971). Induction of the zona reaction in golden hamster eggs by cortical granule material. *Nature (London)* **233**, 268–69.

Birr, C., and Hedrick, J. L. (1979). Immunological identification of the jelly coat ligand for *Xenopus laevis* cortical granule lectin. *Fed. Proc.* **38**, 466a.

Blaxter, J. H. S. (1969). Development eggs and larvae. *In* "Fish Physiology" (W. S. Hoar and D. J. Randall, eds.), Vol. 3, pp. 177–252. Academic Press, New York.

Bleil, J. D., and Wassarman, P. M. (1980a). Synthesis of zona pellucida proteins by denuded and follicle-enclosed mouse oocytes during culture *in vitro*. *Proc. Natl. Acad. Sci. U.S.A.* **77**, 1029–1033.

Bleil, J. D., and Wassarman, P. M. (1980b). Structure and function of the zona pellucida: Identification and characterization of the proteins of the mouse oocyte's zona pellucida. *Dev. Biol.* **76**, 185–202.

Bleil, J. D., and Wassarman, P. M. (1980c). Mammalian sperm–egg interaction: Identification of a glycoprotein in mouse egg zona pellucida possessing receptor activity for sperm. *Cell (Cambridge, Mass.)* **20**, 873–882.

Bleil, J. D., and Wassarman, P. M. (1983). Sperm–egg interactions in the mouse: Sequence of events and induction of the acrosome reaction by a zona pellucida glycoprotein. *Dev. Biol.* **95**, 317–324.

Bleil, J. D., Beall, C. F., and Wassarman, P. M. (1981). Mammalian sperm–egg interaction: Fertilization of mouse eggs triggers modification of the major zona pellucida glycoprotein ZP2. *Dev. Biol.* **86**, 189–197.

Bryan, J. (1970a). The isolation of a major structural element of the sea urchin fertilization membrane. *J. Cell Biol.* **44**, 635–644.

Bryan, J. (1970b). On the reconstitution of the crystalline components of the sea urchin fertilization membrane. *J. Cell Biol.* **45**, 606–614.

Campanella, C., and Andreuccetti, P. (1977). Ultrastructural observations on cortical endoplasmic reticulum and on residual cortical granules in the egg of *Xenopus laevis*. *Dev. Biol.* **56**, 1–10.

Carroll, E. J., Jr. (1976). Cortical granule proteases from sea urchin eggs. *In* "Methods in Enzymology" (L. Lorand, ed.), Vol. 45, pp. 343–353. Academic Press, New York.

Carroll, E. J., Jr., and Epel, D. (1975). Isolation and biological activity of the proteases released by sea urchin eggs following fertilization. *Dev. Biol.* **44**, 22–32.

Chabonneau, M., and Grey, R. D. (1984). The onset of activation responsiveness during maturation coincides with the formation of the cortical enoplasmic reticulum in oocytes of *Xenopus laevis*. *Dev. Biol.* **102**, 90–97.

Chandler, D. E., and Heuser, J. (1980). The vitelline layer of the sea urchin egg and its modification during fertilization. *J. Cell Biol.* **84**, 618–632.

Citkowitz, E. (1971). The hyaline layer: Its isolation and role in echinoderm development. *Dev. Biol.* **24**, 348–362.

Cuthbertson, K. S. R., and Cobbold, P. M. (1985). Phorbal ester and sperm activate mouse oocytes by inducing sustained oscillations in cell Ca^{2+}. *Nature (London)* **316**, 541–542.

Cuthbertson, K. S. R., Whittingham, D. G., and Cobbold, P. M. (1981). Free Ca^{2+} increases in exponential phases during mouse oocyte activation. *Nature (London)* **294**, 754–757.

Deits, T., and Shapiro, B. M. (1985). pH-induced hysteretic transitions of ovoperoxidase. *J. Biol. Chem.* **260**, 7882–7888.

Deits, T., and Shapiro, B. M. (1986). Conformational control of ovoperoxidase catalysis in the sea urchin fertilization membrane. *J. Biol. Chem.* **261**, 12159–12165.

Deits, T., Farrance, M., Kay, E., Turner, E., Weidman, P., and Shapiro, B. M. (1984). Purification and properties of ovoperoxidase, the enzyme responsible for hardening the fertilization membrane of the sea urchin egg. *J. Biol. Chem.* **259**, 13525–13533.

Derbes, M. (1847). Observations sur le mécanisme et les phenomenes qui accompagnent la formation de l'embryon chezl'oursin comestible. *Ann. Sci. Nat. Zool.* **8**, 80–98.

Dumont, J. N., and Brummett, A. K. (1985). Egg envelopes invertebrates. *In* "Developmental Biology: A Comprehensive Synthesis" (L. W. Browder, ed.), Vol. 1, pp. 235–288. Plenum Press, New York.

Dunbar, B. S. (1983). Morphological, biochemical, and immunochemical characterization of the mammalian zona pellucida. *In* "Mechanism and Control of Animal Fertilization (J. F. Hartmann, ed.), pp. 139–175. Academic Press, New York.

Dunbar, B. S., and Wolgemuth, D. J. (1984). Structure and function of mammalian zona pellucida, a unique extracellular matrix. *Mod. Cell Biol.* **3**, 77–111.

Dunbar, B. S., Wardrip, N. J., and Hedrick, J. L. (1980). Isolation, physiochemical properties, and macromolecular composition of zona pellucida from porcine oocytes. *Biochemistry* **19**, 356–365.

Dunbar, B. S., Liu, C., and Sammons, D. W. (1981). Identification of three major proteins of porcine and rabbit zona pellucida by high resolution two-dimensional gel electrophoresis: Comparison with serum, follicular fluid, and ovarian cell proteins. *Biol. Reprod.* **24**, 1111–1124.

Eisen, A., and Reynolds, G. T. (1985). Source and sinks for the calcium released during fertilization of a single sea urchin egg. *J. Cell Biol.* **100**, 1522–1527.

Epel, D. (1970). Methods for the removal of the vitelline layer of sea urchin eggs. II. Controlled exposure to trypsin to eliminate post-fertilization clumping of embryos. *Exp. Cell Res.* **61**, 69–70.

Epel, D., Weaver, A. M., Muchmore, A. V., and Schimke, R. T. (1969). β-1,3-glucanase of sea urchin eggs: Release from particals at fertilization. *Science* **163**, 294–296.

Epel, D., Weaver, A. M., and Mazia, D. (1970). Methods for the removal of the vitelline layer of sea urchin eggs. I. Use of dithiolthreitol (Clevland reagent). *Exp. Cell Res.* **61**, 64–68.

Florman, H. M., and Wassarman, P. M. (1985). O-linked oligosaccharides of mouse egg ZP3 account for its sperm receptor activity. *Cell (Cambridge, Mass)* **41**, 313–324.

Florman, H. M., Bechtol, K. B., and Wassarman, P. M. (1984). Enzymatic dissection of the functions of the mouse egg's receptor for sperm. *Dev. Biol.* **106**, 243–255.

Foerder, C. A., and Shapiro, B. M. (1977). Release of ovoperoxidase from sea urchin eggs hardens the fertilization membrane with tyrosine crosslinks. *Proc. Natl. Acad. Sci. U.S.A.* **74**, 4214–4218.

Foerder, C. A., Klebanoff, S. J., and Shapiro, B. M. (1978). Hydrogen peroxide production, chemiluminescence and the respiratory burst of fertilization: Interrelated events in early sea urchin development. *Proc. Natl. Acad. Sci. U.S.A.* **75**, 3183–3187.

Gerton, G. L., and Hedrick, J. L. (1986). The vitelline envelope to fertilization envelope conversion in eggs of *Xenopus laevis*. *Dev. Biol.* **116**, 1–7.

Gilkey, J. C. (1981). Mechanisms of fertilization in fishes. *Am. Zool.* **21**, 359–375.

Gilkey, J. C., Jaffe, L. F., Ridgway, E. B., and Reynolds, G. T. (1978). A free calcium wave traverses the activating egg of the medaka oryzias latipes. *J. Cell Biol.* **76**, 448–466.

Ginsburg, A. S. (1961). The block to polyspermy in sturgeon and trout with special reference to the role of cortical granules (alveoli). *J. Embryol. Exp. Morphol.* **9**, 173–190.

Ginsburg, A. S. (1972). Fertilization in fishes and the problem of polyspermy. *In* "Academy of Science of the USSR Institute of Developmental Biology (translated from Russian)" (T. A. Detlaf, ed.). Israel Programs for Scientific Translations, Jerusalem.

Greve, J. M., and Wassarman, P. M. (1985). Mouse egg extracellular coat is a matrix of interconnected filaments possessing a structural repeat. *J. Mol. Biol.* **181**, 253–264.

Greve, J. M., Salzman, G. S., Roller, R. J., and Wassarman, P. M. (1982). Biosynthesis of the major zona pellucida glycoproteins secreted by oocytes during mammalian oogenesis. *Cell (Cambridge, Mass.)* **31**, 749–753.

Greve, L. C., and Hedrick, J. L. (1978). An immunocytochemical localization of the cortical granule lectin in fertilized and unfertilized eggs of *Xenopus laevis*. *Gamete Res.* **1**, 13–18.

Greve, L. C., Prody, G. A., and Hedrick, J. L. (1985). N-acetyl-β-D-glucosaminidase activity in the cortical granules of *Xenopus laevis* eggs. *Gamete Res.* **12**, 305–312.

Grey, R. D., Wolf, D. P., and Hedrick, J. L. (1974). Formation and structure of the fertilization envelope in *Xenopus laevis*. *Dev. Biol.* **36**, 44–61.

Grey, R. D., Working, P. K., and Hedrick, J. L. (1976). Evidence that the fertilization envelope blocks sperm entry in eggs of *Xenopus laevis:* Interaction of sperm with isolated envelopes. *Dev. Biol.* **54**, 52–60.

Gulyas, B. J. (1980). Cortical granules of mammalian eggs. *Int. Rev. Cytol.* **63**, 357–392.

Gulyas, B. J., and Schmell, E. D. (1980). Ovoperoxidase activity in ionophore treated mouse eggs. I. Electron microscopic localization. *Gamete Res.* **3**, 267–277.

Guraya, S. S. (1983). Recent progress in the structure, composition and function of cortical granules in the animal egg. *Int. Rev. Cytol.* **78**, 257–360.

Gwatkin, R. B. L. (1977). "Fertilization Mechanisms in Man and Mammals." Plenum, New York.

Gwatkin, R. B. L., Williams, D. T., and Meyenhofer, M. (1979). Isolation of bovine zona pellucida from ovaries with collagenase: Antigenic and sperm receptor properties. *Gamete Res.* **2**, 187–192.

Gwatkin, R. B. L., Anderson, O. F., and Williams, D. T. (1980). Large scale isolation of bovine and pig zona pellucida: Chemical, immunological, and receptor properties. *Gamete Res.* **3,** 217–231.

Hall, H. B. (1978). Hardening of the sea urchin fertilization envelope by peroxidase-catalyzed phenolic coupling of tyrosines. *Cell (Cambridge, Mass.)* **15,** 343–355.

Hedrick, J. L., and Wardrip, N. J. (1986). Isolation of the zona pellucida and purification of its glycoprotein families from pig oocytes. *Anal. Biochem.* **157,** 63–70.

Inoue, M., and Wolf, D. P. (1975). Fertilization-associated changes in the murine zona pellucida: A time sequence study. *Biol. Reprod.* **11,** 558–565.

Ishida, J. (1936). An enzyme dissolving the fertilization membrane of sea urchin eggs. *Annot. Zool. Jpn.* **15,** 453–459.

Jaffe, L. F. (1983). Sources of calcium in egg activation: A review and hypothesis. *Dev. Biol.* **99,** 265–276.

Kapur, R. P., and Johnson, L. V. (1986). Selective sequestratioon of an oviductal fluid glycoprotein in the perivitelline space of mouse oocytes and embryos. *J. Exp. Zool.* **238,** 249–260.

Kay, E. S., and Shapiro, B. M. (1985). The formation of the fertilization membrane of the sea urchin egg. *Biol. Fertil.* **3,** 45–80.

Kay, E. S., and Shapiro, B. M. (1987). Ovoperoxidase localization effects dityrosine cross-linking of specific polypeptides after assembly of the sea urchin fertilization envelope. *Dev. Biol.* **121,** 325–334.

Kay, E. S., Eddy, E. M., and Shapiro, B. M. (1982). Assembly of the fertilization membrane of the sea urchin: Isolation of a divalent cation-dependent intermediate and its cross-linking *in vitro*. *Cell (Cambridge, Mass.)* **29,** 867–875.

Kidd, P. (1978). The jelly and vitelline coats of the sea urchin egg: New ultrastructural features. *J. Ultrastruct. Res.* **64,** 204–215.

Kudo, S., Sato, A., and Inoue, M. (1988). Chorionic peroxidase activity in the eggs of the fish *Tribolodon hakonensis*. *J. Exp. Zool.* **245,** 63–70.

Laale, H. W. (1980). The perivitelline space and egg envelopes of bony fishes: A review. *Copeia* **2,** 210–226.

Loeb, J. (1905). Artificial membrane formation and chemical fertilization in a starfish. Univ. of Calif. Publ., Berkeley Publ. Physiol. II, **2,** 154.

Loeb, J. (1913). "Artificial Parthenogenesis and Fertilization," p. 218. Univ. of Chicago Press, Chicago, Illinois.

Lonning, S., Kjorsvik, E., and Davenport, J. (1984). The hardening process of the egg chorion of the cod, *gadus morhualo* and lumpsucker, *cyclopterus lumpus* L. *J. Fish Biol.* **24,** 505–522.

Majack, R. A., and Bornstein, P. (1987). Thrombospondin: A multifunctional platelet and extracellular matrix glycoprotein. *In* "Cell Membranes: Methods and Reviews" (E. Elson, W. Frazier, and L. Glaser, eds.), Vol. 3, pp. 55–78. Plenum, New York.

Miyazaki, S., Hashimoto, N., Yoshimoto, Y., Kishimoto, T., Igusa, V., and Hiramoto, Y. (1986). Temporal and spatial dynamics of the periodic increase in intracellular free calcium at fertilization of golden hamster eggs. *Dev. Biol.* **118,** 259–267.

Monk, B. C., and Hedrick, J. L. (1986). The cortical reaction in *Xenopus laevis* eggs: Cortical granule lectin release as determined by radioimmunoassay. *Zool. Sci.* **3,** 459–466.

Nakano, E. (1956). Changes in the egg membrane of the fish egg during fertilization. *Embryologia* **3,** 89–103.

Nakano, E. (1969). Fishes. *In* "Fertilization" (C. B. Metz and A. Monroy, eds.), Vol. 2 pp. 295–324. Academic Press, New York.

Nishihara, T., Gerton, G. L., and Hedrick, J. J. (1983). Radioiodination studies of the envelopes from *Xenopus laevis* eggs. *J. Cell. Biochem.* **22,** 235–244.

Ohtsuka, E. (1957). On the hardening of the chorion in the fish egg after fertilization. I. Role of the cortical substance in chorion hardening of the egg of *orzyias latipes. Sibeboldia Acta Biol.* **2**, 19–29.

Ohtsuka, E. (1960). On the hardening of the chorion in the fish egg after fertilization. III. The mechanism of chorion hardening in oryzias latipes. *Biol. Bull. (Woods Hole, Mass.)* **118**, 120–128.

Ohtsuka (1964). Studies on the invisible cortical change of the fish egg. *Embryologia* **2**, 101–114.

Okada, A., Yanagimachi, R., and Yanagimachi, H. (1986). Development of a cortical granule-free area of cortex and the perivitelline space in the hamster oocyte during maturation and following ovulation. *J. Submicrosc. Cytol.* **18**, 233–247.

Perona, R. M., and Wassarman, P. M. (1986). Mouse blastocysts hatch *in vitro* by using a trypsin like proteinase associated with cells of mural trophectoderm. *Dev. Biol.* **114**, 42–52.

Phillips, D. M., and Shalgi, R. (1980). Surface architecture of the mouse and hamster zona pellucida and oocyte. *J. Ultrastruct. Res.* **72**, 1–12.

Pikó, L. (1969). Gamete structure and sperm entry in mammals. *In* "Fertilization" (C. B. Metz and A. Monroy, eds.), Vol. 2, pp. 325–404. Academic Press, New York.

Prody, G. A., Greve, L. C., and Hedrick, J. L. (1985). Purification and characterization of an N-acetyl-β-D-glucosaminidase from cortical granules of *Xenopus laevis* eggs. *J. Exp. Zool.* **235**, 336–340.

Ridgeway, E. B., Gilkey, J. C., and Jaffe, L. F. (1977). Free calcium increases explosively in activating medaka eggs. *Proc. Natl. Acad. Sci. U.S.A.* **74**, 623–627.

Roller, R. J., and Wassarman, P. M. (1983). Role of asparagne linked oligosaccharides in secretion of glycoproteins of the mouse eggs extracellular coat. *J. Biol. Chem.* **258**, 13243–13249.

Rossignol, D. P., Earles, B. J., Decker, G. L., and Lennarz, W. J. (1984). Characterization of the sperm receptor on the surface of eggs of *Strongylocentrotus purpuratus Dev. Biol.* **104**, 308–321.

Sakai, Y. T. (1961). Method for removal of chorian and fertilization of the naked egg in oryzias latipes. *Embryologia* **5**, 357–368.

Schmell, E. D., and Gulyas, B. J. (1980). Ovoperoxidase activity in ionophore treated mouse eggs. II. Evidence for the enzymes role in the hardening of the zona pellucida. *Gamete Res.* **3**, 279–290.

Schmell, E. D., Gulyas, B. J., and Hedrick, J. L. (1983). Egg surface changes during fertilization and the molecular mechanism of the block to polyspermy. *In* "Mechanism and Control of Animal Fertilization" (J. F. Hartmann, ed.), pp. 365–414. Academic Press, New York.

Shapiro, B. M. (1975). Limited proteolysis of some egg surface components is an early event following fertilization of the sea urchin, *Strongylocentrotus purpuratus. Dev. Biol.* **46**, 88–102.

Silverstein, R. L., Leung, L. K., and Nachman, R. L. (1986). Thrombospondin: A versatile multifunctional platelet glycoprotein. *Arteriosclerosis* **6**, 245–253.

Steinhardt, R. A., Epel, D., Carroll, E. J., and Yanagimachi, R. (1974). Is calcium ionophore a universal activator for unfertilized eggs? *Nature (London)* **252**, 41–43.

Stephens, R. E., and Kane, R. E. (1970). Some properties of hyalin, the calcium insoluble protein of the hyaline layer of sea urchin eggs. *J. Cell Biol.* **44**, 611–617.

Truschel, M. R., Chambers, S. A., and McClay, D. R. (1986). Two antigenically distinct forms of β-1, 3,-glucanase in sea urchin embryonic development. *Dev. Biol.* **117**, 277–285.

Turner, E., Somers, C. E., and Shapiro, B. M. (1985). The relationship between a novel NAD(P)H oxidase activity of ovoperoxidase and the CN^- resistant respiratory burst that follows fertilization of sea urchin eggs. *J. Biol. Chem.* **260**, 13163–13171.

Turner, E., Klevit, R., Hopkins, P. B., and Shapiro, B. M. (1986). Ovothiol: A novel thiohistidine compound from sea urchin eggs that confers NAD(P)H-O_2 oxidoreductase activity on ovoperoxidase. *J. Biol. Chem.* **261**, 13056–13063.

Turner, E., Klevit, R., Hager, L. J., and Shapiro, B. M. (1987). The ovothiols: A family of redox-active thiohistidine compounds from marine invertebrate eggs. *Biochemistry* **26**, 4028–4036.

Urch, U. A., and Hedrick, J. L. (1981a). The hatching enzyme from *Xenopus laevis:* Limited proteolysis of the fertilization envelope. *J. Supramol. Struct. Cell. Biochem.* **15**, 111–117.

Urch, U. A., and Hedrick, J. L. (1981b). Isolation and characterization of the hatching enzyme from the amphibian, *Xenopus laevis. Arch. Biochem. Biophys.* **206**, 424–431.

Vacquier, V. D., Tegner, M. J., and Epel, D. (1973). Protease released from sea urchin eggs at fertilization alters the vitelline layer and aids in preventing polyspermy. *Exp. Cell Res. 80*, 111–119.

Veron, M., Foerder, C., Eddy, E. M., and Shapiro, B. M. (1977). Sequential biochemical and morphological events during assembly of the fertilization membrane of the sea urchin. *Cell (Cambridge, Mass.)* **10**, 321–328.

Wassarman, P. M., Greve, J. M., Perona, R. M., Roller, R. J., and Salzmann, G. S. (1984). How mouse eggs put on and take off their extracellular coat. *In* "Molecular Biology of Development" (K. Yamada, ed.), pp. 213–225. Alan R. Liss, New York.

Wassarman, P. M., Bleil, J. D., Florman, H. M., Greve, J. M., Roller, R. J., Salzmann, G. S., and Samuels, F. G. (1985a). The mouse egg's receptor for sperm: What is it and how does it work? *Cold Spring Harbor Symp. Quant. Biol.* **50**, 11–20.

Wassarman, P. M., Florman, H. M., and Greve, J. M. (1985b). Receptor mediated sperm–egg interactions in mammals. *In* "Biology of Fertilization" (C. B. Metz and A. Monroy, eds.), Vol. 2 pp. 341–360. Academic Press, New York.

Weidman, P. J., and Kay, E. S. (1986). Egg and embryonic extracellular coats: Isolation and purification. *Methods Cell Biol.* **27**, 111–138.

Weidman, P. J., and Shapiro, B. M. (1986). Regulation of extracellular matrix assembly: *In vitro* reconstitution of a partial fertilization envelope from isolated components. *J. Cell Biol.* **105**, 561–567.

Weidman, P. J., Kay, E. S., and Shapiro, B. M. (1985). Assembly of the sea urchin fertilization membrane: Isolation of proteoliaisin, a calcium-dependent ovoperoxidase binding protein. *J. Cell Biol.* **100**, 938–946.

Weidman, P. J., Teller, D. C., and Shapiro, B. M. (1987). Purification and characterization of proteoliaisin, a coordinating protein in fertilization envelope assembly. *J. Biol. Chem.* **262**, 15076–15084.

Whitaker, M. J., and Steinhardt, R. A. (1985). Ionic signaling in the sea urchin egg at fertilization. *Biol. Fertil.* **3**, 168–222.

Wolf, D. P. (1974a). The cortical granule reactions in living eggs of the toad, *Xenopus laevis. Dev. Biol.* **36**, 62–71.

Wolf, D. P. (1974b). On the contents of the cortical granules from *Xenopus laevis* eggs. *Dev. Biol.* **38**, 14–29.

Wolf, D. P., Nishihara, T., West, D. M., Wyrick, R. E., and Hedrick, J. L. (1976). Isolation, physicochemical properties, and the macromolecular composition of the vitelline and fertilization envelopes from *Xenopus laevis* eggs. *Biochemistry* **15**, 3671–3678.

Wyrick, R. E., Mishihara, T., and Hedrick, J. L. (1974). Agglutination of jelly coat and cortical granule components and the block to polyspermy in the amphibian *Xenopus laevis*. *Proc. Nat. Accad. Sci. U.S.A.* **71**, 2067–2071.

Yamagami, K. (1981). Mechanisms of hatching in fish: Secretion of hatching enzyme and enzymatic choriolysis. *Am. Zool.* **21**, 459–471.

Yamamoto, M., and Yamagami, K. (1975). Electron microscopic studies on choriolysis by the hatching enzyme of the teleost oryzias latipes. *Dev. Biol.* **43**, 313–321.

Yamamoto, T. S. (1955). The physiology of fertilization in the medaka (oryzias latipes). *Exp. Cell Res.* **10**, 387–393.

Yamamoto, T. S. (1961). Physiology of fertilization in fish eggs. *Int. Rev. Cytol.* **12**, 361–405.

Yamamoto, T. S. (1962). The mechanism of breakdown of cortical alveoli during fertilization in the medaka, oryzias latipes. *Embryologia* **7**, 228–251.

Yurewicz, E. C., Oliphant, G., and Hedrick, J. L. (1975). The macromolecular composition of *Xenopus laevis* egg jelly coat. *Biochemistry* **14**, 3101–3107.

Zotin, A. I. (1958). The mechanism of hardening of the salmonid egg membrane after fertilization or spontaneous activation. *J. Embryol. Exp. Morphol.* **6**, 546–568.

11

Dispermic Human Fertilization: Violation of Expected Cell Behavior

ISMAIL KOLA AND ALAN TROUNSON

Centre for Early Human Development
Monash Medical Centre
Monash University
Clayton, Victoria, Australia 3168

I. THE FERTILIZATION PROCESS IN HUMANS

After the penetration of the zona pellucida by a spermatozoon, fusion with the oolemma occurs. The initial reaction of the human oocyte to fusion with the sperm head is a rapid and progressive release of the cortical granule contents from under the oolemma (Sathananthan and Trounson, 1982). This appears to progress from the site of sperm fusion and may be accelerated by the rapid efflux of calcium ions from the oocyte. Rapid movement of molecules between closely apposed membranes such as those surrounding the cortical granules and the plasma membrane result in their fusion, releasing cortical contents into the perivitelline space. Cortical granules may fuse with one another at the same time. The cortical contents contain a zona-hardening substance which diffuses into the inner zona causing the zona reaction. Alteration

277

THE CELL BIOLOGY OF
FERTILIZATION

occurs to the chemical structure of the inner zona which becomes more electron dense, traps any sperm in the vicinity, and prevents them from penetrating through the zona pellucida. This is the mechanism which normally blocks polyspermy, and any substantial delay of the cortical reaction or a break in the zona will usually result in fertilization by more than one sperm.

Sperm fusion with the oolemma also initiates the completion of meiosis in the oocyte. The daughter chromatids held in the second metaphase stage are separated by the microtubules of the anaphase spindle, and one complete set of chromosomes is expelled from the oocyte in the extrusion of the second polar body (Sathananthan *et al.*, 1986a). Nuclear membranes begin to form around the set of chromosomes retained by the oocyte in close proximity to the second polar body. The sperm head decondenses in the ooplasm and detaches from the flagellum. Specific cytoplasmic factors (proteins) are required for this, which are generated during oocyte maturation (Thibault, 1977). A nuclear membrane is also formed around the sperm chromosomes.

Decondensing sperm heads can be located in the ooplasm within 3 hr of inseminating human oocytes *in vitro* (McMaster *et al.*, 1978; Sathananthan *et al.* 1986a,b). Pronuclei are visible by 6 hr and move from peripheral locations to the center of the oocyte by 12 hr after insemination. The male and female pronuclei in the human are about equal in size and remain separate but very close together until 20–24 hr after insemination at which time the nuclear membrane disappears, and a bipolar spindle forms with apices at opposite poles of the pronuclei. The condensing chromosomes are mixed on the metaphase plate of the first cleavage division at syngamy. The pronuclei are no longer visible during syngamy and the time between disappearance of the pronuclei and the first cleavage division is 3–6 hr.

II. DISPERMIC FERTILIZATION IN ANIMALS AND HUMANS

In certain instances a delay in the cortical granule reaction or a breach in the zona pellucida allows more than one spermatozoon to penetrate the oocyte and gives rise to a polyspermic zygote. Dispermic fertilization is the most common form of polyspermia. The penetration of more than one spermatozoon and the formation of more than one male pronucleus is a normal occurrence in some animal eggs, notably those laden with yolk. Only one of the male pronuclei undergoes syngamy with the female pronucleus, and the other apparently offer no interference to normal development (Austin and Walton, 1960). This condition is referred to as physiological polyspermy and has been recorded in many insects, the polyzoa, some mollusks, the earthworm, the lamprey, urodeles, reptiles, and birds (see Austin and Walton, 1960). In other species, the presence of supernumerary pronuclei precludes normal development, and this condition has been referred to as pathological polyspermy. Polyspermy in mammals belongs to the latter class.

The subdivision of polyspermy into two categories reflects the two mechanisms that exist in the animal egg in response to such fertilization. The first mechanism, Rothschild's (1954) type I inhibition, is the block to polyspermy in the vitelline surface or the zona reaction in mammalian eggs. The second mechanism, type II inhibition, prevents the approach and syngamy of more than one male pronucleus with the female pronucleus. In animals where pathological polyspermy occurs the type I inhibition appears to be operational, whereas in animals with physiological polyspermy type II inhibition is operational (Austin and Walton, 1960). Neither mechanism is fully effective in preventing polyspermy.

The processes involved in physiological polyspermy are well exemplified in the urodele, *Triturus helveticus*. All the male pronuclei are maintained in a normal fashion until one of them comes into contact with the female pronucleus fusing with it. The supernumerary male pronuclei then degenerate, beginning first in the nearer pronuclei, and later involving the others (Fankhauser, 1948). In another urodele, *Diemictylus viridescens*, all the supernumerary pronuclei degenerate simultaneously. Austin and Walton (1960) have suggested that in *T. helveticus*, an inhibitory substance diffuses from the fusion nucleus, or its immediate cytoplasmic surroundings, and that this substance progressively suppresses the supernumerary male pronuclei.

The sea urchin egg has been extensively studied and is a good model of pathological polyspermy. Here, several, but not necessarily all, male pronuclei enter syngamy with the female pronucleus. The male pronuclei may also fuse with each other at syngamy. In the frog egg, type I inhibition evidently operates and only one male pronucleus usually fuses with the female pronucleus (Austin and Walton, 1960).

We (Kola *et al.*, 1987) and others (Sathananthan and Trounson, 1985; Angell *et al.*, 1986) have demonstrated that both sperm in dispermic human oocytes decondense and form individual pronuclei. Consequently, a tripronuclear human zygote results (Fig. 1). These pronuclei do migrate toward one another and come together at the center of the oocyte before syngamy. We have also shown (Kola *et al.*, 1987) that such tripronuclear zygotes do have a triploid complement of chromosomes before the first cleavage division. It is thus clear that human oocytes employ Rothschild's type I inhibition, and that polyspermy is pathological.

A. Incidence of Dispermic Fertilization

Although the incidence of dispermic human fertilization *in vivo* is unknown, 1% of all clinically recognized pregnancies *in vivo* are triploid. The majority of these triploid conceptuses are derived from dispermic fertilization (Jacobs

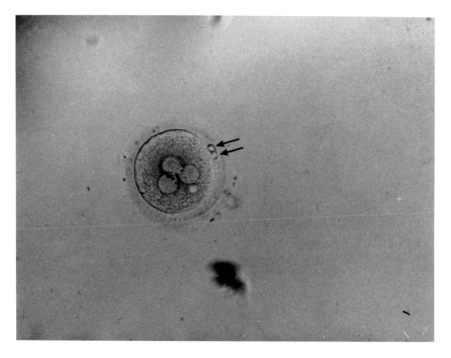

Fig. 1. Photomicrograph of a tripronuclear human zygote ($\times$ 176). The photomicrograph clearly shows three pronuclei that have migrated to the center of the oocyte. The first and second polar bodies (arrows) are also seen, demonstrating that the tripronuclear state of the oocyte is due to dispermic fertilization. (Reproduced with permission from *Biology of Reproduction*.)

et al., 1978). On the other hand, the incidence of triploid zygotes in *in vitro* fertilized human oocytes is approximately 5%. In excess of 80% of these triploid zygotes are due to dispermic fertilization, the remainder being digynic in origin (Trounson *et al.*, 1988). A question which could be posed is, is the incidence of dispermic fertilization (and triploidy) higher in *in vitro* fertilized human oocytes when compared to those fertilized *in vivo*? This question will be examined later (Section II,C,2).

B. The First Cleavage Division of Dispermic Human Zygotes

We (Kola *et al.*, 1987) and others (Rudak *et al.*, 1984) have studied the tripronuclear human zygotes just before the first cleavage division and have demonstrated that these zygotes are karyotypically triploid. Caution should be exercised, however, in that some monospermic fertilized eggs can have a cytoplasmic vacuole present which may have a gross morphological appearance and dimension of a pronucleus. This has been referred to as a pseudopronucleus (Van Blerkom et al., 1987). These vacuolated oocytes can be distinguished from true dispermic eggs by the absence of nucleoli in the pseudopronucleus.

Although it has been demonstrated that dispermic zygotes are karyotypically triploid, there have been reports in the literature (Rudak et al., 1984; Angell et al., 1986) of cleavage-stage dispermic embryos being karyotypically either diploid or triploid, or having severely abnormal chromosome compositions. We have focused on the pattern of the first cleavage division in dispermic zygotes in order to clarify the origins of the variation in chromosome numbers. Of 29 oocytes studied, we found (Kola et al., 1987) that 18 cleaved directly and synchronously from 1- to 3-cells, (Figs. 2 and 3), 7 cleaved to morphologically normal 2-cell embryos (Fig. 4), and 4 cleaved to 2-cells plus a variable size extrusion (Fig. 5).

1. Cleavage to 3-Cells

The finding of dispermic zygotes cleaving simultaneously from 1- to 3-cells was confirmed by us using time-lapse cinematography. The regularity of the three blastomeres in such embryos varied. Some embryos had regular-sized 3-cells (Fig. 2), but others had three irregular-sized blastomeres with one large and two smaller-size blastomeres (Fig. 3). The cleavage of 1- to 3-cells is a relatively rare phenomenon in mammalian cells, and we have only found one other such report. Austin and Braden (1953) studied dispermic rat eggs prior to the first cleavage division and showed that all the chromosomes from the haploid sets in dispermic rat eggs participated in the first cleavage division. Although most of the first cleavage division spindles were observed to be bipolar, a tripolar spindle was observed on one occasion (Austin and Walton, 1960).

In sea urchins, however, two or more sperm nuclei conjugate with the egg nucleus to form a triploid or polyploid fusion-nucleus. Such eggs usually produce a quadripolar or tripolar spindle and divide directly and synchronously to 3-cells (Wilson, 1928). O. Hertwig and F. Fol first reported (see Wilson, 1928) that sea urchin eggs fertilized by two spermatozoa result in three or four-cornered spindles being formed and multipolar cleavage occurring. The embryo consequently divides directly to a 3-or 4-cell embryo. After the initial multipolar division, cleavage proceeds by regular bipolar divisions and a morphologically normal looking blastula is produced. In a small percentage of cases these blastulae produce normal larvae. The majority produce a range of abnormal or "monstrous" larvae (Wilson, 1928) and death ensues.

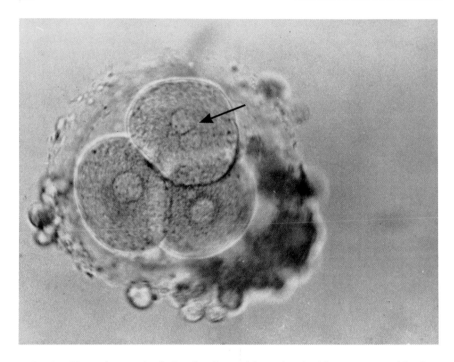

Fig. 2. Photomicrograph of a 3-cell embryo with regular-size blastomeres resulting from the direct cleavage of a tripronuclear zygote ($\times 348$). One blastomere (arrow) is binucleated.

The four asters in dispermic sea urchin eggs arise by the formation of two centers in connection with each spermatozoon. Boveri (see Wilson, 1928) found that triasters are readily produced by shaking the dispermic eggs, which often prevents the division of one center. In our study on dispermic human oocytes, we did not find any oocytes cleaving to 4-cells. It is possible that, because of the constant movement of the dispermic oocytes for microscopic evaluation, the division of one of the centers was prevented, and consequently no 4-cell embryo was found after the first cleavage division. However, it is clear that dispermic human oocytes, like sea urchin oocytes but unlike rat oocytes, result in a multiple cleavage after the first division. This is contrary to the suggestion of Austin and Walton (1960), who proposed that the unique feature of polyspermy in mammalian eggs (inferred chiefly from observations on rat eggs) is that although polyandrous syngamy occurs, it does not usually result in the formation of a multipolar spindle. They further suggested, on the basis that polyspermy did not result in multiple cleavage, that the sperm aster is of less importance in cleavage in mammals than in nonmammals (Austin and Braden,

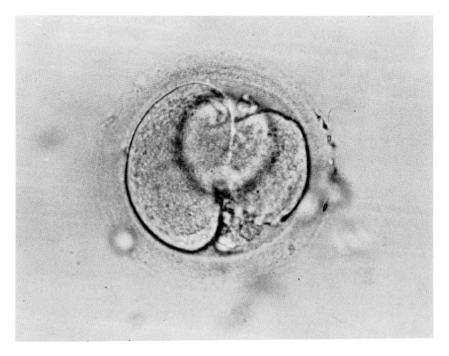

Fig. 3. Photomicrograph of a 3-cell embryo (one large, two smaller blastomeres) derived from the direct cleavage of a tripronuclear human zygote ($\times 348$).

1953; Austin and Walton, 1960). Our data on dispermic human eggs violates this expectation and suggests that the sperm aster and/or centrosome may play an important role in the organization of the spindle at the first cleavage division. Schatten *et al.* (1986) have recently demonstrated that the mouse spermatozoon lacks centrosomes, and that centrosomes in the mouse are maternally inherited. Further studies on the inheritance of centrosomes (i.e., whether their inheritance is maternal or paternal) are certainly indicated in humans.

In the frog egg, only one male pronucleus fuses with the female pronucleus, but the asters that form in conjunction with the supernumerary male pronuclei cause a multiple first cleavage division. A progressive cleavage may occur, each nucleus (including the fusion nucleus) with its attendant astral system, giving rise to a separate bipolar mitotic figure. The cleavage thus produced is more or less regular, according to the number of spermatozoa that have entered the egg, and many of the resulting blastomeres are binucleate. It is interesting that one of the dispermic human eggs which cleaved to a 3-cell embryo with regular blastomeres had one binucleate blastomere (Fig. 2). It is thus possible that this is the result of a tetraster egg which cleaved to 3- instead of 4 cells, with one blastomere thus being binucleate. Obviously this suggestion is speculative and the aetiology or significance of the binucleate blastomere has not been firmly elucidated.

2. Cleavage to 2-Cells

The cleavage of dispermic eggs to regular-looking, 2-cell embryos (Fig. 4) could have been expected to occur on the basis of previous observations on lower mammals (Austin and Braden, 1953). These workers demonstrated that the extra male pronucleus in dispermic rat eggs come into contact with the female pronucleus and all contribute to the first cleavage spindle. They found all but one of the eggs formed normal bipolar spindles and demonstrated that these eggs cleave to normally appearing 2-cell embryos. They also demon-

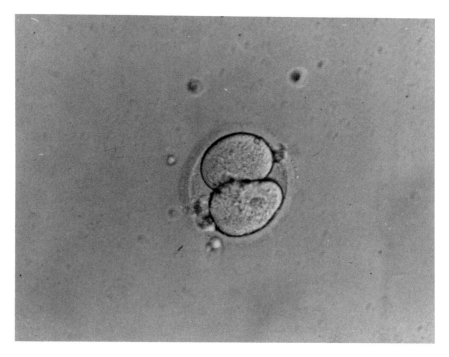

Fig. 4. Photomicrograph of a morphologically normal 2-cell embryo derived from the direct cleavage of a tripronuclear human zygote ($\times$ 174). (Reproduced with permission from *Biology of Reproduction.*)

strated that dispermic rat and mouse eggs did not have nuclei which were significantly larger than those of normal (monospermic) eggs (Austin and Braden, 1955). They concluded, on the basis of the above evidence, that the dispermic rat egg develops into a triploid embryo. The finding that *only* 24% of dispermic human eggs cleaved to 2-cells was thus unexpected.

3. Cleavage to 2-Cells plus an Extrusion

Four of twenty-nine dispermic oocytes cleaved to 2-cells plus an extrusion (Fig. 5). It appears from the karyotype data (discussed later) that the extrusion represents an entire haploid set of chromosomes. Whether the extrusion represents a maternal or paternal genomic complement cannot be elucidated on the basis of current data. We have been unable to find reports in other animal species of the two sperm pronuclei and the egg pronucleus coming together

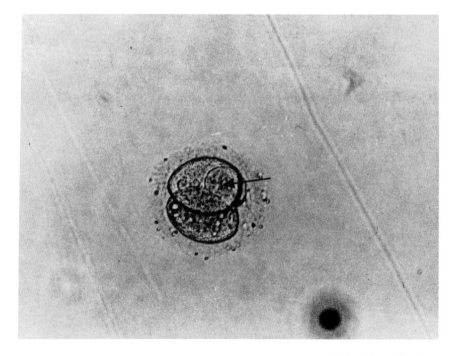

Fig. 5. Photomicrograph of a 2-cell-plus-extrusion embryo, derived from the direct cleavage of a tripronuclear human zygote (× 174). The two blastomeres are clearly visible. The extrusion is indicated by an arrow. (Reproduced with permission from *Biology of Reproduction.*)

and then one entire haploid set of chromosomes not participating in mitotis. What has been reported, however, is that in dispermic sea urchin eggs, on occasions, one of the sperm may remain separate in the protoplasm and does not conjugate with the egg pronucleus. Furthermore, in the hen's egg, only one sperm pronucleus conjugates with the egg nucleus, and the supernumerary ones distribute themselves near the margins of the blastodisc where they remain during the later phases of fertilization and the initial cleavage stages. They undergo a certain amount of division, forming small groups of daughter nuclei, and these divisions are frequently accompanied by partial cleavage of the surrounding cytoplasm on the margin of the blastodisc (Olsen, 1942). At a later period, when the egg has reached the 32-cell stage, they have usually undergone degeneration and have disappeared. However, one important distinguishing feature in the dispermic human egg is that all the pronuclei conjugate, and thereafter an entire set of chromosomes is extruded. This third situation, of dispermic human eggs cleaving to 2-cells plus an extrusion, is a further example of the violation of expected cell behavior by these dispermic human oocytes.

C. Chromosomal Composition of Dispermic Eggs before the Second Cleavage Division

Although dispermic fertilization occurs at a frequency of some 5% in human *in vitro* fertilization, very little information exists about the subsequent chromosomal composition of embryos derived from such oocytes, and the information that does exist suggests a complex picture. As discussed in Section II,B, it has been demonstrated that dispermic eggs are karyotypically triploid before the first cleavage division. After the first cleavage division, however, reports of various chromosomal compositions of embryos derived from dispermic oocytes exist. Van Blerkom et al. (1984) reported that a tripronuclear zygote had developed to the 12-cell stage and that on the basis of DNA microfluorimetry, one of the blastomeres was diploid. Rudak et al. (1984), on the other hand, karyotyped a cleavage-stage embryo which was derived from a tripronuclear zygote and found one cell in mitosis which consisted of 33 chromosomes. Angell et al. (1986) karyotyped 5 such embryos and found one triploid, one diploid, and three with severely abnormal chromosomal constitutions. Two more embryos were found to have an approximately diploid DNA content on the basis of microfluorimetric measurements (Angell et al., 1986).

We (Kola et al., 1987) karyotyped embryos that had cleaved to 3-cells, 2-cells plus an extrusion, and 2-cells just before the second cleavage division. The following results were obtained.

1. Karyotype of Embryos That Cleave to 3 Cells

Of 18 3-cell embryos, 13 were karyotyped, and 1 or more mitotic cells were obtained in 10 of these embryos. None of the cells had an exact triploid or a diploid karyotype. They all had chromosomal compositions of between 34 and 57 chromosomes/cell. Furthermore, in those embryos that had more than one cell in mitosis, the numbers of chromosomes in each cell varied. Two embryos had three cells in mitosis, and the total number of chromosomes in these 3-cells was 138, thus indicating that the 69 chromosomes/cell at the zygote stage had replicated itself.

Interestingly, the karyotype of these 3-cell embryos further parallels that of the sea urchin. As discussed earlier in this chapter (Section II,B,1), in dispermic sea urchin eggs, both sperm pronuclei conjugate with the egg pronucleus producing a tripolar or quadripolar spindle and cleave at once into 3- or 4-cells. In these cases the chromosomes are randomly distributed and almost always irregularly to the three or four poles. The resulting nuclei, therefore, receive varying number or combinations of chromosomes. The results are plainly apparent to the eye in the cells of the resulting larvae, which always show conspicuous variations in the size of the nuclei (Wilson, 1928). After the initial tripolar or quadripolar division, the dispermic eggs continue to segment by bipolar mitosis, and often produce free-swimming larvae; but in the great majority of cases, development sooner or later becomes abnormal and death ensues. Interestingly, Boveri (1902, 1907) demonstrated that the pathological effect is not due to the abnormal chromosome numbers but to their false combinations and, thus as far back as 1902, proposed that the individual chromosomes must be qualitatively different.

Our data (Kola et al., 1987) indicates that the dispermic human egg forms a tripolar spindle and thus cleaves to 3 cells (Fig. 8). Furthermore, the karyotype data discussed above suggests that chromosome segregation on such a tripolar spindle is random and irregular as in the sea urchin. The question as to how far such embryos will continue to develop has not yet been elucidated, but it is reasonable to assume that, as in the sea urchin embryo, the developmental potential of most such embryos is restricted. It is possible, however, that by chance the normal chromosome combination can be achieved in the 3-cells of one embryo. In dispermic sea urchin eggs of 719 triaster eggs, 58 perfect plutei were obtained as compared to 1 (or possibly 2) of 1500 tetraster eggs (see Wilson, 1928). It is claimed that these figures approximate fairly well to the theoretic expectation according to the theory of probabilities (Wilson, 1928). Obviously, the probability of the correct combination resulting from the random segregation of chromosomes on a tripolar spindle in humans is much lower than that of the sea urchin, because of the higher number of chromosomes in the human.

2. Karyotype of Embryos That Cleave to 2-Cells

Out of seven embryos that cleaved to 2-cells, a karyotype was attempted on six of these. Four of the six embryos had 1 or more cells in mitosis. All four embryos were chromosomally triploid (Fig. 6). Evidence presented by Beatty (1957) and Austin and Braden (1953) demonstrated that dispermic fertilization in other mammals results in triploidy. Beatty (1957) has suggested that triploid embryos are most unlikely to survive to birth.

In humans the vast majority of triploid conceptuses are spontaneously aborted. However, Uchida and Freeman (1985) have reported the live birth of seven triploid infants, most of whom died in the neonatal period. Borgaonkar (1984) also reported on five live-born triploid infants who survived from 2 to 7 months. Recently a triploid child, who survived for 10½ months, was documented (Sherard et al., 1986). The child had multiple congenital abnormalities, and it has been claimed to be the longest survival of a triploid infant.

Our data thus demonstrates that only 24% of triploid zygotes are karyotypically triploid after the first cleavage division (Kola et al., 1987). This suggests that the incidence of triploidy (and dispermic fertilization) may be similar in oocytes fertilized in vitro or in vivo. The basis for this suggestion is that after the first cleavage division, only an approximate one fourth of the original 5% (equal to 1.25%) of tripronuclear zygotes in in vitro fertilized oocytes remain triploid, compared to 1% of all in vivo clinically recognized conceptions.

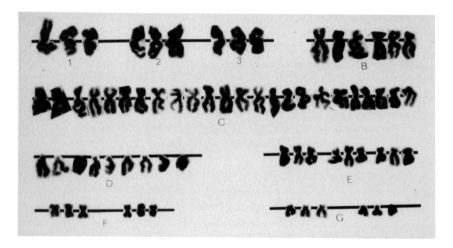

Fig. 6. Karyotype of a cell from a tripronuclear human zygote that had cleaved to a morphologically normal 2-cell embryo (×870). The karyotype demonstrates that the cell is triploid (69XXX). (Reproduced with permission from *Biology of Reproduction*.)

Our data also suggests that in this second case, in which the oocytes cleave to morphologically normal-looking, 2-cell embryos which are karyotypically triploid, a bipolar spindle is formed, and all the chromosomes gather on the metaphase plate (Fig. 8). This situation is what appears to be operating in other mammals (Austin and Walton, 1960).

3. Karyotype of Embryos That Cleave to 2-Cells plus an Extrusion

A karyotype was obtained in all four embryos that cleaved to 2-cells plus an extrusion. Three of the four embryos had both cells in metaphase, and each of the 2-cells had a diploid karyotype (Fig. 7). In the remaining embryo, 2 interphase nuclei and 23 chromosomes were obtained. This data suggests that in this third case, where embryos cleave to 2-cells plus an extrusion, a bipolar spindle is formed and only a diploid set of chromosomes gather on the metaphase plate (Fig. 8). The reasons why an entire haploid set of chromosomes is excluded from the metaphase plate at the first cleavage division is unclear. One possibility is that the chromosomes of one pronucleus have a different degree of condensation from the other two, and they consequently do not gather on the metaphase plate.

Murray et al. (1985) found four sheep embryos that were 1N–2N mosaics.

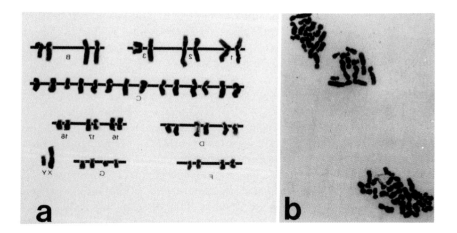

Fig. 7. (a) Karyotype and (b) metaphase spread of a cell from a 2-cell-plus-extrusion embryo (×870). These photomicrographs demonstrate that the cell has a diploid (46XY) karyotype. (Reproduced with permission from *Biology of Reproduction*.)

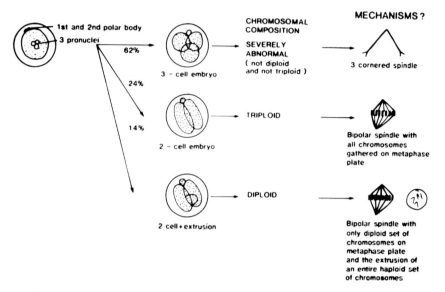

Fig. 8. Diagram demonstrating the cleavage patterns and karyotypic analysis, chromosomal composition, and possible mechanisms of tripronuclear human zygote development.

One of the embryos contained one haploid and two diploid sets. Murray *et al.* (1985) ascribed these haploid–diploid mosaics to polyspermic fertilization. Long and Williams (1980) also obtained a 1N–2N mosaic in their studies on sheep embryos and similarly ascribed this to polyspermic fertilization. These two situations are analogous to the human where the 2-cells are karyotypically diploid and the extrusion appears to be haploid; thus giving rise to a haploid–diploid mosaic. It may thus be possible that dispermic sheep oocytes behave like dispermic human oocytes.

The fate of the extrusion is unclear from our studies. It is unknown whether the extrusion degenerates later during embryogenesis, or even whether it fuses with one of the blastomeres. Further studies which investigate the development of these embryos later in preimplantation embryogenesis is warranted.

It is tempting to speculate that if the maternal set of chromosomes were extruded from the metaphase plate, and that if the division only involved two paternal sets of chromosomes, that this may be a mechanism in which the hydatidiform mole is derived. This may be an etological mechanism for the 46XY hydatidiform mole. However, the finding that 46XX hydatidiform moles are homozygous at all loci would militate against the suggestion that such a mechanism operates in the etiology of the 46XX (entirely homozygous) hydatidiform mole.

III. FUTURE RESEARCH DIRECTIONS

Future research should be directed at an elucidation of the importance of the centrosomes in human fertilization. Similar studies have recently been performed in the mouse (G. Schatten *et al.*, 1985; H. Schatten *et al.*, 1986). Fertilization in the sea urchin is dependent on the paternal contribution of centrosomes. In the mouse, the centrosomes are maternally inherited (Schatten *et al.*, 1986). The data discussed in this chapter, i.e., the multiple cleavage of dispermic human oocytes, suggests that the paternal centrosomes may be important in organizing the spindle in the human.

Although we have proposed certain mechanisms that operate at the first cleavage division of dispermic human zygotes, further research is required to prove or disprove whether or not these mechanisms are operational. The question as to why three different mechanisms operate also needs to be addressed.

IV. CONCLUSIONS

It is clear from the evidence produced in this chapter that the majority of dispermic human oocytes fail to develop into triploid embryos. The oocytes that fail to develop into triploid embryos either cleave directly to 3-cells or to 2-cells plus an extrusion. The multiple cleavage of polyspermic oocytes violates the expected cellular behavior of such oocytes on the basis of data on lower mammals, such as rats, and is suggestive of a role for paternal centrosomes in the organization of the first mitotic spindle. In this regard the dispermic human zygote behaves in a similar manner to the dispermic sea urchin zygote. The cleavage of some of these dispermic oocytes to 2-cells (which are karyotypically diploid) plus an extrusion further violates the expected cell behavior of dispermic human oocytes based on observations in rats in which all such oocytes develop into triploid embryos.

REFERENCES

Angell, R. R., Templeton, A. A., and Messinis, I. E. (1986). Consequences of polyspermy in men. *Cytogenet. Cell Genet.* **42,** 1–7.

Austin, C. R., and Braden, A. W. H. (1953). An investigation of polyspermy in the rat and rabbit. *Aust. J. Biol. Sci.* **6**, 674–696.

Austin, C. R., and Braden, A. W. H. (1955). Observations on nuclear size and form in living rat and mouse eggs. *Exp. Cell Res.* **8**, 163–168.

Austin, C. R., and Walton, A. (1960). Fertilization. *In* "Marshall's Physiology of Reproduction" (A. S. Parkes, ed.), pp. 310–416. Longmans, London.

Beatty, R. A. (1957). "Parthenogenesis and Polyploidy in Mammalian Development." Cambridge Univ. Press, London and New York.

Borgaonkar, D. S. (1984). "Chromosomal Variation in Man—A Catalog of Chromosomal Variants and Anomalies," 4th ed., pp. 778–785. Alan R. Liss, New York.

Boveri, T. (1902). "Ueber mehrpolige Mitosen als Mittel zur Analyse des Zellkerns: V.P.M.G. XXXV."

Boveri, T. (1907). "Zellen-Studien VI." Fischer, Jena.

Fankhauser, G. (1948). The organization of the amphibian egg during fertilization and cleavage. *Ann. N.Y. Acad. Sci.* **49**, 684–694.

Jacobs, P. A., Angell, R. R., Buchanan, J. M., Hassold, T. J., Matsuyama, A. M., and Manuel, B. (1978). The origins of human triploids. *Ann. Hum. Genet.* **42**, 49–57.

Kola, I., Trounson, A. O., Dawson, G., and Rogers, P. (1987). Tripronuclear human oocytes: Altered cleavage patterns and subsequent karyotypic analysis of embryos. *Biol. Reprod.* **37**, 395–401.

Long, S. E., and Williams, C. V. (1980). Frequency of chromosomal abnormalities in early embryos of the domestic sheep (Ovis aries). *J. Reprod. Fertil.* **58**, 197–201.

McMaster, R., Yanagimachi, R., and Lopata, A. (1978). Penetration of human eggs by human spermatozoa *in vitro. Biol. Reprod.* **19**, 212–216.

Murray, J. D., Boland, M. P., Moran, C., Sutton, R., Nancarrow, C. D., Scaramuzzi, R. J., and Hoskinson, R. M. (1985). Occurrence of haploid and haploid/diploid mosaic embryos in untreated and androstenedione-immune Australian Merino sheep. *J. Reprod. Fertil.* **74**, 551–555.

Olsen, M. W. (1942). Maturation, fertilization and early cleavage in the hen's egg. *J. Morphol.* **70**, 513–518.

Rothschild, L. (1954). Polyspermy. *Q. Rev. Biol.* **29**, 332–354.

Rudak, E., Dor, J., Mashiach, S., Nebel, L., and Goldman, B. (1984). Chromosome analysis of multipronuclear human oocytes fertilized *in vitro. Fertil. Steril.* **41**, 538–545.

Sathananthan, A. H., and Trounson, A. O. (1982). Cortical granule release and zona interaction in monospermic and polyspermic human ova fertilized *in vitro. Gamete Res.* **6**, 225–234.

Sathananthan, A. H., and Trounson, A. O. (1985). Human pronuclear ovum: Fine structure of monospermic and polyspermic fertilization *in vitro. Gamete Res.* **12**, 385–398.

Sathananthan, A. H., Ng, S. C., Edirisinghe, R., Ratnam, S. S., and Wong, P. C. (1986a). Sperm–oocyte interaction in the human during polyspermic fertilization *in vitro. Gamete Res.* **15**, 317–326.

Sathananthan, A. H., Trounson, A. O., and Wood, C. (1986b). "Atlas of Fine Structure of Human Sperm, Eggs and Embryos Cultured *in vitro.*" Praeger, Philadelphia, Pennsylvania.

Schatten, G., Simerly, C., and Schatten, H. (1985). Microtubule configurations during fertilization, mitosis, and early development in the mouse and the requirement for egg microtubule-mediated motility during mammalian fertilization. *Proc. Natl. Acad. Sci. U.S.A.* **82**, 4125–4156.

Schatten, H., Schatten, G., Mazia, D., Balczon, R., and Simerly, C. (1986). Behavior of centrosomes during fertilization and cell division in mouse oocytes and in sea urchin eggs. *Proc. Natl. Acad. Sci. U.S.A.* **83**, 105–109.

Sherard, J., Bean, C., Bove, B., Delduca, V., Esterly, K. L., Kavish, H. J., Munshi, G., Reamer, J. F., Suazo, G., Wilmoth, D., Dahlke, M. B., Weiss, C., and Borgaonkar, D. S. (1986). Long survival in a 69XXY triploid male. *Am. J. Med. Genet.* **25,** 305–312.

Thibault, C. (1977). Are follicular maturation and oocyte maturation independent processes? *J. Reprod. Fertil.* **51,** 1–15.

Trounson, A. O., Rogers, P., Kola I., and Sathananthan, A. H. (1988). Fertilization, development and implantation. *In* "Textbook of Obstetrics" (A. Turnbull and G. Chamberlain, eds.). Churchill-Livingstone, Edinburgh and London (in press).

Uchida, I. A., and Freeman, V. C. (1985). Triploidy and chromosomes. *Am. J. Obstet. Gynecol.* **151,** 65–69.

Van Blerkom, J., Henry, G., and Porreco, R. (1984). Preimplantation human embryonic development from polypronuclear eggs after in vitro fertilization. *Fertil. Steril.* **41,** 686–696.

Van Blerkom, J., Bell, H., and Henry, G. (1987). The occurrence, recognition and developmental fate of pseudo-multipronuclear eggs after in vitro fertilization of human oocytes. *Hum. Reprod.* **2,** 217–225.

Wilson, E. B. (1928). "The Cell in Development and Hereditary," pp. 916–979. Macmillan, New York.

III

Ionic Regulation and Its Controls

12

G-Proteins and the Regulation of Oocyte Maturation and Fertilization

PAUL R. TURNER[*] AND LAURINDA A. JAFFE[†]

[*]Department of Zoology
University of California, Berkeley
Berkeley, California 94720

[†]Department of Physiology
The University of Connecticut Health Center
School of Medicine
Farmington, Connecticut 06032

297

I. INTRODUCTION

Guanine nucleotide binding proteins, or G-proteins, are a class of membrane proteins that act as intermediates between membrane receptors and effector proteins in signal transduction pathways. Some receptors are sensitive to chemicals (hormones, neurotransmitters, odorants), while in the visual system, the receptor is sensitive to light. The effector proteins can be enzymes, such as adenylate cyclase, cGMP phosphodiesterase, the phosphodiesterase which hydrolyzes phosphatidylinositol 4,5-bisphosphate (PIP_2), or ion channels such as the potassium channel in cardiac muscle (Yatani et al., 1987). Current research is beginning to show that G-proteins serve a variety of functions, including regulation (stimulation or inhibition of an effector protein), integration (summation of inputs from different types of receptors), diversification of the response (activation of several different types of effectors by a single type of receptor), amplification (activation of many molecules of a particular effector by one receptor molecule), and possibly also adaptation (variation of the magnitude of the amplification step). For excellent reviews concerning G-proteins, see Stryer and Bourne (1986) and Gilman (1987).

G-proteins consist of three subunits, α, β, and γ. The α subunit is believed to be the component that activates the effector protein, and the diversity of G-proteins arises primarily from diversity in this subunit. The α subunit of G_s, (the G-protein that stimulates adenylate cyclase), has a mass of 45–52 kDa; the α subunit of G_i (the G-protein that inhibits adenylate cyclase), has a mass of 41 kDa. The α subunits of both G_o (a G-protein of unknown function) and transducin (the G-protein in visual transduction) are 39 kDa. The β subunits (35–36 kDa) are highly conserved among different G-proteins, whereas the γ subunits (8 kDa), show some variability. The α subunit can bind and hydrolyze GTP to form GDP. With GTP bound, the α subunit dissociates from $\beta\gamma$, and can interact with effector proteins. Once the GTP is hydrolyzed to GDP, the α subunit associates with $\beta\gamma$ again to form an inactive complex. The exchange of GDP for GTP is slow, but the presence of an excited receptor accelerates the exchange of bound GDP for GTP, thereby catalyzing the activation of the G-protein.

The participation of G-proteins in a cellular process can be identified by a number of means.

1. A system involving a G-protein requires GTP. This is because the activation of the G-protein involves the exchange of GDP (bound to the inactivated G-protein) for GTP (the binding of which causes activation of the G-protein and dissociation of α from $\beta\gamma$).

2. Membranes prelabeled with radioactive guanine nucleotides usually show an increased release of label when the G-protein is stimulated.

3. Guanosine-5'-O-(3-thiotriphosphate) (GTP-γ-S) and guanyl-5'-imidodiphosphate [Gpp(NH)p], hydrolysis-resistant analogs of GTP, produce an irreversible activation of G-proteins. This is because GTP-γ-S or Gpp(NH)p become bound to the active site of the α subunit, but are not hydrolyzed to GDP.

4. Guanosine-5'-O-(2-thiodiphosphate)(GDP-β-S), a metabolically stable analog of GDP, inactivates G-proteins by competing with GTP for the guanine nucleotide binding site.

5. Cholera toxin (CTX) and pertussis toxin (PTX) catalyze the ADP-ribosylation of the α subunits of G-proteins; CTX is specific for G_s and transducin, PTX is specific for G_i, G_o and transducin (however, see Owens *et al.*, 1985). If radioactive NAD is present with the toxins, labeling of the G-proteins results, allowing their detection. The toxins also affect the function of the G-proteins; CTX causes irreversible activation by preventing GTP hydrolysis, while PTX causes inactivation by stabilizing the G-protein in the inactive GDP-bound state.

6. The presence of G-proteins can also be detected using antibodies to G-protein subunits (see Mumby *et al.*, 1986).

This chapter concerns the function of G-proteins in oocyte maturation, and in the activation of sperm and eggs at fertilization.

II. OOCYTE MATURATION

A. Hormone Interaction with a Membrane Receptor Inhibits Adenylate Cyclase

In many animals, including amphibians, fish, mammals, and starfish, the fully grown oocytes in the ovary are arrested at the first meiotic prophase, and contain a large nucleus or germinal vesicle. Hormonal stimulation leads to many biochemical and morphological changes, including germinal vesicle breakdown (GVBD) and resumption of meiosis (see Masui and Clarke, 1979). In the frog, meiotic maturation is believed to be reinitiated by progesterone (Masui and Clarke, 1979); a related steroid induces maturation in the fish oocyte (Nagahama and Adachi, 1985), and it is likely that a steroid functions to stimulate meiosis in mammalian oocytes as well (see Tsafriri, 1985; Eppig, 1986). In the starfish, the maturation-inducing hormone is 1-methyladenine (1-MA) (Kanatani et al., 1969).

Progesterone and 1-MA are thought to act at the oocyte surface. Microinjection of progesterone (Masui and Markert, 1971), or of 1-MA (Kanatani and Hiramoto, 1970) does not result in maturation, and entry of these hormones into the oocyte cytoplasm does not appear to be required to stimulate maturation. Desoxycorticosterone bound to agarose beads (Ishikawa et al., 1977) and another progesterone analog, androsta-4-ene-3-one-17β-carboxylic acid, bound to a soluble polymer (Baulieu et al., 1978) both promote GVBD in frog oocytes at similar concentrations to unbound hormone. Similarly, when uptake of 1-MA by starfish oocytes is decreased by use of 1,9-dimethyladenine, the response to 1-MA is not modified (Dorée and Guerrier, 1975). These results suggest that progesterone and 1-MA interact with receptors in the oocyte plasma membrane.

To identify a receptor for progesterone, the frog oocyte surface complex (a plasma membrane and vitelline envelope preparation that might also have contained some residual follicle cells and cytoplasm) was photoaffinity labeled with a progesterone analog. This identified a single steroid binding component of M_r 110,000, with a K_d for steroid binding similar to the EC_{50} for induction of oocyte maturation (Sadler and Maller, 1982). These findings suggest that the 110,000 M_r binding protein is the progesterone receptor; however, the effects of progesterone on binding of the labeled analog are complex, so this conclusion is tentative. There is also evidence for a receptor for 1-MA in the starfish oocyte plasma membrane (Yoshikuni et al., 1988).

The intracellular signaling system that couples activation of the progesterone receptor to meiotic maturation appears to involve inhibition of adenylate cyclase (AC). Applying progesterone to the frog oocyte surface complex decreases the activity of AC by about 50% (Sadler and Maller, 1981), and oocytes exposed to progesterone undergo a 10–50% drop in cAMP; this decrease is hypothesized to lead to the reinitiation of meiosis (Speaker and Butcher, 1977; Maller and Krebs, 1977; Maller et al., 1979; Sadler and Maller, 1981; Schorderet-Slatkine et al., 1982; Cicirelli and Smith, 1985). In support of this hypothesis, injection of the regulatory subunit of cAMP-dependent kinase promotes maturation, while the catalytic subunit inhibits it (Maller and Krebs, 1977). In addition, increasing oocyte AC activity by injecting cholera toxin (Sadler and Maller, 1981) or applying forskolin (Schorderet-Slatkine and Baulieu, 1982) inhibits progesterone-induced maturation (Maller et al., 1979). A similar pathway involving cAMP may function in mammals (Bornslaeger et al., 1986) and in starfish (Dorée et al., 1981; Meijer and Zarutskie, 1987).

Since G-proteins are well known to link surface receptors to the AC system in other cells (Gilman, 1984), it was an exciting possibility that a G-protein might also be involved in this signaling sequence in the oocyte. This led to studies examining the possibility that G-proteins might regulate AC in frog oocytes. Other studies examined the role of G-proteins in initiating GVBD, independent of the question of whether the pathway did or did not involve AC. These two approaches will be considered below.

B. Evidence That G-Proteins Regulate Adenylate Cyclase in Frog Oocytes

As in most other cells (see Gilman, 1984), AC in the frog oocyte surface complex is regulated by two G-proteins, G_s and G_i. When a hormone or transmitter binds to a receptor that is coupled to G_s, G_s is activated and stimulates AC; this increases the production of cAMP (Fig. 1). When a hormone or

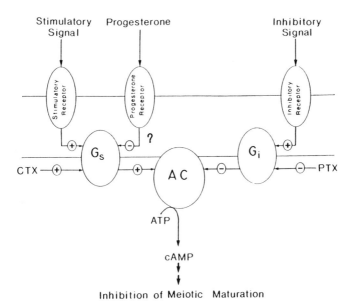

Fig. 1.   G-protein regulation of adenylate cyclase, showing the proposed inhibition of G_S by progesterone in the frog oocyte membrane, and the proposed pathway leading to meiotic maturation. Acetylcholine is known to inhibit AC activity of the frog oocyte surface complex, by way of G_i (Sadler *et al.*, 1984), and there is some evidence to suggest that epinephrine may stimulate frog oocyte AC activity of the oocyte surface complex by way of G_S (Kusano *et al.*, 1982; Van Renterghem *et al.*, 1985). The available evidence does not definitively demonstrate that this regulatory system is located entirely in the oocyte plasma membrane, as opposed to follicle cell membranes. The arrow connecting G_i to AC does not imply that the mechanism of inhibition is direct (see Gilman, 1987, for details).

transmitter binds to a receptor that is coupled to G_i, G_i is activated and inhibits AC; this decreases the production of cAMP (Fig. 1). The discovery that these two G-proteins are present and functional in the frog oocyte surface complex came largely from the use of PTX and CTX. Pertussis toxin catalyzes the ADP-ribosylation of the α subunit of G_i, causing inhibition of G_i, whereas CTX catalyzes the ADP-ribosylation of the α subunit of G_s, causing stimulation of G_s (Fig. 1).

When membranes from *Xenopus* oocytes are [^{32}P]ADP ribosylated by incubation with PTX and ^{32}P-labeled NAD, label is incorporated into a single substrate of 40 kDa (Olate *et al.*, 1984) or 41 kDa (Sadler *et al.*, 1984; Goodhardt *et al.*, 1984). This PTX-sensitive polypeptide comigrates with the α subunit of human erythrocyte G_i (Olate *et al.*, 1984). Incubation of oocyte membranes with CTX results in label incorporation into a 42-kDa substrate (Olate *et al.*, 1984), or 45- and 52-kDa (and several other) substrates (Sadler *et al.*, 1984; Goodhardt *et al.*, 1984). These data indicate that at least two G-proteins are present in frog oocytes. It is conceivable, however, that some of the G-proteins are actually in residual follicle cells on the oocyte. The observation that injection of CTX into the oocyte cytoplasm stimulates AC (Sadler and Maller, 1981) argues that at least some of the G-proteins are in the oocyte plasma membrane and not in follicle cells. However, due to the technical difficulty in removing follicle cells from frog oocytes, subsequent references to "oocyte" G-proteins, adenylate cyclase, or receptors should be understood to mean proteins associated with oocytes after removal of most of the follicle cells.

The oocyte G-proteins function like G_s and G_i to regulate adenylate cyclase. GTP, Gpp (NH)p, the fluoride ion, and CTX, which are known stimulators of G_s (Howlett *et al.*, 1979; Moss and Vaughan, 1979), all activate oocyte AC (Jordana *et al.*, 1981; Sadler and Maller, 1981). Evidence for a functional G_i in the oocyte surface complex comes from the finding that a 35% inhibition of oocyte AC by acetylcholine can be prevented by PTX (Sadler *et al.*, 1984). Pertussis toxin may or may not have an effect on basal AC activity (Sadler *et al.*, 1984; Olate *et al.*, 1984). In conclusion, both G_s-mediated activation and G_i-mediated inhibition appear to be involved in regulating the level of cAMP in the oocyte (see Fig. 1).

C. Does a G-Protein Mediate Progesterone-Induced Inhibition of Adenylate Cyclase in Frog Oocytes?

Since progesterone causes a decrease in AC activity of the oocyte surface complex, and since the oocyte AC activity can be regulated by G-proteins, it seemed logical to infer that progesterone's action on AC is mediated by way of a G-protein. Surprisingly, this inhibition appears not to be mediated by G_i.

1. Pretreatment of oocytes with PTX does not prevent the inhibition of AC by progesterone (Sadler *et al.*, 1984; Olate *et al.*, 1984; Goodhardt *et al.*, 1984; Mulner *et al.*, 1985), even though the PTX treatment is sufficient to ADP-ribosylate most, if not all, of the G_i (Sadler *et al.*, 1984; Olate *et al.*, 1984). In contrast, PTX prevents the inhibition of AC by hormones and transmitters that act through G_i (e.g., Goodhardt *et al.*, 1984), including a 35% inhibition of oocyte AC by acetylcholine (Sadler *et al.*, 1984).
2. Applying progesterone to the oocyte surface complex, preloaded with [^{3}H]GTP, results in a reduced guanine nucleotide release or exchange rate (Sadler and Maller, 1983, 1985). This result differs from results in G_i-mediated systems; hormones which act through G_i, to reduce AC activity, increase the guanine nucleotide exchange rate (Michel and Lefkowitz, 1982).
3. GDP-β-S might be expected to inhibit G_i function, yet GDP-β-S does not prevent the progesterone inhibition of AC (Jordana *et al.*, 1984).

If progesterone does not act by way of G_i, how might it reduce adenylate cyclase activity? The observation that progesterone *reduces* the guanine nucleotide exchange rate of the oocyte surface complex (Sadler and Maller, 1983, 1985) has been interpreted as an indication that progesterone may act by slowing the rate of turnover of guanine nucleotides bound to G_s, and therefore preventing G_s activation (Fig. 1). Alternatively, progesterone might inhibit oocyte AC by way of a different, as yet unidentified, G-protein (Olate *et al.*, 1984), or by a pathway independent of a G-protein.

D. Does a G-Protein Mediate Progesterone-Induced Germinal Vesicle Breakdown in Frog Oocytes?

Irrespective of whether a G-protein regulates progesterone-induced inhibition of AC, studies with PTX suggest that a G-protein may, in some way, be involved in mediating progesterone-induced GVBD in *Xenopus* oocytes. Since PTX inhibits certain G-proteins, PTX might be expected to inhibit progesterone-induced GVBD, if a G-protein was involved in the process. Consistent with this idea, one study (Sadler *et al.*, 1984) reported that PTX, at a concentration outside the oocyte of 0.4 μg/ml, slowed GVBD in response to 0.1 μM progesterone. Pertussis toxin did not slow GVBD in response to 10 μM progesterone. However, another study (Mulner *et al.*, 1985) reported that

PTX *increased* the rate of GVBD in response to progesterone (1 μM). The PTX was applied either externally (2 $\mu g/ml$) or injected (6 ng/oocyte). Possible reasons for the differing results include differences in the concentration or supplier of PTX and differences in the methods of obtaining the oocytes, such as whether the frogs were hormonally "primed."

These observed effects of PTX involved small modifications in the *rate* of GVBD; in no case did PTX completely block GVBD or stimulate it in the absence of progesterone. Thus the PTX experiments suggest that a G-protein might be part of the coupling between 1-MA and GVBD, but are not conclusive. The best evidence for G-protein involvement in mediating the progesterone response is the observation that progesterone decreases the rate of guanine nucleotide exchange (Sadler and Maller, 1983, 1985). This observation suggests an unusual effect of the receptor for progesterone on the G-protein: the receptor appears to inhibit rather than stimulate the G-protein.

E. A Role for ras Proteins in Maturation?

The ras genes, which encode 21-kDa proteins, are present in most species, and frog oocytes appear to contain a low level of ras proteins (Sadler *et al.*, 1986). The ras proteins are similar to G-proteins in that they have GTPase activity, sequence homology with G-proteins, and are localized on the cytoplasmic face of membranes (see references in Birchmeier *et al.*, 1985). The function of ras proteins in vertebrates is not known, but it has been suggested that in transmembrane signal transduction in meiotic maturation in frog oocytes, a ras protein might play a role analogous to that of a G-protein (Birchmeier *et al.*, 1985; Sadler *et al.*, 1986). In yeast, ras proteins activate AC, but injection of ras proteins into frog oocytes does not stimulate or inhibit AC (Birchmeier *et al.*, 1985). Surprisingly, however, injection of 10 ng of ras protein causes meiosis, although the time course is slower than that induced by progesterone. This work suggests that there might be an alternative pathway triggering meiosis, that bypasses changes in cAMP levels (Birchmeier *et al.*, 1985); however, the oocyte disintegration produced by injection of >20 ng of ras protein makes such a conclusion tentative.

Antibodies to ras protein inhibit the AC activity of the *Xenopus* oocyte surface complex by about 50% (Sadler *et al.*, 1986). In addition, antibody injection accelerates the induction of meiosis by progesterone, and in some cases, induces meiosis in the absence of progesterone (Sadler *et al.*, 1986). These results differ from those of Birchmeier *et al.* (1985) in that they support the hypothesis that ras proteins in the oocyte activate AC. However, cross-reaction of the antibodies with G-proteins (such as G_s) has not been ruled out.

F. Possible Role of G-Proteins in the Regulation of Meiotic Maturation in Other Species

Mechanisms controlling meiotic maturation in mammalian oocytes are not well understood, but cAMP appears to be involved (see Bornslaeger et al., 1986). The effects of CTX on cumulus–oocyte complexes (Dekel and Beers, 1980; Bornslaeger and Schultz, 1985) and on follicle-enclosed oocytes (Freter and Schultz, 1984) suggest that a G-protein in the cumulus or granulosa cells could function in the regulation of maturation. G-proteins may also be present in the oocyte itself (Manejwala et al., 1986), but the possible role of such G-proteins in meiotic maturation is unknown.

Injections of PTX and GDP-β-S into starfish oocytes have provided strong evidence that a G-protein may mediate 1-MA induced maturation of starfish oocytes (Shilling and Jaffe, 1987). Pertussis toxin (2–6 μg/ml) blocks GVBD in response to 1-MA (0.1–10 μM), and GDP-β-S (2–4 mM) partially inhibits the response. Effects of CTX on starfish oocyte maturation are less consistent. One study reported that injection of CTX stimulated GVBD in response to subthreshold concentrations of 1-MA (Dorée and Kishimoto, 1981). However, in another study, injection of CTX partially inhibited GVBD in response to 1-MA (F. Shilling and L. A. Jaffe, unpublished). Further experiments will be required to determine what G-proteins may be present in starfish oocytes, how 1-MA might activate these G-proteins, and how G-protein activation might lead to GVBD.

III. SPERM ACTIVATION

A. Binding of ZP3 to a Receptor in the Mouse Sperm Membrane Initiates the Acrosome Reaction

Activation of sperm motility (Lee and Garbers, 1986), chemotaxis of sperm toward eggs (Ward et al., 1985), and the induction of the acrosome reaction (Bleil and Wassarman, 1986) are all thought to be mediated by receptors in the sperm plasma membrane. Of particular relevance to this chapter, the interaction of mouse sperm with a glycoprotein in the zona pellucida, ZP3, initiates the acrosome reaction (Saling et al., 1979; Bleil and Wassarman, 1983). Studies of the binding of [125]I- labeled ZP3 to the plasma membrane of mouse sperm heads suggest that a receptor for ZP3 is present in the sperm plasma membrane (Bleil and Wassarman, 1986). Binding of ZP3 to this receptor presumably initiates the acrosome reaction by way of a second messenger system.

Does a G-protein link activation of the ZP3-receptor to the initiation of the acrosome reaction?

B. G_i- or G_o-like Proteins Are Present in Sperm, While G_s-like Proteins Appear to Be Absent

G-proteins have been identified in sperm from both invertebrate and vertebrate species (Kopf *et al.*, 1986; Bentley *et al.*, 1986). In all species examined, α subunits of these proteins closely resemble G_i or G_o in molecular weight (39,000–41,000), labeling by PTX, and limited proteolytic digest patterns. The PTX substrate from sea urchin sperm copurifies with GTP-binding activity (Bentley *et al.*, 1986). Antibodies against β subunits of G-proteins have been used to demonstrate the presence of β subunits in sperm (Kopf *et al.*, 1986; Bentley *et al.*, 1986).

Use of cholera toxin, however, has not identified any G_s-like proteins in sperm (Hildebrandt *et al.*, 1985; Kopf *et al.*, 1986; Bentley *et al.*, 1986). This suggests that either G_s is absent in sperm, or that present methods have failed to detect it. Since sperm contain a high level of AC activity, the presence of G_s might be expected; however, the sperm AC is not stimulated by the fluoride ion, Gpp(NH)p, CTX, or added G_s, suggesting that it is not regulated by G_s (Hildebrandt *et al.*, 1985; Kopf *et al.*, 1986; Bentley *et al.*, 1986).

C. Evidence That a G-Protein Mediates Induction of the Mouse Sperm Acrosome Reaction by the Zona Pellucida

To test the hypothesis that a G_i- or G_o-like protein mediates the induction of the acrosome reaction by ZP3, Endo *et al.* (1987) exposed mouse sperm to PTX, known in other systems to inactivate these G-proteins. They found that PTX blocked the zona-induced acrosome reaction, suggesting that the reaction is mediated by a G-protein (Endo *et al.*, 1987). This physiological effect of PTX is accompanied by ADP-ribosylation of a 41-kDa polypeptide; the polypeptide resembles the α subunit of G_i, or possibly G_o. The inhibitory effect of PTX on the acrosome reaction is blocked by GTP-γ-S, supporting the idea that PTX is acting by way of a G-protein, and hence that the zona-induced acrosome reaction is mediated by a G-protein.

Surprisingly though, GTP-γ-S is only a very weak stimulator of the acrosome reaction. The target of the PTX-sensitive G-protein of the mouse sperm is unknown. Possibilities include the regulation of a plasma membrane ion channel (leading to an influx of extracellular calcium), regulation of PIP_2 phosphodiesterase (leading to the release of intracellular calcium), or modulation of adenylate cyclase (Endo *et al.*, 1987).

IV. EGG ACTIVATION

A. Responses of the Sea Urchin Egg to Fertilization

The first response of the sea urchin egg to fertilization is a depolarization of the plasma membrane, which prevents further sperm from fusing with the egg (Jaffe, 1976). The sperm–egg interaction then stimulates the metabolism of polyphosphoinositides (Turner *et al.*, 1984), resulting in the cleavage of PIP_2 by the enzyme PIP_2 phosphodiesterase to produce inositol 1,4,5-tris-phosphate ($InsP_3$) and diacylglycerol (DAG) in the egg (Ciapa and Whitaker, 1986). $InsP_3$ stimulates the release of calcium from intracellular stores (Clapper and Lee, 1985; Oberdorf *et al.*, 1986; Swann and Whitaker, 1986), causing the fusion of thousands of cortical vesicles with the plasma membrane (Hamaguchi and Hiramoto, 1981; Whitaker and Irvine, 1984; Turner *et al.*, 1986). This exocytosis results in the elevation of the fertilization envelope, a permanent barrier to the entry of additional sperm (Schuel, 1978). The other product of PIP_2 phosphodiesterase activity, DAG, has been shown to stimulate the Na^+–H^+ antiport in the sea urchin egg, resulting in an increase in acid efflux (Lau *et al.*, 1986; Shen and Burgart, 1986). DAG is thought to stimulate protein kinase C; direct stimulation of protein kinase C with phorbol esters also results in a stimulation of Na^+–H^+ exchange, and an increase in protein synthesis in sea urchin eggs (Swann and Whitaker, 1985). Other responses activated by fertilization include activation of NAD kinase, increases in reduced nicotinamide nucleotides, increased oxygen consumption, activation of amino acid transport, initiation of DNA synthesis, and stimulation of cell division (see Whitaker and Steinhardt, 1985). Some of these events may also result from $InsP_3$ and DAG production.

How can the sperm, acting at the egg surface, regulate the production of $InsP_3$ and DAG? G-proteins have been shown to regulate production of $InsP_3$ and DAG in other systems (see Stryer and Bourne, 1986), so the possible involvement of G-proteins in egg activation was an attractive hypothesis.

B. Sea Urchin Eggs Contain G-Proteins

To look for G-proteins in sea urchin eggs, the labeling of egg proteins in the presence of CTX and PTX and ^{32}P-labeled NAD was examined. Oinuma *et al.* (1986) found a 39-kDa substrate for PTX in the detergent extracted membranes of whole eggs. This 39-kDa protein copurified with the ability to bind GTP-γ-^{35}S, as would be expected for a G-protein. When a complex of sea urchin egg plasma membranes, cortical vesicles, and associated membranes was incubated with toxins, label was incorporated into two substrates (Turner *et al.*, 1987). Cholera toxin catalyzed the ADP-ribosylation of a 47-kDa substrate, and PTX labeled a 40-kDa substrate. Thus it appears that the sea urchin egg contains a PTX-sensitive G_i- or G_o-like G-protein, with an α subunit of 39–40 kDa, and a CTX-sensitive G_s-like G-protein with an α subunit of 47 kDa. By immunoblotting, β subunits of G-proteins have also been identified in sea urchin eggs (Oinuma *et al.*, 1986).

Further analysis of the sea urchin egg PTX substrate showed similarities with G_o (Oinuma *et al.*, 1986). On an SDS gel, the PTX substrate migrated with α_o from rat brain, and not with α_i. Furthermore, the sea urchin egg protein cross-reacted weakly with an antibody against rat brain α_o, but not with an antibody against α_i.

C. Evidence that the Cholera Toxin-Sensitive G-Protein Regulates Exocytosis in Sea Urchin Eggs

To test the hypothesis that a G-protein was a component of the stimulatory pathway leading to exocytosis of cortical vesicles, sea urchin eggs were microinjected with the hydrolysis-resistant analog of GTP, GTP-γ-S. This caused exocytosis (Turner *et al.*, 1986) (Fig. 2a). [In the series of experiments reported by Turner *et al.* (1986), all of the eggs injected with GTP-γ-S (30 μM) underwent exocytosis. In some subsequent experiments, the response to GTP-γ-S was more variable (P. R. Turner and L. A. Jaffe, unpublished). The reason for this variability is not understood.] Injection of GTP (100 μM) did not cause exocytosis (Swann *et al.*, 1987). The stimulation of exocytosis by GTP-γ-S was accompanied by a rise in intracellular calcium (Swann *et al.*, 1987), and the exocytosis could be blocked by preinjection of the calcium buffer EGTA (Turner *et al.*, 1986).

Microinjection, but not external application, of CTX or CTX subunit A also resulted in exocytosis (Turner *et al.*, 1987). The exocytosis in response to CTX injection was blocked by preinjecting eggs with EGTA, suggesting that CTX was stimulating exocytosis by way of an increase in [Ca^{2+}]$_i$. This hypothesis is supported by the observation that microinjection of cAMP, or a hydrolysis-resistant analog of cAMP, did not cause exocytosis (Turner *et al.*, 1987).

If the activation of a G-protein is required for exocytosis, then inactivation of the G-protein should prevent the stimulation of exocytosis by sperm. Indeed, when eggs were preinjected with GDP-β-S, which acts as a competitive inhibitor of GTP at the regulatory site of G-proteins, sperm were prevented from stimulating exocytosis (Turner *et al.*, 1986; Swann *et al.*, 1987) (Fig. 3a). Nevertheless, sperm entered the egg cytoplasm (Swann *et al.*, 1987). If GDP-β-S–injected eggs were subsequently injected with InsP$_3$, exocytosis re-

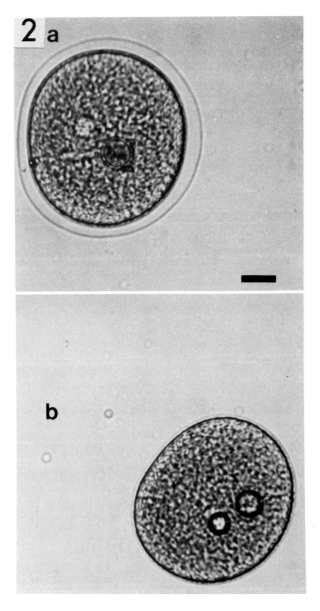

Fig. 2. Calcium-dependent exocytosis stimulated by GTP-γ-S. (a) An egg which was injected with 28 μM GTP-γ-S. Bar, 25 μm. (b) An egg which was first injected with a 1 : 3 mixture of 0.2 M CaEGTA and 0.2 M EGTA ([Ca^{2+}] = 0.1 μM, [EGTA] = 1.6 mM), and then injected with 28 μM GTP-γ-S. [From Turner *et al.* (1986).]

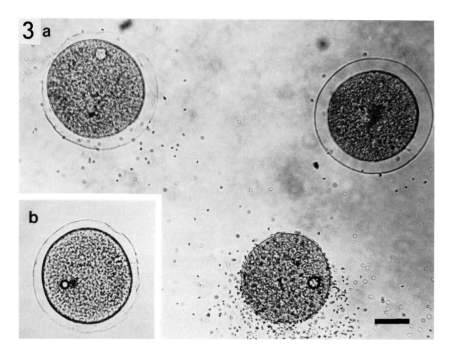

Fig. 3. GDP-β-S effects on exocytosis stimulated by sperm or InsP₃. (a) The lower egg was injected with 3mM GDP-β-S; the two upper eggs are uninjected controls. After insemination, both control eggs elevated fertilization envelopes, but the injected egg did not. The specks around the eggs are sperm heads. Bar, 50 μm. (b) An egg which was injected with 3 mM GDP-β-S and was then injected with 28 nM InsP₃. The fertilization envelope elevated after the injection of InsP₃. [From Turner *et al.* (1986).]

sulted (Turner *et al.*, 1986) (Fig. 3b), indicating that the step involving the CTX-sensitive G-protein precedes the step involving InsP₃ (Fig. 4).

The target of the activated G-protein is unknown. The G-protein may regulate PIP₂ phosphodiesterase, the enzyme which produces InsP₃ and DAG, as has been suggested in other systems (Cockcroft and Gomperts, 1985). That the amounts of the polyphosphoinositides increase following fertilization, but prior to exocytosis (Turner *et al.*, 1984), suggests that certain kinases have been activated, and the G-protein could interact with these kinases (see Pike and Eakes, 1987; Chahwala *et al.*, 1987). It is conceivable that a G-protein could have more than one target.

Injection of PTX into sea urchin eggs did not stimulate or inhibit cortical vesicle exocytosis, but since the PTX was not fully in solution, the amount of PTX injected was not known (Turner *et al.*, 1987). Therefore, the inter-

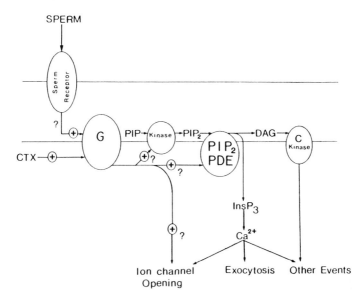

Fig. 4. Proposed role of a G-protein in egg activation. The G-protein is activated by the sperm, possibly by way of a receptor in the egg plasma membrane. The G-protein stimulates the production of InsP₃ and DAG, leading to exocytosis and other developmental events. The G-protein may stimulate PIP₂ phosphodiesterase, or kinases which phosphorylate inositol lipids, or other enzymes; it might also act directly on ion channels. This model is based on experiments with sea urchin, frog, and hamster eggs.

pretation of these experiments is unclear. Furthermore, since PTX action may require time (Shilling and Jaffe, 1987), it will be important to examine the effects of injecting PTX more than 1 hr before insemination. The function of the PTX-sensitive G-protein in the sea urchin egg remains to be determined.

D. G-Proteins and Activation of Other Developmental Events in Sea Urchin Eggs

Observations of GTP-γ-S–injected eggs suggest that, in addition to a rise in calcium and exocytosis, G-protein activation may also result in the activation of Na^+–H^+ exchange, NAD(P) reduction, nuclear envelope breakdown, chromosome condensation, and DNA synthesis (Dargie *et al.*, 1986). Whether some or all of these events result from the production of InsP₃ and DAG (Fig. 4), or whether they result from interaction of the G-protein with other enzymes, is unknown.

E. G-Proteins and Activation of Eggs of Other Species

Injection of GTP-γ-S into *Xenopus* eggs resulted in exocytosis of cortical vesicles and a change in membrane potential like that occurring at fertilization; these responses did not occur if the eggs were preinjected with the calcium chelator BAPTA (Kline and Jaffe, 1987, and unpublished results). Further evidence that a G-protein might be involved in activating frog eggs came from experiments in which exogenous serotonin or muscarinic acetylcholine (M1) receptors were introduced into *Xenopus* egg membranes; when serotonin or acetylcholine was applied, the eggs produced an activation potential and underwent cortical vesicle exocytosis as well as endocytosis and cortical contraction as occur after fertilization (Kline *et al.*, 1988). Since the serotonin and acetylcholine receptors that were introduced are known to act by way of G-proteins (Dascal *et al.*, 1986; Nomura *et al.*, 1987), it was proposed that neurotransmitters activate the egg by interacting with an endogenous G-protein that is normally activated by sperm.

Injection of GTP-γ-S into hamster eggs resulted in a series of periodic rises in intracellular calcium and hyperpolarizations of the membrane; these responses were very much like those occurring at fertilization (Miyazaki, 1988). Injection of GDB-β-S inhibited the hyperpolarizations in response to sperm but not in response to subsequent injection of InsP$_3$ (Miyazaki, 1988). These observations support the conclusion that a G-protein may be involved in the activation of mammalian eggs at fertilization.

V. UNANSWERED QUESTIONS

The study of G-proteins in gametes has identified their function in oocyte maturation, sperm activation, and egg activation, and has posed many additional questions.

With regard to oocyte growth and maturation (Section II): Are the G-proteins of the frog oocyte localized exclusively in the oocyte plasma membrane, or are they also present in follicle cells? Does the model presented in Fig. 1 apply to species other than frogs? Does the type, number, and/or function of G-proteins change during maturation? Why is G$_i$ present? Do G-proteins regulate oocyte growth in some way?

With regard to sperm function (Section III): If sperm do not possess a G$_s$, but do possess a G$_i$, then the regulation of adenylate cyclase is markedly different from that seen in other systems. The target enzyme(s) for the PTX-sensitive G-protein (G$_i$?) is at present unknown. It will also be of great interest to determine if a G-protein regulates other sperm activities, such as motility, chemotaxis, and/or capacitation.

With regard to the activation of development (Section IV): Are G-proteins present in the eggs of mammals and others species? If so, do they function in the activation of development at fertilization? What is the function of the PTX-sensitive G-protein in the sea urchin egg? In frog oocytes, a PTX-sensitive G-protein has been shown to function as G_i, to inhibit adenylate cyclase (Section II). Does this G-protein function (if present) as G_i in the mature frog egg? Does the sperm activate the G-protein via a receptor in the sea urchin egg plasma membrane? It might prove possible to use the G-protein to identify such a receptor. Is the target enzyme for the G-protein PIP_2 phosphodiesterase, or is it a kinase which phosphorylates the lipid precursors of PIP_2? Do G-proteins, activated at fertilization, stimulate other enzymes as well? Do G-proteins directly open or close ion channels? Finally, it will be of great interest to examine whether G-proteins function in the regulation of other aspects of embryonic development, such as cell division, induction, and pattern formation.

ACKNOWLEDGMENTS

We thank Nick Cross, Doug Kline, Greg Kopf, and Clare O'Connor for critical reading of the manuscript. The work from the authors' laboratory was supported by NIH grant HD 14939.

REFERENCES

Baulieu, E. E., Godeau, F., Schorderet, M., and Schorderet-Slatkine, S. (1978). Steroid-induced meiotic division in *Xenopus laevis* oocytes: Surface and calcium. *Nature (London)* 275, 593–598.
Bentley, J. K., Garbers, D. L., Domino, S. E., Noland, T. D., and Van Dop, C. (1986). Spermatozoa contain a guanine nucleotide-binding protein ADP-ribosylated by pertussis toxin. *Biochem. Biophys. Res. Commun.* 138, 728–734.
Birchmeier, C., Broek, D., and Wigler, M. (1985). RAS proteins can induce meiosis in *Xenopus* oocytes. *Cell (Cambridge, Mass.)* 43, 615–621.
Bleil, J. D., and Wassarman, P. M. (1983). Sperm–egg interactions in the mouse: Sequence of events and induction of the acrosome reaction by a zona pellucida glycoprotein. *Dev. Biol.* 95, 317–324.
Bleil, J. D., and Wassarman, P. M. (1986). Autoradiographic visualization of the mouse egg's sperm receptor bound to sperm. *J. Cell Biol.* 102, 1363–1371.
Bornslaeger, E. A., and Schultz, R. M. (1985). Regulation of mouse oocyte maturation: Effect of elevating cumulus cell cAMP on oocyte cAMP levels. *Biol. Reprod.* 33, 698–704.
Bornslaeger, E. A., Mattei, P., and Schultz, R. M. (1986). Involvement of cAMP-dependent kinase and protein phosphorylation in regulation of mouse oocyte maturation. *Dev. Biol.* 114, 453–462.

Chahwala, S. B., Fleischman, L. F., and Cantley, L. (1987). Kinetic analysis of guanosine 5'-O-(3-thiotriphosphate effects on phosphatidylinositol turnover in NRK cell homogenates. *Biochemistry* **26**, 612–622.

Ciapa, B., and Whitaker, M. (1986). Two phases of inositol polyphosphate and diacylglycerol production at fertilization. *FEBS Lett.* **195**, 347–351.

Cicirelli, M. F., and Smith, L. D. (1985). Cyclic AMP levels during the maturation of *Xenopus* oocytes. *Dev. Biol.* **108**, 254–258.

Clapper, D. L., and Lee, H. C. (1985). Inositol trisphosphate induces calcium release from non-mitochondrial stores in sea urchin egg homogenates. *J. Biol. Chem.* **260**, 13947–13954.

Cockcroft, S., and Gomperts, B. D. (1985). Role of guanine nucleotide binding protein in the activation of polyphosphoinositide phosphodiesterase. *Nature (London)* **314**, 534–536.

Dargie, P. J., Agre, M. C., and Lee, H. C. (1986). Parthenogenetic activation of sea urchin egg by microinjection of IP$_3$ and GTP-γ-S. *J. Cell Biol.* **103**, 84a.

Dascal, N., Ifune, C., Hopkins, R., Snutch, T. P, Lübbert, H., Davidson, N., Simon, M. I., and Lester, H. A. (1986). Involvement of a GTP-binding protein in mediation of serotonin and acetylcholine responses in *Xenopus* oocytes injected with rat brain messenger RNA. *Mol. Brain Res.* **1**, 201–209.

Dekel, N., and Beers, W. H. (1980). Development of the rat oocyte in vitro: Inhibition and induction of maturation in the presence or absence of the cumulus oophorus. *Dev. Biol.* **75**, 247–254.

Dorée, M., and Guerrier, P. (1975). Site of action of 1-methyladenine in inducing oocyte maturation in starfish. Kinetic evidence for receptors localized on the cell membrane. *Exp. Cell Res.* **96**, 296–300.

Dorée, M., and Kishimoto, T. (1981). Calcium-mediated transduction of the hormonal message in 1-methyladenine-induced meiosis reinitiation of starfish oocytes. *In* "Metabolism and Molecular Activities of Cytokinins" (J. Guern and C. Peaud-Lenoel, eds.), pp. 338–348. Springer-Verlag, New York.

Dorée, M., Kishimoto, T., Le Peuch, C. J., Demaille, J. G., and Kanatani, H. (1981). Calcium-mediated transduction of the hormonal message in meiosis reinitiation of starfish oocytes. Modulation following injection of cholera toxin and cAMP-dependent protein kinase. *Exp. Cell Res.* **135**, 237–249.

Endo, Y., Lee, M. A., and Kopf, G. S. (1987). Evidence for the role of a guanine nucleotide-binding regulatory protein in the zona pellucida-induced mouse sperm acrosome reaction. *Dev. Biol.* **119**, 210–216.

Eppig, J. J. (1986). Mechanisms controlling mammalian oocyte maturation. *Res. Reprod.* **18**, 1–2.

Freter, R. R., and Schultz, R. M. (1984). Regulation of murine oocyte meiosis: Evidence for a gonadotropin-induced, cAMP-dependent reduction in a maturation inhibitor. *J. Cell Biol.* **98**, 1119–1128.

Gilman, A. G. (1984). G proteins and dual control of adenylate cyclase. *Cell (Cambridge, Mass.)* **36**, 577–579.

Gilman, A. G. (1987). G proteins: Transducers of receptor-generated signals. *Annu. Rev. Biochem.* **56**, 615–649.

Goodhardt, M., Ferry, N., Buscaglia, M., Baulieu, E. E., and Hanoune, J. (1984). Does the guanine nucleotide regulatory protein N$_i$ mediate progesterone inhibition of *Xenopus* oocyte adenylate cyclase? *EMBO J.* **3**, 2653–2657.

Hamaguchi, Y., and Hiramoto, Y. (1981). Activation of sea urchin eggs by microinjection of calcium buffers. *Exp. Cell Res.* **134**, 171–179.

Hildebrandt, J. D., Codina, J., Tash, J. S., Kirchick, H. J., Lipschultz, L., Sekura, R. D., and Birnbaumer, L. (1985). The membrane-bound spermatozoal adenylyl cyclase system does not share coupling characteristics with somatic cell adenylyl cyclases. *Endocrinology (Baltimore)* **116**, 1357–1366.

Howlett, A. C., Sternweis, P. C., Macik, B. A., Van Arsdale, P. M., and Gilman, A. G. (1979). Reconstitution of catecholamine-sensitive adenylate cyclase: Association of a regulatory component of the enzyme with membranes containing the catalytic protein and β-adrenergic receptors. *J. Biol. Chem.* **254**, 2287–2295.

Ishikawa, K., Hanaoka, Y., Kondo, Y., and Imai, K. (1977). Primary action of steroid hormone at the surface of amphibian oocyte in the induction of germinal vesicle breakdown. *Mol. Cell. Endocrinol.* **9**, 91–100.

Jaffe, L. A. (1976). Fast block to polyspermy in sea urchin eggs is electrically mediated. *Nature (London)* **261**, 68–71.

Jordana, X., Allende, C. C., and Allende, J. E. (1981). Guanine nucleotides are required for progesterone inhibition of amphibian oocyte adenylate cyclase. *Biochem. Int.* **3**, 527–532.

Jordana, X., Olate, J., Allende, C. C., and Allende, J. E. (1984). Studies on the mechanism of inhibition of amphibian oocyte adenylate cyclase by progesterone. *Arch. Biochem. Biophys.* **228**, 379–387.

Kanatani, H., and Hiramoto, Y. (1970). Site of action of 1-methyladenine in inducing oocyte maturation in starfish. *Exp. Cell Res.* **61**, 280–284.

Kanatani, H., Shirai, H., Nakanishi, K., and Kurokawa, T. (1969). Isolation and identification of meiosis-inducing substance in starfish *Asterias amurensis*. *Nature (London)* **221**, 273–274.

Kline, D., and Jaffe, L. A. (1987). The fertilization potential of the *Xenopus* egg is blocked by injection of a calcium buffer and is mimicked by injection of a GTP analog. *Biophys. J.* **51**, 398a.

Kline, D., Simoncini, L., Mandel, G., Maue, R., Kado, R. T., and Jaffe, L. A. (1988). Fertilization events induced by neurotransmitters after injection of mRNA in *Xenopus* eggs. *Science* **241**, 464–467.

Kopf, G. S., Woolkalis, M. J., and Gerton, G. L. (1986). Evidence for a guanine nucleotide-binding regulatory protein in invertebrate and mammalian sperm. *J. Biol. Chem.* **261**, 7327–7331.

Kusano, K., Miledi, R., and Stinnakre, J. (1982). Cholinergic and catecholaminergic receptors in the *Xenopus* oocyte membrane. *J. Physiol. (London)* **328**, 143–170.

Lau, A. F., Rayson, T. C., and Humphreys, T. (1986). Tumor promoters and diacylglycerol activate the Na^+/H^+ antiporter of sea urchin eggs. *Exp. Cell Res.* **166**, 23–30.

Lee, H. C., and Garbers, D. L. (1986). Modulation of the voltage-sensitive Na^+/H^+ exchange in sea urchin spermatozoa through membrane potential changes induced by the egg peptide speract. *J. Biol. Chem.* **261**, 16026–16032.

Maller, J. L., and Krebs, E. G. (1977). Progesterone-stimulated meiotic cell division in *Xenopus* oocytes: Induction by regulatory subunit and inhibition by catalytic subunit of adenosine 3′,5′-monophosphate dependent protein kinase. *J. Biol. Chem.* **252**, 1712–1718.

Maller, J. L., Butcher, F. R., and Krebs, E. G. (1979). Early effects of progesterone on levels of cyclic adenosine 3′, 5′-monophosphate in *Xenopus* oocytes. *J. Biol. Chem.* **254**, 579–582.

Manejwala, F., Kaji, E., and Schultz, R. M. (1986). Development of activatable adenylate cyclase in the preimplantation mouse embryo and a role for cyclic AMP in blastocoel formation. *Cell (Cambridge, Mass.)* **46**, 95–103.

Masui, Y., and Clarke, H. G. (1979). Oocyte maturation. *Int. Rev. Cytol.* **57**, 185–282.

Masui, Y., and Markert, C. L. (1971). Cytoplasmic control of nuclear behavior during meiotic maturation of frog oocytes. *J. Exp. Zool.* **177**, 129–146.

Meijer, L., and Zarutskie, P. (1987). Starfish oocyte maturation: 1-methyladenine triggers a drop of cAMP concentration related to the hormone-dependent period. *Dev. Biol.* **121**, 306–315.

Michel, T., and Lefkowitz, R. J. (1982). Hormonal inhibition of adenylate cyclase. α_2 adrenergic receptors promote release of [^{3}H] guanylimidodiphosphate from platelet membranes. *J. Biol. Chem.* **257**, 13557–13563.

Miyazaki, S. (1988). Inositol 1,4,5-trisphosphate-induced calcium release and GTP-binding protein-mediated periodic calcium rises in golden hamster eggs. *J. Cell Biol.* **106**, 345–353.

Moss, J., and Vaughan, M. (1979). Activation of adenylate cyclase by choleragen. *Annu. Rev. Biochem.* **48**, 581–600.

Mulner, O., Megret, F., Alouf, J. E., and Ozon, R. (1985). Pertussis toxin facilitates the progesterone-induced maturation of *Xenopus* oocyte. *FEBS Lett.* **181**, 397–402.

Mumby, S. M., Kahn, R. A., Manning, D. R., and Gilman, A. G. (1986). Antisera of designed specificity for subunits of guanine nucleotide-binding regulatory proteins. *Proc. Natl. Acad. Sci. U.S.A.* **83**, 265–269.

Nagahama, Y., and Adachi, S. (1985). Identification of maturation-inducing steroid in a teleost, the amago salmon *(Oncorhynchus rhodurus)* *Dev. Biol.* **109**, 428–435.

Nomura, Y., Kaneko, S., Kato, K., Yamagishi, S., and Sugiyama, H. (1987). Inositol phosphate formation and chloride current responses induced by acetylcholine and serotonin through GTP-binding proteins in *Xenopus* oocyte after injection of rat brain messenger RNA. *Mol. Brain Res.* **2**, 113–123.

Oberdorf, J. A., Head, J. F., and Kaminer, B. (1986). Calcium uptake and release by isolated cortices and microsomes from the unfertilized egg of the sea urchin *Strongylocentrotus droebachiensis*. *J. Cell Biol.* **102**, 2205–2210.

Oinuma, M., Katada, T., Yokosawa, H., and Ui, M. (1986). Guanine nucleotide-binding protein in sea urchin eggs serving as the specific substrate of islet-activating protein, pertussis toxin. *FEBS Lett.* **207**, 28–34.

Olate, J., Allende, C. C., Allende, J. E., Sekura, R. D., and Birnbaumer, L. (1984). Oocyte adenylyl cyclase contains N_i, yet the guanine nucleotide-dependent inhibition by progesterone is not sensitive to pertussis toxin. *FEBS Lett.* **175**, 25–30.

Owens, J. R., Frame, L. T., Ui, M., and Cooper, D. M. F. (1985). Cholera toxin ADP-ribosylates the islet-activating protein substrate in adipocyte membranes and alters its function. *J. Biol. Chem.* **260**, 15946–15952.

Pike, L. J., and Eakes, A. T. (1987). Epidermal growth factor stimulates the production of phosphatidylinositol monophosphate and the breakdown of polyphosphoinositides in A431 cells. *J. Biol. Chem.* **262**, 1644–1651.

Sadler, S. E., and Maller, J. L. (1981). Progesterone inhibits adenylate cyclase in *Xenopus* oocytes. Action on the guanine nucleotide regulatory protein. *J. Biol. Chem.* **256**, 6368–6373.

Sadler, S. E., and Maller, J. L. (1982). Identification of a steroid receptor on the surface of *Xenopus* oocytes by photoaffinity labeling. *J. Biol. Chem.* **257**, 355–361.

Sadler, S. E., and Maller, J. L. (1983). Inhibition of *Xenopus* oocyte adenylate cyclase by progesterone and 2′,5′-dideoxyadenosine is associated with slowing of guanine nucleotide exchange. *J. Biol. Chem.* **258**, 7935–7941.

Sadler, S. E., and Maller, J. L. (1985). Inhibition of *Xenopus* oocyte adenylate cyclase by progesterone: A novel mechanism of action. *Adv. Cyclic Nucleotide Protein Phosphorylation Res.* **19**, 179–194.

Sadler, S. E., Maller, J. L., and Cooper, D. M. F. (1984). Progesterone inhibition of *Xenopus* oocyte adenylate cyclase is not mediated via the *Bordatella pertussis* toxin substrate. *Mol. Pharmacol.* **26**, 526–531.

Sadler, S. E., Schechter, A. L., Tabin, C. J., and Maller, J. L. (1986). Antibodies to the ras gene product inhibit adenylate cyclase and accelerate progesterone induced cell division in *Xenopus laevis* oocytes. *Mol. Cell. Biol.* **6**, 719–722.

Saling, P. M., Sowinski, J., and Storey, B. T. (1979). An ultrastructural study of epididymal mouse spermatozoa binding to zonae pellucidae in vitro: Sequential relationship to the acrosome reaction. *J. Exp. Zool.* **209**, 229–238.

Schorderet-Slatkine, S., and Baulieu, E. -E. (1982). Forskolin increases cAMP and inhibits progesterone induced meiosis reinitiation in *Xenopus laevis* oocytes. *Endocrinology (Baltimore)* **111**, 1385–1387.

Schorderet-Slatkine, S., Schorderet, M., and Baulieu, E. -E. (1982). Cyclic AMP-mediated control of meiosis: Effects of progesterone, cholera toxin and membrane-active drugs in *Xenopus laevis* oocytes. *Proc. Natl. Acad. Sci. U.S.A.* **79**, 850–854.

Schuel, H. (1978). Secretory functions of egg cortical granules in fertilization and development: A critical review. *Gamete Res.* **1**, 299–382.

Shen, S. S., and Burgart, L. J. (1986). 1,2-diacylglycerols mimic phorbol 12-myristate 13-acetate activation of the sea urchin egg. *J. Cell. Physiol.* **127**, 330–340.

Shilling, F., and Jaffe, L. A. (1987). Evidence that a G-protein mediates 1-methyladenine induced maturation of starfish oocytes. *Biol. Bull. (Woods Hole, Mass.)* **173**, 427.

Speaker, M. G., and Butcher, F. R. (1977). Cyclic nucleotide fluctuations during steriod-induced meiotic maturation of frog oocytes. *Nature (London)* **267**, 848–849.

Stryer, L., and Bourne, H. R. (1986). G proteins: A family of signal transducers. *Annu. Rev. Cell Biol.* **2**, 391–419.

Swann, K., and Whitaker, M. (1985). Stimulation of the Na/H exchanger of sea urchin eggs by phorbol ester. *Nature (London)* **314**, 274–277.

Swann, K., and Whitaker, M. (1986). The part played by inositol trisphosphate and calcium in the propagation of the fertilization wave in sea urchin eggs. *J. Cell Biol.* **103**, 2333–2342.

Swann, K., Ciapa, B., and Whitaker, M. (1987). Cellular messengers and sea urchin egg activation. *In* "Molecular Biology of Invertebrate Development" (J. D. O'Connor, ed.), pp. 45–69. Alan R. Liss, New York.

Tsafriri, A. (1985). The control of meiotic maturation in mammals. *In* "Biology of Fertilization" (C. B. Metz and A. Monroy, eds.), Vol. 1, pp. 221–252. Academic Press, New York.

Turner, P. R., Sheetz, M. P., and Jaffe, L. A. (1984). Fertilization increases the polyphosphoinositide content of sea urchin eggs. *Nature (London)* **310**, 414–415.

Turner, P. R., Jaffe, L. A., and Fein, A. (1986). Regulation of cortical vesicle exocytosis in sea urchin eggs by inositol 1,4,5-trisphosphate and GTP-binding protein. *J. Cell Biol.* **102**, 70–76.

Turner, P. R., Jaffe, L. A., and Primakoff, P. (1987). A cholera toxin-sensitive G-protein stimulates exocytosis in sea urchin eggs. *Dev. Biol.* **120**, 577–583.

Van Renterghem, C., Penit-Soria, J., and Stinnakre, J. (1985). β-adrenergic induced K^+ current in *Xenopus* oocytes: Role of cAMP, inhibition by muscarinic agents. *Proc. R. Soc. London, Ser. B* **223**, 389–402.

Ward, G. E., Brokaw, C. J., Garbers, D. L., and Vacquier, V. D. (1985). Chemotaxis of *Arbacia punctulata* spermatozoa to resact, a peptide from the egg jelly layer. *J. Cell Biol.* **101**, 2324–2329.

Whitaker, M., and Irvine, R. F. (1984). Inositol 1,4,5-trisphosphate microinjection activates sea urchin eggs. *Nature (London)* **312**, 636–639.

Whitaker, M. J., and Steinhardt, R. A. (1985). Ionic signaling in the sea urchin egg at fertilization. *In* ''Biology of Fertilization'' (C. B. Metz and A. Monroy, eds.), Vol. 3, pp. 167–221. Academic Press, New York.

Yatani, A., Codina, J., Brown, A. M., and Birnbaumer, L. (1987). Direct activation of mammalian atrial muscarinic potassium channels by GTP regulatory protein G_K. *Science* **235**, 207–211.

Yoshikuni, M., Ishikawa, K., Isobe, M., Goto, T., and Nagahama, Y. (1988). Characterization of 1-methyladenine binding in starfish oocyte cortices. *Proc. Natl. Acad. Sci., U.S.A.* **85**, 1874–1877.

13

The Relaxation State of Water in Unfertilized and Fertilized Sea Urchin Eggs

SELMA ZIMMERMAN,[*] IVAN L. CAMERON,[†] AND
ARTHUR M. ZIMMERMAN[‡]

[*]Division of Natural Sciences
Glendon College
York University
Toronto, Ontario, Canada, M4N 3M6

[†]Department of Cellular and Structural Biology
The University of Texas Health Science Center at San Antonio
San Antonio, Texas 78229

[‡]Department of Zoology
University of Toronto
Toronto, Ontario, Canada M5S 1A1

I. INTRODUCTION

The nature of water in cells, and its relationship to cell structure and function is a subject of increasing research interest (Drost-Hansen and Clegg, 1979; Beall, 1980; Ling, 1984; Clegg, 1984; Beall *et al.*, 1984). Because of the het-

319

erogenous nature of cellular interfaces, regions of greater and lesser restrictions of the mobility of water molecules might be expected to exist. Two extreme models of cellular water have been proposed by Clegg (1984). The first model considers the cytoplasm as consisting of an aqueous solution in which the water is thought to be similar to pure or bulk water with the exception of a small fraction of bound water (water of hydration) located proximally to intracellular surfaces. The second model of cellular water contends that most if not all the cellular water interacts with macromolecules such that it differs (physically) in its rotational and translocational properties from that of pure water. Since water plays a central biochemical role in all cellular activity we cannot hope to fully understand molecular and cellular function without further information on the properties of cellular water. Such information has recently been gained in studies on changes in the physical–chemical properties of water in unfertilized and fertilized sea urchin eggs.

Nuclear magnetic resonance (NMR) spectroscopy is a method which has been used to study the physical properties of water molecules in various cells and tissues. In this method, energy is transferred to water proton nuclear particles (from outside sources such as magnets and radiofrequency electrical circuits) which disturbs their equilibrium state. One of the measurements made of the time which occurs between the initiation of equilibrium perturbation and the return to an equilibrium condition is called the spin-lattice relaxation which has a characteristic time constant, designated as T_1. The T_1 relaxation time is a measure of the average of relaxation rates of the bulk and hydration water fractions.

Recently, a Fast Proton Diffusion (FPD) analytical approach has been introduced (Fullerton *et al.*, 1983, 1986; Merta *et al.*, 1986) which proposes that multiple fast exchanging hydration compartments constitute the physical status of water in biological systems. According to the FPD model these compartments (or phases) consist of four distinct masses of water which are defined on the basis of their freedom of molecular motion, as dictated by the interactions between water molecules and other chemical species. A bulk water compartment resembling pure water consists of water molecules whose molecular motion depends on interactions only with other water molecules. On the other hand, water of hydration phases, i.e., structured, bound and superbound water compartments consist of water molecules that are motionally perturbed by macromolecules. The FPD analysis of the globular protein, lysozyme, suggests that water molecules in the structured compartment are motionally perturbed but not directly bound to macromolecules; water molecules in the bound phase are directly hydrogen bonded to macromolecular polar sites; and water molecules in the superbound water compartment are bound to ionic sites on macromolecules. Each of these phases can be isolated and characterized by stepwise dehydration and NMR analysis of proton relaxation times (Fullerton *et al.*, 1986; Merta *et al.*, 1986).

II. SEA URCHIN STUDIES

A. NMR Analysis

NMR spectroscopy was used to measure the water proton spin lattice relaxation times (T_1) of unfertilized and fertilized sea urchin eggs of *Strongylocentrotus purpuratus* (Zimmerman *et al.*, 1985, 1987) and *Lytechinus variegatus* (Cameron *et al.*, 1987). Since these measurements relate to water molecules, they can be influenced by differences in the egg sample fluid (seawater). To eliminate this possibility the egg cells were uniformly packed to a constant volume in 1 ml NMR tubes and the free water above the packed eggs was removed prior to T_1 measurements. In addition, since jelly coats were found to affect T_1 relaxation time, they were routinely removed from the sea urchin eggs with acidified seawater (pH 5) prior to packing cells in the NMR sample tubes. T_1 (spin-lattice) relaxation times were measured immediately after sample preparation using a Praxis model II instrument (Praxis Corp., San Antonio, Texas) equipped with a 0.25 tesla permanent magnet, a sample coil, and R. F. pulser tuned to 10.7 MHz. The pulsed proton NMR relaxation analyses enlisted a saturation recovery pulse sequence of 90°-τ-90°. An interfaced microcomputer provided rapid data acquisition and analysis. The T_1 decay curve is the product of the resultant analysis of 30 free-induction-decay (FID) peak heights with a sequence of increasing interpulse delay times (Fullerton *et al.*, 1986).

In order to assess the possible influence of differences in intracellular water content between unfertilized and fertilized sea urchin eggs on T_1 relaxation time measurements, it was necessary to determine the water content and cell volume of the cells at the developmental stages studied. Water content was assessed by weighing samples from NMR tubes in pretared weighing pans followed by dehydration in a vacuum oven at 90°C until a stable weight was reached. The difference between the initial wet weight and the final dry weight of the samples was used to determine the percentage of water in the cells. There were no significant differences found in water content during the first cell cycle of fertilized eggs (Fig. 1). Cell volume calculations were made from diameter measurements which were made with the aid of an ocular micrometer in conjunction with a Zeiss microscope (100× magnification). For volume measurements, egg flattening was averted by using a microslide without a coverslip, and by using a water immersion lens (Fig. 2). Volume calculations made from the egg diameter measurements yielded the following values: (1) unfertilized eggs averaged $3.159 \times 10^5 \pm 0.054$ μm³ ($n = 30$), and (2) fertilized eggs proper possessed an average volume of $3.35 \times 10^5 \pm 0.09$ μm³ ($n = 30$). The volume of the unfertilized egg and the volume of the fertilized egg proper was not significantly different ($p > .05$). Following fertilization the elevation

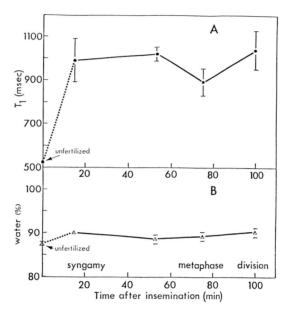

Fig. 1. Water proton T_1 (A) and the percentage of water (B) in eggs of the sea urchin *S. purpuratus* are shown at various stages of the cell cycle. Each measurement was made on a separate freshly packed sample of eggs. The T_1 at metaphase is statistically lower than the other T_1 values in the fertilized eggs. The water content of the fertilized and unfertilized eggs did not change at each of the stages. [From Zimmerman *et al.* (1985).]

of the fertilization membrane was responsible for a large overall increase in egg volume. The photomicrograph in Fig. 2 illustrates this change in overall volume at fertilization.

NMR measurements of unfertilized and fertilized sea urchin eggs revealed a major increase in T_1 water proton relaxation time upon fertilization. In *S. purpuratus* (Zimmerman *et al.*, 1985) the mean water proton relaxation time of unfertilized eggs was 520 msec. Following fertilization, the T_1 time rose to 991 msec. *Lytechinus variegatus* eggs showed an increase in T_1 time from 385 msec to 929 msec at fertilization. This increase was in large measure accounted for by the accumulation of extracellular water in the perivitelline space (Zimmerman *et al.*, 1985, 1987; Merta *et al.*, 1986; Cameron *et al.*, 1987). Analyses of the contribution of water in the perivitelline space compartment to T_1 relaxation measurements of fertilized eggs were carried out by chemical dissection of the fertilization membrane and hyaline layer. In experiments on *S. purpuratus* (Zimmerman *et al.*, 1987) the fertilization membrane was prevented from forming by treating unfertilized eggs with 0.01% panprotease for 10 min which enzymatically removes the vitelline membrane from which the fertil-

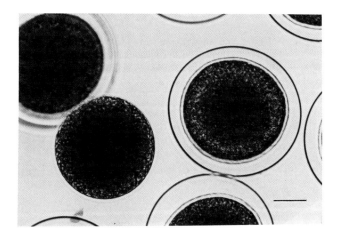

Fig. 2.   Photomicrograph of unflattened, unfertilized and fertilized sea urchin eggs. Thirty unfertilized eggs like that on the left average $3.16 \times 10^5 \pm 0.05$ μm^3 in volume. After fertilization, the volume of the fertilized egg proper was $3.35 \times 10^5 \pm 0.09$ $((m) = 30)$ μm^3 (not significantly changed from the unfertilized egg), but with the elevation of the fertilization membrane (right), the overall volume of the egg increased about 2-fold. Bar, 20 μm. [From Merta *et al.* (1986).]

ization membrane is formed. Another method used to remove the fertilization membrane was dithioerythritol treatment which interferes with disulfide bonding in the newly forming fertilization membrane (Epel *et al.*, 1970). Yet another method of fertilization membrane removal was used on *L. variegatus* eggs. These eggs were treated with 1 μM 3-amino-1,2,4-triazole for 3 min at 6 min postfertilization and then passed through a 4× nylon screen to remove the membranes (Cameron *et al.*, 1987). Sea urchin eggs subjected to any one of the above treatments do not develop a fertilization membrane but do develop an intact hyaline layer following insemination with sperm. The hyaline layer was prevented from forming by culturing such eggs with Ca^{2+}-free sea water.

In experiments with *S. purpuratus* (Zimmerman *et al.*, 1987), proton T_1 relaxation time measurements as well as cell volume measurements were made on (1) unfertilized eggs, (2) fertilized eggs, (3) fertilized eggs with a hyaline layer but with the fertilization membrane removed, and (4) fertilized eggs without the fertilization membrane and the hyaline layer (Table I). There was a significant decrease in both T_1 time and volume in the absence of a fertilization membrane. Removal of the hyaline layer caused a further significant decrease in T_1 time and volume. The T_1 time of fertilized eggs without the fertilization membrane and hyaline layer did not differ significantly from the T_1 of unfertilized eggs. Thus, each of the extraneous coats contributes to the

TABLE I

Analysis of the Proton T_1 Relaxation Time (msec) and of Cell Volume of (1) the Unfertilized Egg, (2) the Fertilized Egg within the Raised Fertilization Membrane (FM), (3) Fertilized Egg plus the Hyaline Layer but less the Fertilization Membrane, and (4) Fertilized Egg less both the Hyaline Layer and Fertilization Membrane[a,b]

		Extracellular membrane		
	(1)	(2)	(3)	(4)
FM		+	−	−
Hyaline		+	+	−
Treatments			Fertilized after PP or DTE treatment	Washed 3× in Ca^{2+}-free seawater
No treatment	307	644		
	325	692		
	387	611		
	342	580		
	335	538		
Panprotease (PP)	277		423	456
treated prior to	344		489	371
insemination	357		443	284
Dithioerythritol	327		440	394
(DTE) treated	336		533	297
	362		445	293
Mean ± SE	336.3 ± 8.3	613.0 ± 23.6	462.2 ± 15.3	390.8 ± 21.0

[a]The justification for pooling T_1 values from different treatments in each column is that the different treatments did not cause significantly different mean T_1 values as tested by ANOVA.

[b]Results of ANOVA: F ratio = 47; p of F value = <.001. S/N/K multiple range test showed that all column means except (1) and (4) to be significantly different ($p < .05$).

Volume in μm^3 ± SE: (1) $3.17 \times 10^5 \pm 0.10$; (2) $6.27 \times 10^5 \pm 0.11$; (3) $3.60 \times 10^5 \pm 0.04$; (4) $2.57 \times 10^5 \pm 0.15$. Results of ANOVA: F ratio = <0.0001, S/N/K multiple range test shows all means to be significantly different; $p < .05$; n = 22–58 eggs for each condition.

overall proton T_1 relaxation time in the fertilized eggs. Similar results were obtained with *L. variegatus* eggs (Cameron *et al.*, 1987).

B. FPD Analyses

The water in unfertilized and fertilized *Strongylocentrotus* eggs has been further characterized with the FPD model to define the behavior of the intracellular water compartments (Merta *et al.*, 1986). With this method each fast-exchanging water compartment was sequentially removed by dehydration, and the T_1 relaxation rates and water content of each compartment was determined. The NMR titration series, conducted at sequential hydration states, consistently yielded single exponential T_1 relaxation behavior throughout the process for both unfertilized and fertilized egg samples. The T_1 relaxation rates $1/T_1(sec^{-1})$, are plotted against corresponding concentration levels (grams solid per 100 g water) (Figs. 3 and 4). In this context, "solid" refers to dry mass, which consists of all cellular components except water.

The NMR titration data for the unfertilized and fertilized sea urchin eggs is graphically illustrated in Fig. 3A, and B, respectively. Linear regression and exponential curve analyses for each set of data were conducted to ascertain the best possible fit. The linear curve "fits" gave higher correlation coefficients. The linear correlation values (as shown on Table II) reveal three linear segments. Each line corresponds to a distinct water compartment, which is characterized by the slope and intercepts of this line. The initial, second, and third line segments demarcate the bulk, structured, and bound water phases, respectively. Points of intersection enable one to determine the mass of water in each fraction. Table II summarizes these analyses and lists the amount of water in each of the compartments. The range of values shown in Fig. 3 did not require specimen heating which could cause denaturation and FPD analysis is therefore applied only to data in this range.

To gain information on possible subcompartments within the bound water compartment, Merta *et al.*, (1986) collected NMR titration data over a wider range of hydrations than could be obtained by vacuum dehydration at room temperature. Upon acquisition of a constant mass, dehydration was further promoted by a daily 10°C stepwise increase in temperature up to 90°C. The extended range of T_1 relaxation values obtained by this procedure is shown in Fig. 4A and B. The horizontal line segment $1/T_1 = 10.686 \pm 0.18 \ sec^{-1}$ represents the T_1 relaxation rate of the pooled, dry solid fractions after lipid extraction. Intersection of this line with the third line segment from the left gives an estimate of the superbound water compartment. In the unfertilized eggs (Fig. 4A) the intersection is 26.307, 10.686. In the fertilized eggs (Fig. 4B) the intersection is 27.386, 10.686. The superbound water compartment is therefore considered to be a subcompartment of the bound water compartment determined by the FPD model analysis (Table II). The superbound water compartment amounts to 3.80 and 3.65 g of water/100 grams of dry solids for the unfertilized and fertilized eggs, respectively.

Table III summarizes the amount of water in each of the water compartments in unfertilized and fertilized eggs.

According to the FPD model, the bulk water compartment is made up of unperturbed bulk water molecules which comprise the largest phase in the eggs. Unfertilized and fertilized eggs contained 238 and 470 g of water/100 g of solids, respectively, by gravimetric measurements (Table III). Of this water,

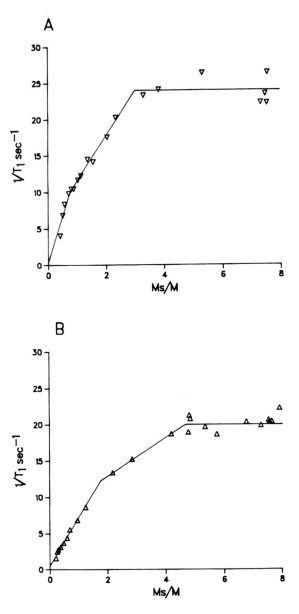

Fig. 3. Fast proton diffusion (FPD) model plots of unfertilized (A) and fertilized (B) sea urchin eggs. Each point represents an independent T_1 (spin lattice) relaxation rate measurement, $1/T_1(sec^1)$, at its respective hydration level (Ms/M, gram dry mass per gram water). Both $1/T_1$ and Ms/M values can be readily converted to T_1 relaxation time (sec) and gram water per gram dry mass respectively, by taking the reciprocal ($1/x$) of any coordinate value from the plots. Line and curve best fit regression analysis yielded three linear segments on

128 and 56 g of water/100 g of solids are in the respective hydration compartments. Thus 111 and 414 g of water/100 g of solid are in the bulk water phase for the unfertilized and fertilized cells, respectively. The corresponding $1/T_1$ rates of bulk water in the unfertilized and fertilized eggs, as extrapolated from the FPD plot (Fig. 3 and Table 2) were 0.272 sec^{-1} (T_1 = 3861 msec) and 0.570 sec^{-1} (T_1 = 1799 msec), respectively; these relaxation rates in unfertilized and fertilized eggs were not statistically different from the rate for bulk water 0.37 sec^{-1} due to the large relative spread, (SEE) standard error of estimation = 0.913 and 0.259, respectively.

The structured water compartment is made up of water molecules that are motionally perturbed by, but not bound to macromolecules (Fullerton *et al.*, 1986). In unfertilized eggs, this fraction contains 94.5 g of water/100 g dry mass. The T_1 time for the structured water compartment is 191 ± 39 msec for unfertilized eggs and 131 ± 28 msec for fertilized eggs.

The bound water compartment, in which molecules are believed to be either ionically or hydrogen bonded to macromolecules, changed from 33 g for unfertilized to 21 g per 100 g dry mass in fertilized eggs. Extrapolated T_1 time values for this bound water compartment in unfertilized and fertilized eggs were 42 ± 1.2 msec and 50 ± 0.6 msec, respectively (Table II).

Finally, the superbound water compartment, comprised of water molecules that are bound to ionic sites on macromolecules (Fullerton *et al.*, 1986), was roughly estimated for each egg sample. Unfertilized and fertilized eggs yielded 3.80 and 3.65 g/100 grams dry mass, respectively. These quantities of water are subfractions of the bound water compartment.

An analysis of the above data, as determined by the FPD model (Table IV) reveals the contribution that each of the three water fractions make to the overall spin-lattice relaxation rate of unfertilized and fertilized eggs. The results in Table IV indicate that in unfertilized eggs, the relaxation rate is dominated by the structured (38%) and bound (60%) water fractions, while the bulk water compartment contributes only 2%. In fertilized eggs, the contribution of the bulk water fraction increases to 44%, while the contributions from the hydration water fraction (structured plus bound) is proportionately decreased. Can the uptake of 232 g bulk water/100 g dry mass, that occurs upon fertilization, account for the overall spin-lattice rate change? As can be seen in Table IV, the calculated T_1 relaxation rate change due to influx of 232 g water/

---

each of the plots. The values of Ms/M >3.03 for unfertilized eggs and >4.69 for unfertilized eggs were independent of concentration and were averaged. The three segments, from left to right, represent bulk, structured, and bound water compartments. The parameters of each compartment were obtained from points of intersection and *Y*-intercepts of the respective line segment. [From Merta *et al.* (1986).]

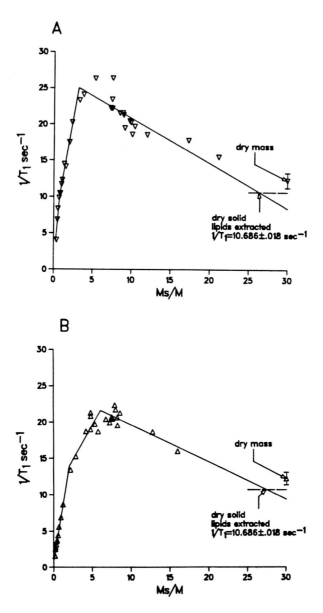

Fig. 4. Graphic representation of the further dehydration of unfertilized (A) and fertilized (B) egg samples by heating. A constant mass was achieved on day 27 of vacuum dehydration. Dehydration was further promoted by temperature elevation (a daily 10°C stepwise increase in temperature up to 90°C) in vacuum oven. After 9 days of heating, a lipid extraction was carried out. An estimate of the mass of superbound water, i.e., that fraction ionically bound to macromolecules, was obtained from the data by calculating the point of intersection between the line segment which decreased in slope and the line $1/T_1 = 10.686 \pm 0.18$ sec^{-1}, which corresponds to the relaxation rate of the pooled, dry solid fraction (lipid extracted). [From Merta *et al.* (1986).]

TABLE II

Characterization of Water Compartments in Sea Urchin Eggs[a,b]

Water compartments	Specifications	Unfertilized eggs	Fertilized eggs
Bulk	Line 1 equation	$1/T_1 = 12.54 \times Ms/M + 0.272^c$	$1/T_1 = 6.62 \times Ms/M + 0.570^c$
	N value	7	10
	Correlation coefficient	0.976	0.995
	Standard error of estimate	0.91260	0.25930
	p value	0.0002	0.000
	Point of intersection with line 2	(0.782, 10.074)	(1.774, 12.313)
	T_1 time of compartment	3676 ± 1716 msec	1754 ± 541 msec
	Mass of water in compartment	110.79 g/100 g	415.33 g/100 g
Structures	Line 2 equation	$1/T_1 = 6.21 \times Ms/M + 5.22^c$	$1/T_1 = 2.64 \times Ms/M + 7.63^c$
	N value	6	6
	Correlation coefficient	0.985	0.971
	Standard error of estimate	0.62589	0.84080
	p value	0.0003	0.0013
	Point of intersection with line 3	(3.027, 23.997)	(4.691, 20.013)
	T_1 time of compartment	191 ± 39 msec	131 ± 28 msec
	Mass of water in compartment	94.48 g/100 g	34.28 g/100 g

(continues)

TABLE II (continued)

Water compartments	Specifications	Unfertilized eggs	Fertilized eggs
Bound	Line 3 equation	$1/T_1 = 23.997\ \text{sec}^{-1}$	$1/T_1 = 20.013\ \text{sec}^{-1}$
	N value	7	7
	Standard error of estimate	0.658	0.256
	T_1 time of compartment	42 ± 1.2 msec	50 ± 0.6 msec
	Mass of water in compartment	33.04 g/100 g	21.32 g/100 g

[a]From Merta et al. (1986).

[b]FPD model characterization of unfertilized and fertilized sea urchin eggs. From the fast proton diffusion (FPD) model plot (Fig. 3) of the NMR titration data, individual water compartments in the unfertilized and fertilized egg samples were evaluated according to previously reported methods (Fullerton et al., 1983, 1986). The mass of water in each compartment was calculated through the isolation of each region (as defined by a line segment) by way of the respective points of intersection with neighboring line segments.

[c]Ms/M, gram dry mass per gram water.

TABLE III

Water Compartments in Unfertilized and Fertilized Sea Urchin Eggs (g water/
100 g dry mass)[a]

Compartment	Unfertilized	Fertilized	Change on fertilization
Superbound	3.80[b]	3.65[b]	−0.15 (−4%)[c]
Polar bound	29.2	17.7	−11.5 (−39%)
Total bound	33.0	21.3	−11.7 (−35%)
Structured	94.5	34.3	−60.2 (−64%)
Total water of hydration	127.5	55.6	−71.9 (−56%)
Bulk intracellular	111.0	182.4[d]	+71.4 (+64%)
Bulk perivitelline	0.0	232.0[d]	+230.0
Total bulk	111.0	414.4[d]	+303.4 (+273%)
Grand Total	238.0	470.0	+232.0 (+98%)

[a]From Merta et al. (1986).
[b]From the data reported in Fig. 4 (see text).
[c]Percentage change.
[d]Calculated from the fact that the volume of the egg proper did not show significant differences between fertilized and unfertilized eggs (see text).

100 g dry mass is 2.92. This calculated rate differs from the measured rate (1.58) with an error of + 84.8%. Thus Merta et al. (1986) verified that most of the change in relaxation rate at fertilization can be accounted for by uptake of bulk water as suggested by Zimmerman et al., (1985). However the calculated relaxation rate of the fertilized eggs using the FPD model (1.15) differs from the measured rate (1.58; Zimmerman et al., 1985) with an error of only −27.2%. Thus use of the FPD model which takes into account the relaxation rate of each water compartment gives a more accurate explanation of the overall relaxation rate of fertilized eggs than does the assumption that all of the relaxation rate change at fertilization is due to uptake of bulk water.

C. Changes Accompanying Fertilization That May Be Related to Hydrational Modifications

As shown previously dramatic alterations in the cellular hydration compartments coincide with fertilization in the sea urchin egg (Tables II and III). This hydration modification may be associated with other intracellular changes which occur at fertilization. For example, shifts in concentration of various intracellular ions are known to occur at fertilization (Schmidt et al., 1982). A transient rise in Ca^{2+} concentration follows the sperm-induced release of Ca^{2+}

TABLE IV

Relative Fraction and Spin-Lattice Relaxation Rate Data for Unfertilized and Fertilized Sea Urchin Eggs[a,b]

	Bulk water			Structured water			Bound water			R_1 calculated[c]	R_1 measured[d]	Precentage of error calculated to measured
	fw	Rw	fw · Rw	fs	Rs	fs · Rs	fb	Rb	fb · Rb			
Unfertilized	0.466	0.272	0.127 (2%)	0.397	5.24	2.08 (38%)	0.138	23.8	3.28 (60%)	5.49	4.05	+35.5%
Fertilized	0.882	0.570	0.503 (44%)	0.073	7.63	.557 (48%)	0.045	20.0	0.090 (8%)	1.15	1.58	−27.2%
Calculated[c] change due to influx of 232 g water/100 g dry mass	0.730[e]	0.272	0.199 (7%)	0.201[e]	5.24	1.053 (36%)	0.70[e]	23.8	1.666 (57%)	2.92	1.58	+84.8%

[a]From Merta et al. (1986).

[b]Numbers in parenthesis represent the percentage of contribution to the overall calculated relaxation rate.

[c]$R_1 = 1/T_1 = fw \cdot Rw + fs \cdot Rs + fb \cdot Rb$, where fw, fs, and fb and Rw, Rs, and Rb are the relative fractions of water and relaxation rates for the bulk water (w), structured water (s), and bound water (b) fractions, respectively.

[d]Measured within 30 min after the eggs were centrifuged.

[e]Total relative water mass = 238 + 232 = 470 g H_2O/100 g dry mass; fw = 343/470 = 0.730; fs = 94.5/470 = 0.201; fb = 33.0/470 = 0.070.

stores. This event is accompanied by an influx of Na^+ and a rapid depolarization of the egg plasma membrane (Schmidt *et al.*, 1982). A Na^+-dependent efflux of H^+ results in a rise in the internal pH from approximately 0.4 to 0.6 pH units to about pH 7.2 (Lee and Epel, 1983). The increase in intracellular pH triggers an increase in protein synthesis as preexisting messenger RNAs are unmasked and ribosomes become activated. In addition, nucleotide and amino acid transport is activated (Rebhun *et al.*, 1982). Thus numerous ionic changes may be related to the observed alterations of intracellular water at fertilization. Events, such as changes in concentration of ions might profoundly affect the conformation-dependent water binding and structuring capacities of macromolecules such as proteins.

Extensive reorganization of the cytoskeletal framework also accompanies fertilization. For example, cortical reserves of G-actin are rapidly polymerized (Spudich *et al.*, 1982) and soluble actin drops from 0.58 ng/egg in the unfertilized egg to 0.40 ng/egg in the fertilized egg (Otto *et al.*, 1980). At the same time, cortical actin goes from an average of 0.056 ng/egg in the unfertilized egg to 0.24 ng/egg in the fertilized egg (Otto *et al.*, 1980). Thus a correlation may exist between actin polymerization and a decreased hydration fraction. Although actin is only 1.4% of total egg protein, it constitutes 7% of the soluble protein in unfertilized egg extracts (Bryan and Kane, 1982) and may still play a major role in changes in the cellular hydration fraction. For example, F-actin in the cytoplasmic environment readily binds proteins (Clegg, 1984; Schliwa *et al.*, 1981; Moon *et al.*, 1983) and polyribosomes (Moon *et al.*, 1983). Hence by decreasing the interactive availability of surfaces of some intracellular macromolecules, actin could effectively limit their ability to perturb water motion.

D. Changes in Water Proton Relaxation Time during the Cell Cycle

During the first cell cycle, the T_1 relaxation time shortened at mitosis and lengthened at cytokinesis with no change in water content (Fig. 1) (Zimmerman *et al.*, 1985, 1987; Cameron *et al.*, 1987). Zimmerman *et al.*, (1987) and Cameron *et al.*, (1987) examined the possibility that events associated with the perivitelline space were responsible for T_1 time fluctuations during mitosis and cytokinesis. They found no differences in the cell cycle pattern of T_1 relaxation time changes between normal eggs and eggs that lacked a fertilization membrane and hyaline layer. This indicates that the fluctuations in T_1 relaxation time in sea urchin eggs during mitosis and cytokinesis are due to intracellular events.

Cameron *et al.* (1985, 1987) demonstrated that it is possible to examine changes in the state of hydration water in the cytoplasm by analyzing changes in the growth of intracellular ice crystals upon rapid freezing. The growth of ice crystals in biological materials is dependent on the water content of the cell and its macromolecular composition. Ice crystal imprint size is limited by the number of water molecules that can migrate to an ice crystal nidation site prior to cooling to the point where complete freezing of the cell occurs. Cellular water is structured by its interactions with protein surfaces.

Water molecules in the structured water state diffuse more rapidly within the layer along a protein surface than they can diffuse away from the protein surface into the unstructured water surroundings. Thus, cells with long filamentous protein assemblages provide longer surfaces along which water can migrate to an ice crystal nidation site. It would then appear that two cells of equal water content may show different sized ice crystals depending upon the globular or filamentous nature of the macromolecules in the cell. In this way, ice crystal size is a measure of the amount of hydration water and length of surface along which water can move in the cell (Cameron *et al.*, 1985).

Ice crystal imprint size was examined in thin sections of *L. variegatus* eggs during the first cell cycle (Cameron *et al.*, 1987). These eggs were frozen in liquid propane at regular intervals during the first cell cycle; the frozen samples were sectioned to a thickness of 1–2 μm, cryoabsorbed, examined, and photographed in the scanning transmission electron microscope. The ice crystal imprint size in the cytoplasm of the eggs showed cell cycle-dependent changes in radius size which paralleled those seen with measurements of T_1 relaxation time (Fig. 5). Thus, there was a drop in ice crystal size during mitosis and a rise to premitosis levels during cleavage.

E. Relationship of Cytoskeletal Changes to Changes in Water Proton Relaxation Time during the Cell Cycle

What macromolecular events might account for the pattern of water ordering observed during the cell cycle? Cytoskeletal changes such as tubulin polymerization occur at mitosis, and actin polymerization accompanies fertilization (Spudich *et al.*, 1982) and is involved in the contractile ring and cortical microfilaments which form at cleavage. In addition, the ice crystal data discussed previously suggests that T_1 time measurements during the cell cycle reflect the state of the macromolecules. Thus, changes in the egg cytoskeleton at fertilization and during the cell cycle may be responsible for changes in the water proton NMR signal observed.

The possible role that the cytoskeletal protein tubulin might play in water proton relaxation time changes was examined with the use of taxol, a microtubule inducing agent, and colchicine and low temperature which depolymerize microtubules (Zimmerman *et al.*, 1985, 1987). Treatment of unfertilized sea

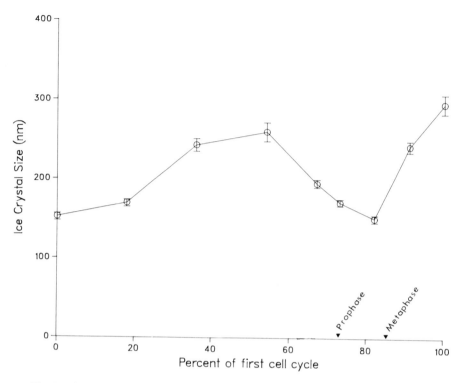

Fig. 5. Cell cycle-dependent changes in cytoplasmic ice crystal size during the first cell cycle of *L. variegatus* eggs. A small sample of concentrated eggs was frozen in liquid propane cooled in liquid nitrogen. Frozen 1-µm sections of frozen eggs were examined in a scanning electron microscope, and the radii of cytoplasmic ice crystal imprints were measured at various times after insemination. The data is presented as the percentage of first cell cycle to account for variability in the cleavage time among the different batches of eggs. Error bars indicate the standard error of the mean. [From Cameron *et al.* (1987).]

urchin eggs with 10 µM taxol for 120 min caused the formation of numerous cytasters containing microtubules. When 10 µM colchicine was added to such taxol treated cells, the cytasters disappeared. The data in Table V shows that the T_1 relaxation time of the unfertilized eggs was not significantly changed due to drug treatment (Zimmerman *et al.*, 1987). In other experiments, taxol (10 µM) was added to cultures of fertilized eggs 55 min postfertilization. At 75 min postfertilization, the mitotic spindles of the taxol-treated eggs were observed to be much larger than that of the nontaxol treated eggs. The T_1 relaxation time of the eggs with a taxol enlarged mitotic spindle did not vary significantly from that of the normal-size mitotic spindle. In addition, eggs in

TABLE V

Analysis of the Effects of Taxol and Colchicine Treatment on the Proton T_1 Relaxation Time and on Cell Volume of Unfertilized Sea Urchin Eggs

Treatment and duration	T_1 relaxation time		Cell volume	
	Number of runs (n)	Mean ± SE (msec)	Number of eggs (n)	Mean ± SE (μm^3)
Taxol (120 min)	4	297 ± 7.3	21	$3.17 \times 10^5 \pm 0.5$
Colchicine (30 min)	4	284 ± 4.2	48	$3.39 \times 10^5 \pm 0.4$
Taxol (120 min) + Colchicine (last 30 min)	4	306 ± 6.5	48	$3.48 \times 10^5 \pm 0.3$
Untreated	4	279 ± 9.0	21	$3.35 \times 10^5 \pm 0.9$
Results of ANOVA		F ratio = 0.216 (not significant)		F ratio = 1.17 (not significant)

mitosis were treated with colchicine to examine the role of microtubule assembly (associated with the mitotic apparatus) in the increase in water order observed at metaphase. The colchicine-induced disassembly of the mitotic spindle in sea urchin eggs did not affect the proton T_1 relaxation time. Thus, neither taxol-induced assembly nor colchicine-induced disassembly of cellular microtubules could be linked to measurable changes in the proton T_1 relaxation time in fertilized sea urchin eggs.

The involvement of the cytoskeletal protein actin in changes in the physical properties of water during the cell cycle was studied with the use of cytochalasin B which disorganizes actin filaments. Sea urchin eggs were treated with 25 μM cytochalasin B prior to the beginning of mitosis (Cameron *et al.*, 1987; Zimmerman *et al.*, 1987). The results of these experiments showed that the fall in T_1 relaxation time during mitosis was not inhibited by cytochalasin B, but the rise in T_1 relaxation time at cleavage was completely abolished (Fig. 6). Thus the correlation between G- and F-actin transition and the extent of cellular hydration water which occurs shortly after fertilization may also occur at mitosis and cytokinesis.

This postulated pattern of polymerization of actin after fertilization, depolymerization at mitosis, and polymerization at cleavage is consistent with the observations of a number of investigators. After fertilization in *S. purpuratus*, there is a burst of microvillar elongation, followed by a second burst that ceases prior to mitosis, but at no time during the first cell cycle is there

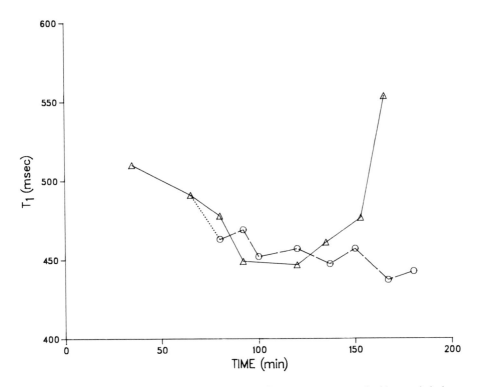

Fig. 6. Water proton T_1 relaxation time in eggs of *S. purpuratus* treated with cytochalasin B. Fertilized eggs (at 15°C) were treated with 25 μm cytochalasin B at 70 min after insemination. T_1 values of cytochalasin-treated (o--o) and control cells (Δ———Δ) are plotted as a function of time after insemination. At 120 min, 65% of control cells displayed furrows. The cytochalasin-treated cells showed a decrease in T_1 values at mitosis, similar to control cells. At cytokinesis, the control cells displayed an increase in T_1 values; however, the T_1 values of cytochalasin-treated cells remained low. [From Zimmerman *et al.* (1987).]

any evidence for a shortening of microvilli (Schroeder, 1979). Otto *et al.* (1980) observed that prior to fertilization in the Hawaiian sea urchin, *Tripneustes gratilla,* 9.2% of the total actin in the egg polymerized. After fertilization, 35% of the total actin was polymerized. They attributed this change in the amount of actin polymerized to microvillar elongation. There are two cytological events at the time of cleavage involving the polymerization of actin. Usui and Yoneda (1982) reported the appearance of a meshwork of actin filaments in the egg cortex at anaphase of the first cell cycle which increased in density during telophase. The other event involving actin at the time of cleavage is the assembly of the contractile ring. The source of contractile-ring actin is not known. It has not been ascertained whether the contractile ring assembles from a pool of monomeric actin, or whether actin in the cortex depolymerizes to provide actin subunits (Mabuchi, 1986). Another physical property of the egg that is related to the state of actin polymerization is protoplasmic viscosity. The pattern of changes in the consistency of the protoplasm (Hiramoto, 1970) shows similar cell cycle-dependent changes as seen with T_1 relaxation time and ice crystal imprint size. While all these observations are consistent with the changes in T_1 relaxation time and ice crystal imprint size being related to the state of polymerization of actin, they are far from conclusive evidence. There is little evidence to indicate that there is a depolymerization of actin at mitosis in a large enough proportion of cellular actin to cause the fall in T_1 relaxation time and ice crystal imprint size. However, the studies cited were concerned with cortical actin, and none dealt with actin in the rest of the egg, the cytomatrix actin. Events involving the state of cytomatrix actin polymerization may be more important to changes in T_1 relaxation time and ice crystal imprint size than events involving cortical actin.

Since the polymerization and depolymerization of actin may be an explanation for the changes in T_1 relaxation time and ice crystal imprint size *in vivo*, the changes in the physical properties of water upon polymerization of purified actin were investigated as an *in vitro* correlate to the changes in T_1 relaxation time and ice crystal imprint size observed in *L. variegatus* eggs. To do this, T_1 relaxation times and ice crystal radii were measured for an actin solution at various concentrations of potassium chloride, a salt known to promote the polymerization of actin (Bryan and Kane, 1982). The polymerization of actin was evident macroscopically as an increase in viscosity and gelation of the solution and microscopically as the presence of fibrous mats. The ice crystal radius was seen to increase upon polymerization of the actin solution. This was attributed to the change of the actin from the globular monomer state to a filamentous state because there was no change in water content.

The results of the *in vitro* experiments with actin are consistent with an interpretation of the changes in the properties of intracellular water, during the first cell cycle, being related to actin polymerization and depolymerization. However, the transition of a portion of the total cellular actin from a globular to a filamentous form cannot explain completely the T_1 relaxation time and ice crystal imprint size changes.

It has been shown that there are cell cycle-dependent changes in the motional properties of cellular water in sea urchin eggs, as reflected by changes in proton NMR T_1 relaxation time and ice crystal imprint size. The state of polymerization of cytoplasmic actin is suggested as the most likely macromolecular feature of the egg that could account for these changes, based on experiments by Cameron *et al.* (1987) and Zimmerman *et al.* (1987) showing that cytochalasin B treatment disrupts the pattern of changes in T_1 time in the egg. The postulated pattern of cell cycle-dependent changes in the state of actin polymerization is not inconsistent with observations in actin *in vivo*, and changes in the motional properties of water due to this postulated pattern would be consistent with the characteristics of water associated with purified actin polymerized *in vitro*. The polymerization or depolymerization of cytoplasmic actin could not account for the magnitude of T_1 relaxation time changes, suggesting that in addition to the state of actin polymerization, other macromolecular events must occur to make this hypothesis tenable. Clegg (1984) has suggested that the association of cytoplasmic proteins with actin filaments in the cytomatrix changes the extent of hydration water in the cytoplasm. If this is true, the changes in T_1 relaxation time and ice crystal imprint size during the first cell cycle of *L. variegatus* eggs not accounted for by actin polymerization could conceivably be due to the "docking" of cytoplasmic proteins to filamentous actin in the cytomatrix. Thus one might envision that each globular protein molecule has a layer of hydration water and that the polymerization of such globular proteins releases hydration water from between the docked molecules to become bulk water. Such a model can be used to explain the water relaxation and hydration water changes which occur at fertilization and during the first cell cycle in sea urchin eggs.

REFERENCES

Beall, P. T. (1980). Water–molecular interactions during the cell cycle. *In* "Nuclear-Cytoplasmic Interactions in the Cell Cycle" (G. L. Whitson, ed.), pp. 223–247. Academic Press, New York.

Beall, P. T., Amtey, S. R., and Kasturi, S. R. (1984). "NMR Data Handbook for Biomedical Applications." Pergamon, Oxford.

Bryan, J., and Kane, R. E. (1982). Actin gelation in sea urchin egg extracts. *Methods Cell Biol.* **25,** 175–199.

Cameron, I. L., Hunter, K. E., Ord, V. A., and Fullerton, G. D. (1985). Relationships between ice crystal size, water content and protein NMR relaxation times in cells. *Physiol. Chem. Phys. Med. NMR* **17,** 371–386.

Cameron, I. L., Cook, K. R., Edwards, D., Fullerton, G. D., Schatten, G., Schatten, H., Zimmerman, A. M., and Zimmerman, S. (1987). Cell cycle changes in water properties in sea urchin eggs. *J. Cell Physiol.* **133,** 14–24.

Clegg, J. S. (1984). Properties and metabolism of the aqueous cytoplasm and its boundaries. *Am. J. Physiol.* **246,** R133–R151.

Drost-Hansen, W., and Clegg, J. S., eds. (1979). "Cell-Associated Water." Academic Press, New York.

Epel, D., Weaver, A. M., and Mazia, D. (1970). Methods for removal of the vitelline membrane of sea urchin eggs. *Exp. Cell Res.* **61,** 64–68.

Fullerton, G. D., Seitz, P. K., and Hazlewood, C. F. (1983). Application of the fast proton diffusion model to evaluation of water in artemia cysts. *Physiol. Chem. Phys. Med. NMR* **15,** 489–499.

Fullerton, G. D., Ord, V. A., and Cameron, I. L. (1986). An evaluation of lysozyme hydration by an NMR titration method. *Biochim. Biophys. Acta* **869**, 230–246.

Hiramoto, Y. (1970). Rheological properties of sea urchin eggs. Biorheology **6**, 201–234.

Lee, H. C., and Epel, D. (1983). Changes in intracellular acidic compartments in sea urchin eggs after activation. *Dev. Biol.* **98**, 446–454.

Ling, G. N. (1984). "Search of the Physical Basis of Life." Plenum, New York.

Mabuchi, I. (1986). Biochemical aspects of cytokinesis. *Int. Rev. Cytol.* **101**, 175–216.

Merta, P. A., Fullerton, G. D., and Cameron, I. L. (1986). Characterization of water in unfertilized and fertilized sea urchin eggs. *J. Cell. Physiol.* **127**, 439–447.

Moon, R. T., Nicosa, R. F., Olsen, C., Hill, M. B., and Jeffery, W. R. (1983). The cytoskeletal framework of sea urchin eggs and embryos: Developmental changes in the association. *Dev. Biol.* **95**, 447–458.

Otto, J. J., Kane, R. E., and Bryan, J. (1980). Redistribution of actin and fascin in sea urchin eggs after fertilization. *Cell Motil.* **1**, 31–40.

Rebhun, L. I., Begg, D. A., and Fisher, G. (1982). Reorganization of the cortex of sea urchin eggs as a function of activation. *Cell Differ.* **11**, 271–276.

Schliwa, M., Van Blerkom, J., and Porter, K. R. (1981). Stabilization of the cytoplasmic ground substance in detergent opened cells and a structural and biochemical analysis of its composition. *Proc. Natl. Acad. Sci. U.S.A.* **78**, 4329–4333.

Schmidt, T., Patton, C., and Epel, C. (1982). Is there a role for the Ca^{2+} influx during fertilization of the sea urchin egg? *Dev. Biol.* **90**, 284–290.

Schroeder, T. E. (1979). Surface area changes at fertilization: Resorption of the mosaic membrane. *Dev. Biol.* **70**, 306–326.

Spudich, A., Giffard, R. G., and Spudich, J. A. (1982). Molecular aspects of cortical actin filament formation upon fertilization. *Cell Differ.* **11**, 281–284.

Usui, N., and Yoneda, M. (1982). Ultrastructural basis of the tension increase in sea urchin eggs prior to cytokinesis. *Dev., Growth Differ.* **24**, 453–465.

Zimmerman, S., Zimmerman, A. M., Fullerton, G. D., Luduena, R. F., and Cameron, I. L. (1985). Water ordering during the cell cycle: NMR studies of the sea urchin egg. *J. Cell Sci.* **79**, 247–257.

Zimmerman, S., Zimmerman, A. M., Cameron, I. L., Fullerton, G. D., Schatten, H., and Schatten, G. (1987). Effects of cytoskeletal inhibitors on water proton relaxation time changes in unfertilized and fertilized sea urchin eggs. *Cell Biol. Int. Rep.* **11**, 605–614.

14

Calcium and Mitosis: A Mythos?

CHRISTIAN PETZELT AND MATHIAS HAFNER

Institute of Cell and Tumor Biology
German Cancer Research Center
D-6900 Heidelberg 1, Federal Republic of Germany

It is almost taken for granted that mitosis is controlled and regulated by Ca ions. They are especially suited to perform locally and spatially such a function given by their unique ionic properties. For an extensive discussion of this topic the reader is referred to Blaustein (1984) and Rasmussen (1984). New methods have become available in recent years to locate calcium and even to visualize free Ca ions, but not infrequently the interpretation of the data was based on the dogma of Ca ions governing the course of mitosis. In this review, we will critically examine the data so far accumulated, and we will ask how strongly the data really support the idea of mitosis being governed by Ca ions. It is purposefully intended to be provocative and does not claim to present a complete overview of the relevant literature. For another recent review of the topic, the reader is referred to Poenie and Steinhard (1987).

Mitosis can be defined as a sequence of events leading to the equal distribution of the chromatin into the two daughter nuclei after which the cell undergoes cytokinesis to give rise to two daughter cells. The beginning of

THE CELL BIOLOGY OF
FERTILIZATION

mitosis may be arbitrarily set to the visual onset of chromatin condensation and the accompanying separation of the centrosomes.

To study the role of calcium during this process, one can examine components of the mitotic process for their Ca sensitivity, search for calcium in the mitotic apparatus, look for Ca-sequestering systems at mitosis, follow qualitative and quantitative changes in the intracellular calcium, or, by analogy with other known Ca-dependent processes, try to identify components with similar functions and properties as in the known systems.

I. MICROTUBULES ARE SENSITIVE TO CALCIUM

Microtubules form the main part of the mitotic apparatus (reviewed by Inoué, 1981). At the beginning of mitosis the intracellular microtubular network undergoes a complete reorganization giving rise to the bipolar mitotic apparatus (cf. Harris, 1982; Hollenbeck and Cande, 1985; Balczon and Schatten, 1983; Schatten et al., 1985). Since the pioneering work of Weisenberg (1972), microtubules have been known to depend on the absence of more than micromolar amounts of calcium in order to undergo assembly.

Two ways of achieving such a sensitivity have been suggested: Tubulin itself is thought to bind calcium thereby changing from a polymerizable into a nonpolymerizable state (Berkowitz and Wolff, 1981; Serrano et al., 1986). The latter authors describe tubulin as having a high-affinity Ca-binding site with a dissociation constant in the order of $10^{-6}\ M$ and another class containing several low-affinity sites that have a dissociation constant of $10^{-4}\ M$. At present, it is unclear if this property is common to tubulin from all sources or is restricted only to tubulin of neuronal origin. As long as assembly competent tubulin is not isolated from mitotic apparatus and shown to bind calcium, one should probably refrain from considering a calcium sensitivity of the mitotic apparatus as a direct consequence of a calcium–tubulin interaction (cf. Suprenant and Rebhun, 1983). Such a property is most probably conveyed to tubulin by Ca-sensitive, microtubule-associated proteins (Vallee and Bloom, 1983; Lee and Wolff, 1984); but in that case, it is not proved that such an observation applies also to mitosis.

II. FROM CALCIUM TO CALMODULIN TO MICROTUBULES

A more coherent picture emerges when the relationship between calmodulin and tubulin is analyzed. Calmodulin is an ubiquitous, small protein that changes its configuration upon binding of calcium in a very characteristic way (reviewed by Kretsinger, 1979; Babu *et al.*, 1985; Williams, 1986; Klee *et al.*, 1986; Cox, 1988). Such a pronounced alteration, its high intracellular concentration, and its solubility properties make it the ideal candidate for conferring effects of calcium to a target. In fact, using indirect immunofluorescence techniques with antibodies to calmodulin, it was shown that calmodulin occurs in the mitotic apparatus of animal cells (Anderson *et al.*, 1978; Welsh *et al.*, 1978; Marcum *et al.*, 1978; De Mey *et al.*, 1980), as well as of plant cells (Vantard *et al.*, 1985). Such results have been confirmed by the injection experiments of fluorescently labeled calmodulin showing the same distribution of the protein as obtained on fixed cells (Zavortink *et al.*, 1983; Welsh *et al.*, 1981; Hamaguchi and Iwasa, 1982; Hamaguchi *et al.*, 1985; Stemple *et al.*, 1988). Keith *et al.* (1983, 1985a) went even one step further when they injected Ca-saturated calmodulin into fibroblasts. Only the Ca–calmodulin complex caused the local dissolution of microtubules; no such effect was observed when calmodulin alone was injected. Moreover, the intracellular mobility of the Ca–calmodulin was sharply reduced in contrast to the Ca-free protein, suggesting a possible local restriction of calcium effects. In order to maintain the skeptical approach, the minimum of arguments for the involvement of Ca–calmodulin in the regulation of mitosis would be: (1) The occurrence of calmodulin in the mitotic apparatus along the kinetochore fibers and at the poles at anaphase appears to be unambiguously acceptable and (2) Ca-saturated calmodulin can cause *in vivo* disassembly of microtubules. It still remains unproved whether the cell itself applies such a system for local depolymerization at the various mitotic steps. Nishida and Kumagai (1980) and also Salmon (1982), on the other hand, have shown that the calcium sensitivity of echinoderm spindles is not altered by the addition of purified calmodulin or its antagonist, trifluoperazine. Therefore, a direct tubulin–calmodulin relationship appears to be rather improbable.

III. EFFECTS OF Ca IONS ON MITOTIC SPINDLES *IN VIVO* AND *IN VITRO*

Naturally, the main interest is always focused on observations obtained with living cells. Unfortunately, only very few biologists were fortunate enough to conceive feasible and meaningful experiments. The already classical experiments by Kiehart (1981) are such an example. Using the sea urchin embryo and the clearly visible property of the mitotic apparatus to show birefringence in polarized light, he injected Ca ions onto various sites of the mitotic apparatus and other locations in the cell. If the injection site was close enough to the mitotic apparatus, the spindle fibers were dissolved. No such effect could be observed after an injection of Ca ions at sites more distant to the mitotic apparatus. Whereas such a result confirms the susceptibility of the spindle fibers to artificially introduced high Ca concentrations ($>10^{-6}$ M), it does not have any direct bearing on the *in vivo* conditions. His second observation appears to be much more important, namely, that the artificially induced high Ca concentration is rapidly taken up by a Ca-sequestering system in or close to the mitotic apparatus. After such an uptake has occurred, the birefringence of the mitotic apparatus is restored and mitosis resumed. From these results one can certainly conclude that a very potent Ca transport system is present, but not necessarily that such a system is actually used for the mitotic process.

Injection of Ca ions has also been used by Izant (1983) in PtK cells. In contrast to Kiehart (1981), Izant found that these cells already are killed by injection of Ca ions above 50 μM (needle concentration), whereas the sea urchin eggs survived injection of millimolar Ca ions. Several results are remarkable in his paper: (1) Injection of micromolar calcium at metaphase accelerates the onset of anaphase compared to the injection of a cell with an unrelated buffer, (2) Injection of a Ca–EGTA buffer with a low Ca concentration ($<10^{-7}$ M) prolongs the onset of anaphase, (3) Anaphase itself is not susceptible to Ca injections and, (4) Chromosome splitting cannot be induced by an increase in calcium as demonstrated by the injection of Ca onto chromosomes of cells arrested in metaphase by nocodazole. These results point to "some" function of Ca ions at the metaphase–anaphase transition, but they seem to exclude a direct connection to the splitting of the chromosomes, provided that nocodazole-arrested chromosomes resemble the *in vivo* conditions. These results concerning the metaphase–anaphase transition cannot give a sufficiently satisfying answer as yet, although evidence for a Ca-dependent essential event comes also from the work of Hepler (1985) showing the extension of metaphase in *Tradescantia* stamen hair cells when calcium is artificially reduced. The insensitivity of anaphase to the Ca injections is intriguing. Only if one assumes the presence of a very powerful Ca-sequestering system along the kinetochore fibers efficiently removing the injected calcium, can one reconcile these results with a regulation of anaphase progress by calcium. Alternatively, it must be postulated that both anaphase A and B can proceed without being under the control of calcium, a rather sacrilegious idea in these times. First support for such a possibility can be found in the recent studies by Gorbsky *et al.* (1987) showing a relatively autonomous kinetochore.

The question of Ca dependence is more indirectly approached in the experiments of Wagenaar (1983). By changing the external Ca concentration and at the same time the permeability properties of the cell (mitotic sea urchin eggs), he could change the rate of chromosome movement and, thereby, speed up the mitotic process. Taken at face value, these results demonstrate clearly that mitosis depends on a highly ordered intracellular ionic environment, and that any alteration especially of Ca ions changes the tick–tack rhythm of the complex cell cycle clock. Do such artificial interferences really indicate the natural function of these ions in the untreated cell?

A similar indirect result is represented by the work of Yoshimoto and Hiramoto (1985) and Yoshimoto *et al.* (1985). They measured the efflux of free calcium at various stages of the cell cycle, in sea urchin eggs as well as in *Medaka* eggs, after permeabilizing the cell with Ca Ionophor A 23187. Whenever a cell population reached the last stage of mitosis, the efflux decreased indicating a minimum in the cytoplasmic level of free calcium. It is not clear if such a holistic approach really provides useful clues regarding the Ca control of mitosis, but from these experiments as well as from others (starting with the early *Xenopus* experiments by Baker and Warner, 1972; Schantz, 1985) it can be accepted that during mitosis Ca levels change continously.

IV. CALCIUM CHANGES AT MITOSIS

A confirmation for this hypothesis was obtained by Bennett and Mazia (1981) when unfertilized sea urchin eggs that had been fused with fertilized embryos underwent only cortical granule breakdown in the unfertilized half when the fertilized half began mitosis, an indication that an increase in free calcium had occurred in the mitotic cell.

Another approach was used by Wolniak and Bart (1985a). They applied quin-2 externally to stamen hair cells from *Tradescantia*. The addition of this Ca chelator indirectly lowered the intracellular Ca concentration and resulted in an arrest of the cells at the metaphase–anaphase transition. It will be seen later that this point is alluded to by some authors as exhibiting a dramatic increase ("burst") of free calcium in other cells. Still, this experiment, as well as others, does not prove conclusively the essential and direct involvement of Ca ions in the control of mitosis. A more basic argument applies to these results: As Wolniak and Bart (1985b) have shown, the Ca channel blocker Nifedipine, when present during mitosis, blocks the stamen hair cells in metaphase. When Nifedipine is removed by inactivation with ultraviolet light, anaphase occurs within minutes but only when calcium is present in the outside medium. This shows clearly an influx of Ca ions from the outside necessary for the entry into anaphase. Similar observations have been reported by Hepler (1985) who showed that treatment of *Tradescantia* stamen hair cells with La^{3+} or D600 extended metaphase transit time, and that the addition of $CaCl_2$ to the medium promoted anaphase onset. Such results have not been obtained with animal cells. It is an established fact, for example, that fertilized sea urchin eggs develop perfectly well in a completely Ca-free environment (cf.

Dube *et al.*, 1985). Recently, Hepler (1986) analyzed the course of calcium fluctuations in *Tradescantia* stamen hair cells during mitosis using arsenazo-III with special emphasis to the metaphase–anaphase transition. He observed an increase at that time, but such an increase occurred *after* the chromosomes had split (Hepler and Callaham, 1987). One is now faced with the rather unsatisfying situation that completely different results are obtained by different authors using differing Ca dyes and cell types.

Keith *et al.* (1985b) used quin-2 and observed plant endosperm cells; they actually found a slight increase of calcium at the poles throughout the duration of anaphase. Ratan *et al.* (1986) using fura-2 and PtK cells found an increase in whole cell calcium but no significant calcium increases inside the mitotic apparatus. No metaphase–anaphase burst was found.

In an extensive and well-documented study, Poenie *et al.* (1985) using fura-2 on sea urchin eggs observed a "burst" in free calcium in all cells at the metaphase–anaphase transition, besides several small increases at nuclear envelope breakdown (NEB). This increase in free calcium at the metaphase–anaphase transition was sometimes obscured by the already high internal Ca level. The same group also found an increase in PtK cells at metaphase and related it to splitting of the chromosomes (Poenie *et al.*, 1986). With regard to the Ca^{2+} dependency of NEB, some more important facts have emerged from the work of Steinhardt and Alderton (1988) and Twigg *et al.* (1988). The former authors analyzed in depth the increased Ca^{2+} requirement for NEB in ammonia-activated sea urchin eggs. They found that blocking the increase of intracellular Ca^{2+} by various chelators immediately before NEB reversibly prevented this event. The actual Ca^{2+} increases described by the authors showed that the concentration of intracellular Ca^{2+} raises by a factor of 2 (from $\sim 1.5 \times 10^{-7}$ to 3×10^{-7} M). Twigg *et al.* (1988) demonstrated in sea urchin eggs the susceptibility of NEB and chromatin condensatin to increases in intracellular calcium. Such a fluctuation was induced intracellularly by inositol 1,4,5-trisphosphate or Ca buffers and could be prevented by Ca chelators. Additionally, the authors found that the nucleus is not continuously sensitive to such fluctuations, but that some "conditioning" and/or synthesis of protein molecules must occur before the Ca effects can take place. These results refine our emerging picture of the way Ca^{2+} ions may exert various regulatory functions during the cell cycle, being dependent on the presence of the corresponding receptor molecule that, in turn, may be synthesized periodically.

Taken at face value, all of these results confirm a calcium requirement for NEB. The Ca^{2+} fluctuations during the second half of mitosis, however, are still pieces of a puzzle searching for their place in the picture. Of course, speculation persists that the continuous disassembly of microtubules at the kinetochores is driven by local increases of calcium, and, in such a case, an artificial increase of calcium would not influence mitosis since the kinetochorial compartments would be protected by the powerful Ca-sequestering system. At the present stage, one cannot yet definitively connect overall calcium changes with an ordered progress through mitosis. We can likewise assume that the Ca dyes with their high affinity for calcium act competitively with Ca-binding proteins. By changing many of them intracellularly into a Ca-free state, one again achieves an unspecific alteration of systems not directly related to mitosis. A good example of such systems is the Ca-activated proteases (reviewed in Suzuki, 1987), which would probably be changed in their activity by the presence of the chelator.

The insensitivity of mitosis to changes in the intracellular Ca concentration is demonstrated convincingly if Ca ions are not introduced to the mitotic apparatus [as extremely "minatory ions" (Rasmussen, 1984)] due to local sequestration in a very fast and efficient manner (cf. the classical experiments by Rose and Loewenstein, 1975, 1976). This can be achieved by the injection of inositol 1,4,5-trisphosphate. If this compound is injected in physiological concentrations ($\sim 10^{-6}$ M final concentration) into mitotic sea urchin eggs, the Ca^{2+} concentration can be shown to rise instantaneously (M. Hafner and C. Petzelt, unpublished). Nevertheless, neither a reduction in birefringence nor a change in the rate of the mitotic progress can be observed.

What other groups of molecules remain to be studied with regard to their Ca susceptibility at mitosis? Chromatin condensation depends to a high degree on the continuous presence of Ca–calmodulin (Chafouleas et al., 1984), whereas a relationship to the splitting of the chromosomes remains to be demonstrated (cf. Izant, 1983). Even less clear is the case for calcium as controlling agent in a possible sol–gel transformation occurring in the mitotic apparatus involving the actin system. Although many years of evidence have shown that actin occurs in the spindle (cf. Sanger, 1978), no involvement of an actin–myosin system in mitosis has been proved. On the contrary, Kiehart et al. (1982) were unable to block mitosis by the injection of antimyosin antibodies. Whether molecules involved in microtubule-dependent movement in other cellular processes, as is kinesin, are actually regulated at mitosis by calcium is at present completely unclear (cf. Scholey et al., 1985; Vale et al., 1985).

What we have now is abundant evidence that during mitosis Ca^{2+} ions always undergo fluctuations, but, besides speculation, there is not the slightest proved causal link to any event in the real untreated mitotic cell. However, in order to strengthen the case for the importance of calcium in mitosis— albeit in an indirect way—we will look at the presence of Ca-regulating systems at the site where mitosis occurs, the mitotic apparatus, as evidence that such a close relationship must have its functional equivalent, otherwise evolution would not have allowed survival. We will therefore examine the "hard facts" for the presence of Ca-regulatory systems in the mitotic apparatus.

V. CALCIUM IN THE MITOTIC APPARATUS

Before such an endeavor is undertaken, one can ask whether there is a more direct way to show that calcium has a regulatory function in the mitotic apparatus than by using Ca injection experiments, as described above. In the age of the dogma of microtubule all importance, it was only a consequence that methods were developed to preserve (or generate?) Ca sensitivity in isolated mitotic apparatuses (Salmon and Segall, 1980; Rebhun et al., 1980; Keller et al., 1982). The same argument can be used here as in all other cases where isolated mitotic apparatus have been described with regard to certain properties, localization of molecules, etc. Since they are unable to perform in vitro as in vivo, the results obtained are not of sufficient proof. In the case of Ca localization using chlorotetracycline in isolated mitotic apparatus of sea urchins, the arguments are much stronger for a true in vivo-like property of the mitotic apparatus. Schatten et al. (1982) showed, however, the distribution of calcium to be dependent on the mode of isolation. Therefore, all of these experiments allow one to conclude that Ca-sensitive mitotic apparatus can be obtained only under certain conditions, and that calcium can be found in mitotic apparatus containing membrane vesicles by using chlorotetracycline.

Ca-containing membranes in the mitotic apparatus have been demonstrated on an ultrastructural level by Hepler and his co-workers. For an extensive description of these and related results, the reader is referred to a review by Hepler and Wolniak (1984).

In short, Ca precipitates are shown to occur inside of membrane cisternae along the chromosomal fibers and appear to engulf the mitotic apparatus. Such results have been obtained in plant cells, whereas an equivalent Ca localization in the mitotic apparatus of animal cells has not been achieved so far.

VI. Ca-SEQUESTERING MEMBRANES IN THE MITOTIC APPARATUS

An elaborate membrane system is found in almost all mitotic cells of animal or plant origin. Moll and Paweletz (1980) demonstrated such a system in mitotic mammalian cells, but it was Harris (cf. 1975, 1978, 1982) who in describing the membranes in the mitotic apparatus of sea urchin eggs linked these systems to the Ca regulation of mitosis. Does such a hypothesis have its support in the existence of an identified Ca transport system? At present, several laboratories have reported on Ca-sequestering membranes active during mitosis. Suprynowicz and Mazia (1985) used electrically permeabilized sea urchin eggs and followed the Ca uptake through the cell cycle. They found cyclic oscillations with a maximum of the uptake at prophase and a second one at telophase. These results show clearly that Ca fluctuations occur during mitosis, but because of the properties of the method used, the Ca uptake cannot be related to a locally defined site in the cell; the desired causal connection to the course of mitosis is not possible.

A different approach was used by Silver *et al.* (1980). They isolated membrane-containing mitotic apparatus from sea urchin eggs and found them capable of sequestering calcium. If one considers the above described abundance of membranes in and around the mitotic apparatus and takes into consideration the method of isolation, i.e., breaking up cells in hypotonic medium containing 10 mM EDTA, it is no surprise that such mitotic apparatuses sequester calcium. In 1986, Silver reported on the identification of an enzyme that had immunological similarities with the well-known Ca–ATPase of the sarcoplasmic reticulum. He took these common epitopes as an indication of common function in a mitosis-specific Ca-sequestering system. At present, it seems to be most desirable to obtain more reliable data for such a hypothesis since the results would have immediate and important implications on our understanding of Ca regulation at mitosis.

There are a few other reports on the isolation of Ca-sequestering vesicles possibly involved in Ca regulation at mitosis. Inoue and Yoshioka (1982) describe Ca-sequestering vesicles from unfertilized and fertilized sea urchin eggs where the uptake capacity is increased five times in the fertilized eggs. In their preliminary findings, Clapper and Lee (1985) reported that sea urchin egg membranes that sequester calcium can be isolated from the endoplasmic reticulum, but that Ca release can be induced by inositol 1,4,5-trisphosphate and by recently discovered other second messengers.

VII. CALCIUM RELEASE BY INOSITOL POLYPHOSPHATES

Much interest is currently focused on the intracellular messenger inositol 1,4,5-trisphosphate (InsP$_3$) and its action in the regulation of cytosolic Ca^{2+} levels (Berridge and Irvine, 1984; Berridge, 1986). The importance of InsP$_3$ as an intracellular secondary messenger is demonstrated by its ability to stimulate a variety of complex physiological processes when injected into living cells (Fein *et al.*, 1984; Turner *et al.*, 1986; Swann and Whitaker, 1986). InsP$_3$, which is released into the cytosol from phosphatidylinositol bisphosphate in the plasma membrane, has been identified as the intermediary messenger between agonist binding and release of calcium. The other product of this hydrolysis is diacylglycerol, which acts on the stimulation of protein kinase C (Nishizuka, 1984). The nature of this transduction mechanism thus results in a bifurcation of the signal pathway (Berridge and Irvine, 1984). Mobilization of calcium by InsP₃ seems primarily responsible for triggering a variety of cellular processes, whereas the diacylglycerol protein kinase C pathway accounts for the modulation of how the Ca signal acts on its cellular targets. There is now considerable evidence that all of the calcium released by InsP₃ originates solely from the endoplasmic reticulum. Much of the evidence is based on using permeabilized cells and microsome fractions. Blocking the uptake of calcium into the mitochondria using various inhibitors has no effect on uptake into the endoplasmic reticulum, nor is the subsequent release by InsP₃ inhibited (Streb *et al.*, 1983, 1985; Biden *et al.*, 1984; Prentki *et al.*, 1984; Hirata *et al.*, 1984; Clapper and Lee, 1985). However, the nonmitochondrial Ca stores *in situ* may be heterogenous with regard to their sensitivity to InsP₃; only a fraction of the endoplasmic reticulum may release calcium upon stimulation by InsP₃ (Muallem et al., 1985; Clapper and Lee, 1985; Oberdorf *et al.*, 1986).

In addition to the early ionic events at fertilization leading to an increase of calcium and pH through activation of the Na^+–H^+ exchanger (Swann and Whitaker, 1983), another intriguing aspect of the phosphatidyl inositol pathway is its possible involvement in the control of mitosis. Ciapa and Whitaker (1986) find a second phase of inositol polyphosphate production after the sperm-induced Ca transient declines, indicating that InsP₃ functions as a second messenger during the cell cycle. In addition, lithium ions have been shown to block mitosis in sea urchin eggs and *Tradescantia* stamen hair cells (Sillers and Forer, 1985; Wolniak, 1986, 1987). It has been suggested that this effect was achieved by blocking the conversion of inositol 1-phosphate to *myo*-inositol to prevent resynthesis of phosphatidylinositol and, subsequently, further release of calcium (Sillers and Forer, 1985; Wolniak, 1986). The ability of exogenuosly applied *myo*-inositol to overcome this block was taken as evidence for a phosphoinositol-mediated release of calcium. However, the effectors of the release mechanism for the local regulation of Ca concentration within the mitotic apparatus remain obscure. Moreover, as the impressive studies by Sherman *et al.* (1985) and Ackermann *et al.* (1987) show, such explanations are far too simple. One is faced with multiple ways of generating the various inositol phosphates whereby the lithium block may be bypassed.

Although all known processes leading to the production of InsP₃ are receptor stimulated through hydrolysis of phosphatidylinositol and require an extracellular agonist, it is conceivable that a coordination between the mitotic apparatus and the cortex may be involved in the breakdown of phosphatidylinositol. In addition to InsP₃, other mediators for Ca release such as GTP and NADP have been demonstrated (Wolf *et al.*, 1987; Clapper *et al.*, 1986; Chueh and Gill, 1986). The latter is known to increase after fertilization (Epel *et al.*, 1981). NADP was added to InsP₃-refractory homogenates and released nearly as much calcium as was induced by the initial InsP₃ (Clapper *et al.*, 1986).

These findings open up exciting new ways of regulating intracellular calcium in addition to the various control mechanisms of the plasma membrane and the membranes of the endoplasmic reticulum, which themselves appear to be regulated by different agents, i.e., inositol 1,4,5-trisphosphate and NADP, respectively. The picture becomes even more complex when the recent results by Irvine and Moore (1986) are taken into consideration. They describe the existence of several inositol phosphates, each capable of functioning as a "second messenger" in its own way. Additionally, one has to realize that an involvement of components of the plasma membrane via G-proteins may reflect the intrinsic interrelationship between the mitotic apparatus and the plasma membrane. Such a connection has not yet been established for mitosis, but there is evidence supporting regulation of the polyphosphoinositide phosphodiesterase by a guanine nucleotide-binding protein (Cockroft, 1987). One, therefore, would not be surprised if the compartmentalized regulation of Ca ions in the mitotic apparatus is orchestrated by a number of control substances and is directed against several, nonidentical Ca regulatory systems.

VIII. FUNCTIONAL ANALYSIS OF THE CALCIUM TRANSPORT SYSTEM

In our laboratory we have developed methods for the isolation and characterization of such a system (Petzelt and Wülfroth, 1984). These membranes show a cell cycle-dependent Ca uptake with a maximum at mitosis. The complex protein composition has been analyzed using monoclonal antibodies that were obtained by *in vitro* immunization of lymphocytes with the isolated membranes. One of these antibodies, reacting with a 46-kDa protein, specifically stains the mitotic apparatus in intact sea urchin egs and inhibits Ca uptake into the isolated vesicles (Petzelt and Hafner, 1986). If these properties are indeed a reflection of a system not only present but also active at mitosis, microinjection of the defined monoclonal antibody should influence the mitotic process (Hafner and Petzelt, 1987). An immediate and profound effect can be observed if mitotic sea urchin eggs are injected with the antibody. As shown in Fig. 1, birefringence of the mitotic apparatus starts to fade away almost immediately, the cell arrested in mitosis. Moreover, a preloading of blastomeres with the Ca dye fura-2 allows the analysis of changes in the actual concentration of intracellular calcium. If indeed the microinjection of the antibody interferes with the Ca regulatory system, Ca^{2+} should rise concomitantly with reduction of the birefringence of the mitotic apparatus. In all cells studied (see Table I), such an increase can be observed, thus establishing a direct line from the isolation of Ca-sequestering vesicles, their analysis and locali-

zation with monoclonal antibodies in intact cells, to the experimental interference with intracellular Ca concentrations and the corresponding effects on
mitosis. That these membranes are not located only in the vicinity of the
mitotic apparatus inside the whole cell could be shown by a new membrane-
preserving method for the isolation of the mitotic apparatus. With certain
modifications, the immunogenicity of the components of the mitotic apparatus
is preserved, thus allowing their immunocytochemical localization (Petzelt *et
al.*, 1987). Figure 2 shows a triple staining of such an isolated mitotic apparatus.
Antitubulin staining reflects structural integrity; Hoechst dye reacting with
the chromosomes allows a correct determination of the mitotic stage; and,
finally, reaction of the 46-kDa antibody presents a picture of the localization
of the calcium transport system in the isolated mitotic apparatus. Distribution
of the vesicles, predominantly localized in the asters, corresponds exactly to
the description given earlier for whole eggs (Petzelt and Hafner, 1986) and
emphasizes the dominance of the mitotic poles for the three-dimensional organization of not only the microtubules but also the membranous elements
(cf. Mazia, 1986). From all of these experiments one can now draw the following conclusions. (1) A Ca-sequestering system assembles at mitosis inside
the mitotic apparatus. It must undergo modifications compared to its state at
interphase, since the 46-kDa antibody does not react with it at interphase. (2)
Microinjection of the antibody immediately blocks mitosis and causes disappearance of the birefringence of the mitotic apparatus. (3) Following the
injection of the antibody, calcium starts to rise immediately in the injected
cell and stays at a submicromolar level; the cell remains arrested, and the
untreated daughter cell develops normally.

IX. SUMMARY AND OUTLOOK

We have presented many facts (and some fictions) accumulated over the
years all pointing to a regulatory function of Ca ions in mitosis. The effects

---

Fig. 1. Effect of the injection of the monoclonal antibody against the 46-kDa protein
into a sea urchin blastomere at mitosis and concomitant observation of the intracellular
fluctuations of free calcium. In order to obtain an equal distribution of the calcium dye fura-
2, a sea urchin egg was injected with ~10 μm fura-2 (free acid), after the second mitosis
had been reached (a, and a'); one blastomere was injected with the antibody (see oil droplet
in the left blastomere in the following sequence). Approximately 60 sec after injection, the
birefringence of the mitotic apparatus has disappeared (b), and the intracellular calcium is
greatly increased (b'). c–f, taken every 3 min, show the continuing process through mitosis
of the uninjected blastomere, whereas the injected one remains blocked. c'–f' demonstrate
that this block is also expressed as a continuation of the elevated calcium level in the injected
cell.

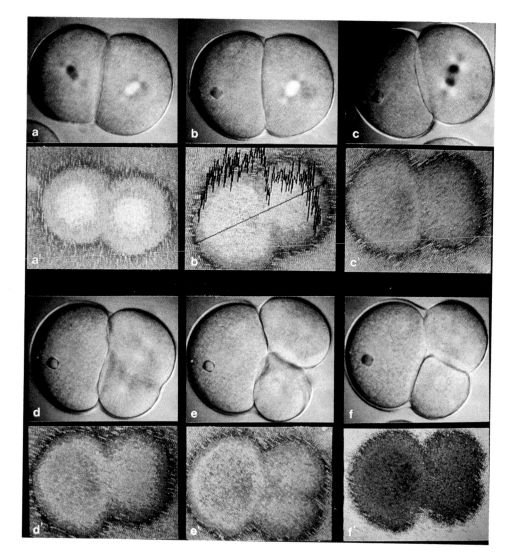

TABLE I

Anti-46-kDa Protein Antibody Injection into *Lytechinus pictus* Embryos[a]

Solution injected	IgM (mg/ml)	Mitosis
Anti-46-kDa protein[b]	2–4	0/14[e]
Denatured anti-46-kDa protein[b]	2–6	4/4
Antihyalin[b]	2–8	5/5
Antichromosal protein[b]	4–15	4/4
BSA	5–15	3/3
Rabbit Ig fraction[c]	20	3/3
Injection buffer[d]	—	4/4

[a] The high specificity and reproducibility of the anti-46-kDa injection experiments are demonstrated. In no case could a cell injected with the antibody overcome the mitotic block.

[b] Affinity purified monoclonal antibody (mAb).

[c] Total protein concentration.

[d] Potassium aspartate (100 mM) and 10 mM HEPES (pH 7.0).

[e] Microinjection of affinity-purified mAb into sea urchin embryos of *L. pictus* caused arrest of mitosis in 14/14 trials. Direct pressure injection of other antibodies or serum proteins and buffer neither affected mitotic apparatus birefringence nor progress through mitosis.

that experimental modifications of the intracellular Ca concentration have on the mitotic process, the emerging picture of a highly active Ca-sequestering system well preserved throughout evolution, and, as a new and exciting field, the discoveries of various second messengers for the Ca-sequestering system(s) make it now highly suggestive that calcium and mitosis are intimately and causally interwoven.

There is only one (but the all-important one) impediment to our progressive understanding of such a function: We do not really have an intellectually satisfying vision of the picture that will emerge from all of these various pieces of the puzzle. In other words, in spite of all the speculations, we do not convincingly know the target for calcium at mitosis. Whether it will be the sum of many Ca-dependent entities or whether one has to assume the existence of completely new mechanisms will be recognized only by a continuation of the intense and fascinating work on this intriguing subject.

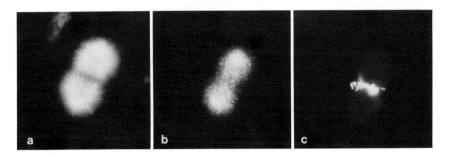

Fig. 2. Mitotic apparatus from the sea urchin *Paracentrotus lividus* was isolated using modifications of a method developed by Mazia and Savin (cf. Petzelt *et al.*, 1987). Inherent components of the mitotic apparatus are visualized using triple staining of one mitotic apparatus with the respective antibodies or dyes. Staining with antitubulin; the microtubules have retained their characteristic distribution and appearance as in the whole cell, indicating the usefulness of the isolation method used. The same mitotic apparatus is stained with the anti-46-kD protein, showing the localization of the Ca-sequestering system; a close correlation to the distribution obtained for whole cells (cf. Petzelt and Hafner, 1986) is found, allowing the conclusion that the Ca-sequestering system is indeed an intrinsic part of the mitotic apparatus. (c) The distribution of the chromosomes as seen with Hoechst dye H-33248, showing the metaphase stage of the isolated mitotic apparatus.

REFERENCES

Ackermann, K. E., Gish, B. G., Honchar, M. P., and Sherman, W. R. (1987). Evidence that inositol 1-phosphate in brain of lithium-treated rats results mainly from phosphatidylinositol metabolism. *Biochem. J.* **242**, 517–524.

Anderson, B., Osborn, M., and Weber, K. (1978). Specific visualization of the distribution of the calcium-dependent regulatory protein of cyclic nucleotide phosphodiesterase (modulator protein) in tissue culture cells by immunofluorescence microscopy: Mitosis and intercellular bridge. *Cytobiologie* **17**, 354–364.

Babu, Y. S., Sack, J. S., Greenhough, T. J., Bugg, C. E., Means, A. R., and Cook, W. J. (1985). Three-dimensional structure of calmodulin. *Nature (London)* **315**, 37–40.

Baker, P. F., and Warner, A. E. (1972). Intracellular calcium and cell cleavage in early embryos of Xenopus laevis. *J. Cell Biol.* **53**, 579–581.

Balczon, R., and Schatten, G. (1983). Microtubule-containing detergent-extracted cytoskeletons in sea urchin eggs from fertilization through cell division. *Cell Motil.* **3**, 213–226.

Bennett, J., and Mazia, D. (1981). Fusion of fertilized and unfertilized sea urchin eggs. Maintenance of cell surface integrity. *Exp. Cell Res.* **134**, 494–498.

Berkowitz, S. A., and Wolff, J. (1981). Intrinsic calcium sensitivity of tubulin polymerisation. The contributions of temperature, tubulin concentrations, and associated proteins. *J. Biol. Chem.* **256**, 11216–11223.

Berridge, M. J. (1986). Regulation of ion channels by inositol trisphosphate and diacylglycerol. *J. Exp. Biol.* **124**, 323–335.

Berridge, M. J., and Irvine, R. F. (1984). Inositol trisphosphate, a novel second messenger in cellular signal transduction. *Nature (London)* **312**, 315–321.

Biden, T. J., Prentki, M., Irvine, R. F., Berridge, M. J., and Wollheim, C. B. (1984). Inositol 1,4,5-trisphosphate mobilizes intracellular Ca2 + from permeabilized insulin secreting cells. *Biochem. J.* **223**, 237–248.

Blaustein, M. P. (1984). Intracellular calcium as a second messenger. What's so special about calcium. *In* "Calcium Regulation in Biological Systems" (S. Ebashi, M. Endo, K. Imahori, S. Kakiuchi, and Y. Nishizuka, eds.), pp 23–33. Academic Press, New York.

Chafouleas, J. G., Lagace, L., Bolton W. E., Boyd, D., III, and Means, A. (1984). Changes in calmodulin and its mRNA accompany reentry of quiescent (Go) cells into the cell cycle. *Cell (Cambridge, Mass.)* **36**, 73–81.

Chueh, S., and Gill, D. L. (1986). Inositol 1,4,5-trisphosphate and guanine nucleotides activate calcium release from endoplasmic reticulum via distinct mechanisms. *J. Biol. Chem.* **261**, 13883–13886.

Ciapa, B., and Whitaker, M. (1986). Two phases of inositol polyphosphate and diacylglycerol production at fertilization. *FEBS Lett.* **195**, 347–351.

Clapper, D. L., and Lee, H. C. (1985). Inositol trisphosphate induces calcium release from nonmitochondrial stores in sea urchin egg homogenates. *J. Biol. Chem.* **260**, 13947–13954.

Clapper, D. L., Dargie, P. J., and Lee, H. C. (1986). Effects of pyridine nucleotides on calcium release from sea urchin egg homogenates. *J. Cell Biol.* **103**, 85a.

Cockroft, S. (1987). Polyphophosphoinositide phosphodiesterase: Regulation by a novel guanine nucleotide binding protein, Gp. *Trends Biochem. Sci.* **12**, 75–78.

Cox, J. A. (1988). Interactive properties of calmodulin. *Biochem. J.* **249**, 621–629.

De Mey, J., Moeremans, M., Geuens, G., Nuydens, R., Van Belle, H., and De Brabander, M. (1980). Immunocytochemical evidence for the association of calmodulin with microtubules of the mitotic apparatus. *In* "Microtubules and Microtubule Inhibitors" (M. De Brabander and J. De Mey, eds.) Elsevier/North-Holland, Amsterdam.

Dube, F., Schmidt, T., Johnson, C. H., and Epel, D. (1985). The hierarchy of requirements for an elevated intracellular pH during early development of sea urchin embryos. *Cell (Cambridge, Mass.)* **40**, 657–666.

Epel, D., Patton, C., Wallace, R. W., and Cheung, W. Y. (1981). Calmodulin activates NAD kinase of sea urchin eggs: An early event of fertilization. *Cell (Cambridge, Mass.)* **23**, 543–549.

Fein, A., Payne, R., Corson, D. W., Berridge, M. J., and Irvine, R. F. (1984). Photorezeptor excitation and adaption by inositol 1,4,5-trisphosphate. *Nature (London)* **311**, 157–160.

Gorbsky, G. J., Sammak, P. J., and Borisy, G. G. (1987). Chromosomes move poleward in anaphase along stationary microtubules that coordinately disassemble from their kinetochore ends. *J. Cell Biol.* **104**, 9–18.

Hafner, M., and Petzelt, C. (1987). Inhibition of mitosis by an antibody to the mitotic calcium transport system. *Nature (London)* **330**, 264–266.

Hamaguchi, Y., and Iwasa, F. (1982). Localization of fluorescent labeled calmodulin in living sea urchin eggs during early development. *Biomed. Res.* **1**, 502–509.

Hamaguchi, Y., Toriyama, M., Sakai, H., and Hiramoto, Y. (1985). Distribution of fluorescently labeled tubulin injected into sand dollar eggs from fertilization through cleavage. *J. Cell Biol.* **100**, 1262–1272.

Harris, P. (1975). The role of membranes in the organization of the mitotic apparatus. *Exp. Cell Res.* **94**, 409–425.

Harris, P. (1978). Triggers, trigger waves, and mitosis: A new model. *In* "Cell Cycle Regulation" (J. R. Jeter, Jr., I. L. Cameron, G. M. Padilla, and A. M. Zimmerman, eds.) pp. 75–104. Academic Press, New York.

Harris, P. (1982). Effects of caffeine on mitosis in eggs of the sea urchin *Strongylocentrotus purpuratus:* The possible role of calcium. *Cell Differ.* **11**, 357–358.

Hepler, P. K. (1985). Calcium restrictions prolongs metaphase in dividing Tradescantia stamen hair cells. *J. Cell Biol.* **100**, 1363–1368.

Hepler, P. K. (1986). Calcium changes during mitosis in Tradescantia hair cells measured with arzenazo. III. *J. Cell Biol.* **103**, 453a.

Hepler, P. K., and Callaham, D. A. (1987). Free calcium increases during anaphase in stamen hair cells of *Tradescantia. J. Cell Biol.* **105**, 2137–2143.

Hepler, P.K., and Wolniak, S. M. (1984). Membranes in the mitotic apparatus: Their structure and function. *Int. Rev. Cytol.* **90**, 169–236.

Hirata, M., Suematsu, E., Hashimoto, T., Hamachi, T., and Koga, T. (1984). Release of Ca2 + from a non-mitochondrial store site in peritoneal macrophages treated with saponin by inositol 1,4,5-trisphosphate. *Biochem. J.* **223**, 229–236.

Hollenbeck, P. J., and Cande, W. Z. (1985). Microtubule distribution and reorganization in the first cell cycle of fertilized eggs of Lytechinus pictus. *Eur. J. Cell Biol.* **37**, 140–148.

Inoue, H., and Yoshioka, T. (1982). Comparison of Ca2 + uptake characteristics of microsomal fractions isolated from unfertilized and fertilized sea urchin eggs. *Exp. Cell Res.* **140**, 283–288.

Inoué, S. (1981). Cell division and the mitotic spindle. *J. Cell Biol.* **91**, 131s–147s.

Irvine, R. F., and Moore, R. M. (1986). Micro-injection of inositol 1,3,4,5-tetrakisphosphate activates sea urchin eggs by a mechanism dependent on external Ca2 + . *Biochem. J.* **240**, 917–920.

Izant, J. G. (1983). The role of calcium ions during mitosis. Calcium participates in the anaphase trigger. *Chromosoma* **88**, 1–10.

Keith, C. H., Dipaola, M., Maxfield, F. M., and Shelanski, M. L. (1983). Microinjection of Ca2 +–calmodulin causes a localized depolymerization of microtubules. *J. Cell Biol.* **97**, 1918–1924.

Keith, C. H., Maxfield, F. R., and Shelanski, M. L. (1985a). Intracellular free calcium levels are reduced in mitotic PtK2 epithelial cells. *Proc. Natl. Acad. Sci. U.S.A.* **82**, 800–804.

Keith, C. H., Bajer, A. S., Ratan, R., Maxfield, F. R., and Shelanski, M. L. (1985b). Calcium and calmodulin in the regulation of the microtubular cytoskeleton. *In* "Microtubules and Microtubule Inhibitors" (M. De Brabander and J. De Mey, eds.), pp. 89–96. Elsevier/North-Holland, Amsterdam.

Keller, T. C. S., Jemiolo, D. K., Burgess, W. H., and Rebhuhn, L. I. (1982). Strongylocentrotus purpuratus spindle tubulin. II. Characteristics of its sensitivity to Ca2 + and the effect of calmodulin isolated from bovine brain and S. purpuratus eggs. *J. Cell Biol.* **93**, 797–803.

Kiehart, D. P. (1981). Studies on the in vitro sensitivity of spindle microtubules to calcium ions and evidence for a vesicular calcium-sequestering system. *J. Cell Biol.* **88**, 604–617.

Kiehart, D. P., Mabuchi, I., and Inoue, S. (1982). Evidence that myosin does not contribute to force production in chromosome movement. *J. Cell Biol.* **94**, 165–178.

Klee, C. B., Newton, D. L., Ni, W. -C., and Haiech, J. (1986). Regulation of the calcium signal by calmodulin. *Ciba Found. Symp.* **122**, 162–170.

Kretsinger, R. H. (1979). The informational role of calcium in the cytosol. *Adv. Cyclic Nucleotide Res.* **11**, 1–26.

Lee, J. C., and Wolff, J. (1984). Calmodulin binds to both microtubule-associated protein 2 and tau proteins. *J. Biol. Chem.* **259**, 1226–1230.

Marcum, J. M., Dedman, J. R., Brinkley, B. R., and Means, A. R. (1978). Control of microtubule assembly–disassembly by calcium-dependent regulator protein. *Proc. Natl. Acad. Sci. U.S.A.* **75**, 3771–3775.

Mazia, D. (1986). The chromosome cycle and the centrosome cycle in the mitotic cycle. *Int. Rev. Cytol.* **100**, 49–92.

Moll, E., and Paweletz, N. (1980). Membranes of the mitotic apparatus of mammalian cells. *Eur. J. Cell Biol.* **21**, 280–287.

Muallem, S., Schoeffield, M., Pandol, S., and Sachs, G. (1985). Inositol trisphosphate modification of ion transport in rough endoplasmic reticulum. *Proc. Natl. Acad. Sci. U.S.A.* **82**, 4433–4437.

Nishida, E., and Kumagai, H. (1980). Calcium sensitivity of sea urchin tubulin in in vitro assembly and the effects of calcium-dependent regulator (CDR) proteins isolated from sea urchin eggs and porcine brain. *J. Biochem. (Tokyo)* **87**, 143–151.

Nishizuka, Y. (1984). The role of protein kinase C in cell surface signal transduction and tumor promotion. *Nature (London)* **308**, 693–697.

Oberdorf, J. A., Head, J. F., and Kaminer B. (1986). Calcium uptake and release by isolated cortices and microsomes from the unfertilized egg of the sea urchin *Strongylocentrotus droebachiensis*. *J. Cell Biol.* **102**, 2205–2210.

Petzelt, C., and Hafner, M. (1986). Visualization of the $Ca2+$-transport system of the mitotic apparatus of sea urchin eggs with a monoclonal antibody. *Proc. Natl. Acad. Sci. U.S.A.* **83**, 1719–1722.

Petzelt, C., and Wülfroth, P. (1984). Cell cycle specific variations in transport capacity of an isolated $Ca2+$-transport system. *Cell Biol. Int. Rep.* **8**, 823–840.

Petzelt, C., Hafner, M., Mazia, D., and Sawin, K. (1987). Microtubules and Ca^{2+}-sequestering membranes in the mitotic apparatus, isolated by a new method. *Eur. J. Cell Biol.* **45**, 268–273.

Poenie, M., and Steinhardt, R. A. (1987). The dynamics of $(Ca^{2+})_i$ during mitosis. *In* "Calcium and Cell Function" (E. Cheung, ed.), pp. 133–157. Academic Press, New York.

Poenie, M., Alderton, J., Tsien, R. Y., and Steinhardt, R. A. (1985). Changes of free calcium levels with stages of the cell division cycle. *Nature (London)* **315**, 147–149.

Poenie, M., Alderton, J., Steinhardt, R., and Tsien, R. (1986). Calcium rises abruptly and briefly throughout the cell at the onset of anaphase. *Science* **233**, 886–889.

Prentki, M., Wollheim, C. B., and Lew, P. D. (1984). $Ca2+$ homeostasis in permeabilized human neutrophils. Characterization of $Ca2+$-sequestering pools and the action of inositol 1,4,5-trisphosphate. *J. Biol. Chem.* **259**, 13777–13782.

Rasmussen, H. (1984). Calcium ion. A synarchic and mercurial but minatory messenger. *In* "Calcium Regulation in Biological Systems" (S. Ebashi, M. Endo, K. Imahori, S. Kakiuchi, and Y. Nishizuka, eds.), pp. 13–22. Academic Press, New York.

Ratan, R. R., Shelanski, M. L., and Maxfield, F. R. (1986). Transition from metaphase to anaphase is accompanied by local changes in cytoplasmic free calcium in PtK2 kidney epithelial cells. *Proc. Natl. Acad. Sci. U.S.A.* **83**, 5136–5140.

Rebhun, L. I., Jemiolo, D., Burgess, W., and Kretsinger, R. (1980). Calcium, calmodulin and control of assembly of brain and spindle microtubules. *In* "Microtubules and Microtubule Inhibitors" (M. De Brabander and J. De Mey, eds.). Elsevier/North-Holland, Amsterdam.).

Rose, B., and Loewenstein, W. R. (1975). Permeability of cell junctions depends on local cytoplasmic activity. *Nature (London)* **254**, 250–252.

Rose, B., and Loewenstein, W. R. (1976). Calcium ion distribution in cytoplasm visualized by aequorin: Diffusion in cytosol restricted by energized sequestering. *Science* **190**, 1204–1206.

Salmon, E. D. (1982). Calcium, spindle microtubule dynamics and chromosome movement. *Cell Differ.* **11**, 353–355.

Salmon, E. D., and Segall, R. R. (1980). Calcium-labile mitotic spindles isolated from sea urchin eggs (Lytechinus variegatus). *J. Cell Biol.* **86**, 355–365.

Sanger, J. W. (1976). The presence of actin during chromosome movement. *Proc. Natl. Acad. Sci. U.S.A.* **70**, 2451–2455.

Schantz, A. R. (1985). Cytosolic free calcium-ion concentration in cleaving embryonic cells of Oryzias latipes measured with calcium-selective microelectrodes. *J. Cell Biol.* **100**, 947–954.

Schatten, G., Schatten, H., and Simerly, C. (1982). Detection of sequestered calcium during mitosis in mammalian cell cultures and in mitotic apparatus isolated from sea urchin zygotes. *Cell Biol. Int. Rep.* **6**, 717–724.

Schatten, G., Simerly, C., and Schatten, H. (1985). Microtubule configurations during fertilization, mitosis, and early development in the mouse and the requirement for egg microtubule-mediated motility during mammalian fertilization. *Proc. Natl. Acad. Sci. U.S.A.* **82**, 4152–4156.

Scholey, J. M., Porter, M. E., Grissom, P., and McIntosh, J. R. (1985). Identification of kinesin in sea urchin eggs, and evidence of its localization in the mitofic spindle. *Nature (London)* **318**, 483–486.

Serrano, L., Valencia, A., Caballero, R., and Avila J. (1986). Localization of the high affinity calcium-binding site on tubulin molecule. *J. Biol. Chem.* **261**, 7076–7081.

Sherman, W. R., Munsell, L. Y., Gish, B. G., and Honchar, M. P. (1985). Effect of systematically administered lithium on phophoinositide metabolism in rat brain, kidney and testis. *J. Neurochem.* **44**, 798–807.

Sillers, P. J., and Forer, A. (1985). Ca2 + in fertilization and mitosis: The phosphatidylinositol cycle in sea urchin gametes and zygotes is involved in control of fertilization and mitosis. *Cell Biol. Int. Rep.* **9**, 275–282.

Silver, R. B. (1986). Mitosis in sand dollar embryos is inhibited by antibodies directed against the calcuim transport enzyme of muscle. *Proc. Natl. Acad. Sci. U.S.A.* **83**, 4302–4306.

Silver, R. B., Cole, R. D., and Cande, W. Z. (1980). Isolation of mitotic apparatus containing vesicles with calcium sequestration activity. *Cell (Cambridge, Mass.)* **19**, 505–516.

Steinhardt, R. A., and Alderton, J. (1988). Intracelluar free calcium rise triggers nuclear envelope breakdown in the sea urchin embryo. *Nature (London)* **332**, 364–366.

Stemple, D. L., Sweet, S. C., Welsh, M. J. and McIntosh, J. R. (1988). Dynamic of a fluorescent calmodulin analog in the mammalian mitotic spindle at metaphase. *Cell Motil. Cytoskeleton* **9**, 231–242.

Streb, H., Irvine, R. F., Berridge, M. J., and Schulz, I. (1983). Release of Ca2 + from a nonmitochondrial intracellular store in pancreatic acinar cells by inositol 1,4,5-tris-phosphate. *Nature (London)* **306**, 67–69.

Streb, H., Heslop, J. P., Irvine, R. F., Schulz, I., and Berridge, M. J. (1985). Relationship between secretagogue-induced Ca2 + release and inositol polyphosphate production in permeabilized pancreatic acinar cells. *J. Biol. Chem.* **260**, 7309–7315.

Suprenant, K. A., and Rebhun, L. I. (1983). Assembly of unfertilized sea urchin egg tubulin at physiological temperatures. *J. Biol. Chem.* **258**, 4518–4525.

Suprynowicz, F., and Mazia, D. (1985). Fluctuation of the Ca2 +-sequestering activity of permeabilized sea urchin embryos during the cell cycle. *Proc. Natl. Acad. Sci. U.S.A.* **82**, 2389–2393.

Suzuki, K. (1987). Calcium activated neutral protease: Domain structure and activity regulation. *Trends Biochem. Sci.* **12**, 103–105.

Swann, K., and Whitaker, M. (1983). Stimulation of the Na/H exchanger of sea urchin eggs by phorbol ester. *Nature (London)* **314**, 274–277.

Swann, K., and Whitaker, M. (1986). The part played by inositol trisphosphate and calcium in the propagation of the fertilization wave in sea urchin eggs. *J. Cell Biol.* **103**, 2333–2342.

Turner, P., Jaffe, L. A., and Fein, A. (1986). Regulation of cortical vesicle exocytosis in sea urchin eggs by inositol 1,4,5-trisphosphate and GTP-binding protein. *J. Cell Biol.* **102**, 70–76.

Twigg, J., Patel, R., and Whitaker, M. (1988). Translational control of InsP$_3$-induced chromatin condensation during the early cell cycles of sea urchin embryos. *Nature (London)* **332**, 366–369.

Vale, R. D., Reese, T. S., and Sheets, M. P. (1985). Identification of a novel force-generating protein kinesin involved in microtubule based motility. *Cell (Cambridge, Mass.)* **42**, 32–50.

Vallee, R. B., and Bloom, G. S. (1983). Isolation of sea urchin egg microtubules with taxol and identification of mitotic spindle microtubule-associated proteins with monoclonal antibodies. *Proc. Natl. Acad. Sci. U.S.A.* **80**, 6259–6263.

Vantard, M., Lambert, A. M., De Mey, J., Picquot, P., and Van Eldik, L. J. (1985). Characterization and immunocytochemical distribution of calmodulin in higher plant endosperm cells: Localization in the mitotic apparatus. *J. Cell Biol.* **101**, 488–499.

Wagenaar, E. B. (1983). Increased free Ca+ levels delay the onset of mitosis in fertilized and artificially activated eggs of the sea urchin. *Exp. Cell Res.* **148**, 73–82.

Weisenberg, R. C. (1972). Microtubule formation in vitro in solutions containing low calcium concentrations. *Science* **177**, 1104–1105.

Welsh, M. J., Dedman, J. R., Brinkley, B. R., and Means, A. R. (1978). Calcium dependent regulator protein: Localization in mitotic apparatus of eukaryotic cells. *Proc. Natl. Acad. Sci. U.S.A.* **75**, 1867–1871.

Welsh, M. J., Johnson, M., Zavortnik, M., and McIntosh, J. R. (1981). The distribution of calmodulin in living mitotic cells. *Exp. Cell Res.* **149**, 375–385.

Williams, R. J. P. (1986). The physics and chemistry of the calcium-binding proteins. *Ciba Found. Symp.* **122**, 145–159.

Wolf, B. A., Florholmen, J., Colca, J. R., and McDaniel, M. L. (1987). GTP mobilization of Ca2+ from the endoplasmic reticulum of islets. Comparison with myo-inositol 1,4,5-trisphosphate. *Biochem. J.* **242**, 137–141.

Wolniak, S. M. (1986). Lithium, calcium and mitotic progression in Tradescantia. *J. Cell Biol.* **103**, 137a.

Wolniak, S. M. (1987). Lithium alters mitotic progression in stamen hair cells of *Tradescantia* in a time-dependent and reversible fashion. *Eur. J. Cell Biol.* **44**, 286–293.

Wolniak, S. M., and Bart, K. M. (1985a). The buffering of calcium with quin2 reversibly forestalls anaphase onset in stamen hair cells of Tradescantia. *Eur. J. Cell Biol.* **39**, 33–40.

Wolniak, S. M., and Bart, K. M. (1985b). Nifedipine reversibly arrests mitosis in stamen hair cells of *Tradescantia*. *Eur. J. Cell Biol.* **39**, 273–277.

Yoshimoto, Y., and Hiramoto, Y. (1985). Cleavage in a saponin model of the sea urchin egg. *Cell Struct. Funct.* **10**, 29–36.

Yoshimoto, Y., Iwamatsu, T., and Hiramoto, Y. (1985). Cyclic changes in intracellular free calcium levels associated with cleavage cycles in echinoderm and medaka eggs. *Biomed. Res.* **6**, 387–394.

Zavortink, M., Welsh, M. J., and McIntosh, J. R. (1983). The distribution of calmodulin in living mitotic cells. *Exp. Cell Res.* **149**, 375–385.

15

Arousal of Activity in Sea Urchin Eggs at Fertilization

DAVID EPEL

Department of Biological Sciences
Stanford University
Hopkins Marine Station
Pacific Grove, California 93950

I. INTRODUCTION

The unfertilized egg is dormant, and its metabolism and activity are aroused upon fertilization. The arousal of this cell and concomitant initiation of development has captured the imagination of developmental biologists since fertilization first began to be studied. One example is seen in the 1916 publication of Jacques Loeb (1916) entitled "How the Sperm Saves the Life of the Egg." Perhaps a (male) chauvinistic title, but one which well defines the problem of egg activation—how does the sperm transform the egg, a dormant cell which will die if not fertilized, into an active cell which eventually yields an embryo?

361

Somehow, this sperm–egg contact results in sperm incorporation, restoration of the diploid genome, and an intermeshed sequence of synthetic and structural changes that culminate in a series of mitoses, coordinated cellular movements, and finally a program of gene activity yielding a new embryo.

In this chapter I will review the current status of our knowledge about egg activation at fertilization, concentrating on the sea urchin egg [see recent reviews by Shapiro *et al* (1981), Trimmer and Vacquier (1986), Whitaker and Steinhardt (1985), and Swann *et al.*, (1987)]. I will begin with a brief summary of what is known about the initial signals that emanate from sperm–egg contact, primarily centering around the changes in two cations, a transient increase in intracellular calcium (Ca_i) and a permanent change in intracellular pH (pH_i). These two changes are complete within a few minutes of sperm–egg contact. Somehow their effects are lasting and their interplay initiates the programmed sequence of development. Work in the last few years has provided important insights into the nature of these first events. The challenge for the future (and the major focus of this chapter) is to discern how these transient ionic changes result in egg activation.

II. SIGNAL TRANSDUCTION MECHANISMS

It has been known for more than a decade that, following sperm–egg contact, there occurs a transient increase in Ca_i followed closely in time by an activation of a $Na^+–H^+$ exchanger which elevates pH_i [see reviews by Epel and Dubé (1987), Whitaker and Steinhardt (1985), and Chapter 12 by Turner and Jaffe in this volume]. Work in the last few years has demonstrated that these two events are preceded by the hydrolysis of polyphosphoinositides (PIP_2). The current view is that following sperm–egg interaction there is an increased turnover of PIP_2 (Turner *et al.*, 1984) and transient increases in the levels of inositol trisphosphate (IP_3) and diacylglycerol (DAG) (Kamel *et al.*, 1985; Ciapa and Whitaker, 1986). This IP_3, in common with the situation seen in other cell types, somehow causes the release of Ca^{2+} from intracellular stores, and this most likely accounts for the increase in Ca_i normally seen following fertilization.

The simultaneous production of diacylglycerol ensuing from the hydrolysis of PIP_2 is apparently the signal for activating $Na^+–H^+$ exchange. Thus, incubation of eggs in phorbol esters, which mimic diacylglycerol in activating protein kinase C, initiates a sodium-dependent alkalinization of the cell (Swann and Whitaker, 1985). This effect can also be seen with synthetic diacylglycerols and with other mimics of diacylglycerol (Shen and Burgart, 1986; Lau *et al.*, 1986).

The initial studies showed that microinjection of IP_3 initiates the Ca^{2+}-dependent postfertilization cortical reaction (Whitaker and Irvine, 1984). Subsequent research has demonstrated a direct increase in Ca_i as a consequence of IP_3 injection into eggs (Swann and Whitaker, 1986) or from addition of IP_3 to the microsome fraction of eggs (Clapper and Lee, 1985) or to isolated cortices (Oberdorf *et al.*, 1986). Finally, ultrastructural studies on calcium localization in eggs, using a newly developed fluoride precipitation procedure, showed that the major calcium store in eggs is in the endoplasmic reticulum (Poenie and Epel, 1987) and that this reticulum is transiently emptied of Ca^{2+} after fertilization.

Imaging of this Ca^{2+} release in single cells, using either aequorin or fura-2, indicates that the calcium release occurs as a propagated wave moving through the egg beginning at the point of sperm–egg contact (Swann and Whitaker, 1986). The timing of the calcium release, which slightly precedes the onset of the cortical reaction as assessed by elevation of the fertilization membrane, is consistent with the known calcium sensitivity of the exocytosis (Vacquier, 1975; Baker *et al.*, 1980) and indeed suggests that the exocytosis can be used as an indirect visualization of the intracellular calcium release.

The propagated release of calcium through the cell may be intimately related to PIP_2 hydrolysis. Since a Ca^{2+}-sensitive phospholipase C has been demonstrated in the egg cortex (Whitaker and Aitchison, 1985), an interesting scenario is that release of calcium at one site in the egg cortex would activate this phospholipase, which would in turn break down more PIP_2 releasing additional IP_3, etc. This autocatalytic sequence could thus account for the observed calcium wave propagation through the egg (see Swann and Whitaker, 1986, for a detailed discussion of this concept).

How does sperm–egg contact initiate calcium release? Work of Jaffe and her collaborators suggests that a classic G-protein is involved. For example, injection of GTPγs will directly activate eggs; injection of GDPβs, an antagonist of G-proteins, prevents the activation by the sperm (Turner *et al.*, 1986, 1987). These results suggest that a sperm-specific plasma membrane receptor is involved, whose action is mediated by a G-protein, probably coupled to phospholipase C. If so, the enzyme might be controlled by two factors, sperm and Ca^{2+}, with the Ca^{2+}-sensitivity of the phospholipase C accounting for the propagation through the egg once the initial Ca^{2+} release occurs at the site of "successful" sperm binding to a plasma membrane receptor.

One problem with any hypothesis about a plasma membrane receptor, however, is that although many sperm bind to the egg surface, the wave of calcium release only begins at the point of contact (fusion?) of the fertilizing or "successful sperm," as opposed to all sperm-binding sites. It appears that simple binding of sperm to the egg does not necessarily lead to receptor occupancy and polyphosphoinositide hydrolysis.

An alternative possibility is that sperm–egg fusion causes some intermediary event which then results in polyphosphoinositide hydrolysis. This latter idea is supported by two observations. The first is that injection of sperm extracts into eggs will cause egg activation (Dale *et al.*, 1985).

The second observation, consistent with the idea that sperm might inject an activator, is that sperm–egg fusion apparently occurs long before the calcium rise can be detected and close to the time of initial sperm–egg contact. For example, eggs were impaled with a microelectrode and then fixed for later ultrastructural observations at various times after the first electrical event (fertilization potential) was seen. The eggs were then serially sectioned and the timing of sperm–egg fusion determined relative to the time when the electrical change was first apparent. These results showed morphological criteria of fusion 5–8 sec after the electrical event (Longo *et al.*, 1986).

A second experiment showing that fusion precedes Ca^{2+} release comes from studies on capacitance changes. Sperm are added to eggs through a micropipette that has formed a tight patch around the surface of the egg. Membrane potential changes and capacitance changes measured simultaneously in the patch area indicate that a capacitance change similar to that expected upon sperm–egg fusion occurs almost simultaneously with the initial depolarization (McCulloh and Chambers, 1987).

The above results then indicate that (1) a sperm extract can activate eggs and (2) that sperm–egg fusion is a very early event. These give credence to the idea that fusion and perhaps injection of an activator may be the trigger for the initial PIP_2 hydrolysis, as opposed to a receptor occupancy mechanism coupled to phospholipase C activity as seen in somatic cells.

III. STRATEGIES FOR STUDYING EGG ACTIVATION

A. The Description of the Activation Process

The analysis of egg activation first requires an understanding of what is activated. What metabolic pathways are turned on? What enzymes become more or less active? What structural changes occur? Finally, are these independent or interconnected pathways? Do causal relationships exist between the varied responses of fertilization?

A listing of some of the changes that take place after fertilization is shown in Table I and Fig. 1. Table I is a listing of the changes and Fig. 1 attempts to put these into a chronological sequence so that one can better analyze possible causal interactions.

Figure 2 is a simplistic schema of the types of interactions that might be

TABLE I

Some Changes That Take Place after Fertilization

Membrane events
   Depolarization and hyperpolarization
   $PIP–PIP_2$ turnover
   Ca^{2+} influx
   Cortical granule exocytosis
   Endocytosis
   $Na^+–H^+$ exchange
   Tyrosine kinase increase
   Increased transport (amino acids, phosphate,
     nucleosides)
Cytoskeletal changes
   Actin polymerization in cortex
   Actin extension/bundling in microvilli
   Actin polymerization in fertilization cone
   Microtubule-mediated pronuclear movements
   Movement of acidic vesicles/pigment granules to cortex
Cytoplasmic–nuclear changes
   NAD kinase activation
   Cytoplasmic alkalinization
   G6PD translocation/change in activity
   Lipoxygenase activity
   Stimulation of many enzyme's activities (global effect?)
   Increased respiration
   Increased rate of protein synthesis
     Faster elongation rate
     Recruitment of mRNA
   Initiation of DNA synthesis

involved; for example, the entire activation could be causally linked to one event, to a "master switch" type of reaction, in which subsequent events are linked in series in a dependent fashion. In this case, switching on the master switch or an early link will lead to induction of the subsequent links; conversely, inhibition of the master switch or an early link will lead to inhibition of the sequence.

Alternatively, fertilization might result in activation through production of one or several activating agents which act independently to turn on various processes in a parallel fashion. The two models depicted in Fig. 2 are, of course, extremes, and the activation *in vivo* may involve a mixture of these two modes.

Let us consider some possible mechanisms for egg activation. Referring to Fig. 2, is there a single enzymatic change, a master switch, which is responsible

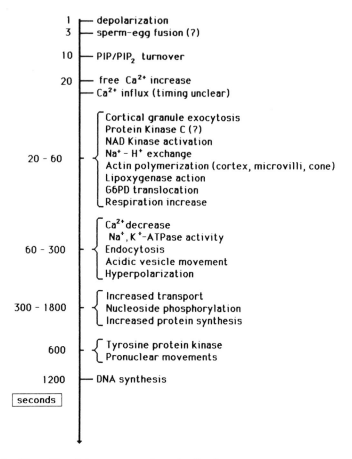

Fig. 1. Timetable of the program of postfertilization events in the zygote of *Strongylocentrotus purpuratus*. Time (in seconds) after sperm–egg contact. The times are meant to give a feeling of the sequence of events and the approximate initiation of each event and its relationship to other postfertilization changes.

for the activation of all pathways in the egg? An example could be changes in the phosphorylation status of many proteins, mediated by protein kinases or protein phosphatases. Another possibility could be a zymogenlike activation, in which a protease is activated early on in the fertilization process (e.g., by calcium) which then turns inactive proenzymes into active enzymes in an irreversible fashion.

Again referring to Fig. 2, another option is for several mechanisms with branch points. For example, covalent modifications such as by phosphorylation

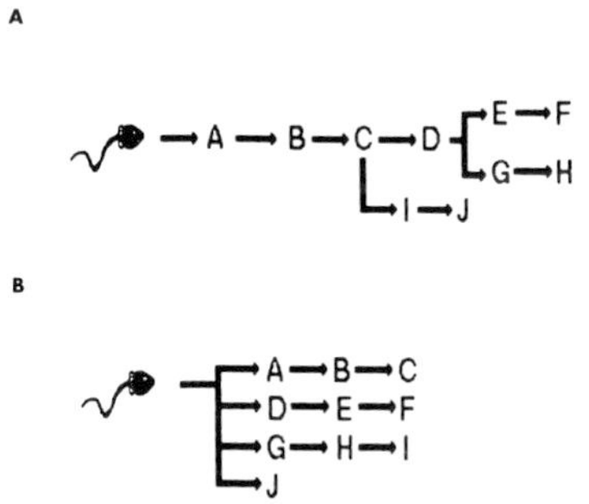

Fig. 2.   Two views of how sperm–egg contact might initiate the postfertilization changes. At one extreme (A) is the "master switch" or "series" or "dependent" model, which assumes one change with all other events emanating from it. The other extreme (B) is a "parallel" or "independent" model in which a number of primary changes occur, which initiate independent sequences of events.

or proteolysis could be the initial change, but the ultimate translation of this phosphorylation or proteolysis into egg activation could then involve cell structural changes. Alternatively, activation could ensue from subtle changes in the cell, such as changes in the state of hydration of proteins associated with changes in the nature of cell water.

B. Experimental Design

Let us next consider experimental strategies for studying egg activation. A critical question is deciding which reactions to study in depth. Should one study a multienzyme system, such as protein and DNA synthesis? Or is it better to study a single enzyme in that pathway? Should one restrict attention only to enzymes involved in metabolic activation? Or is it worthwhile looking at *any* postfertilization change, with the premise that understanding any change will give insights into the general mechanisms of egg activation?

As an example of the utility of this latter approach, assume that one master switch is operative at fertilization. If so, then studying any postfertilization change—whether it be on the main pathway of metabolic activation or on some side track—could lead backward to give insights into this master switch. Similarly, if there are several independent tracks, studying any change could provide insights on that pathway. Also, using this view, one needn't restrict attention to the major synthetic sequellae of activation such as protein synthesis; rather, insights might more easily arise from studies on single enzymes which, although not on the main biosynthetic pathway, are linked to them by some common activation mechanism.

C. Dissection of the Activation Process

Four major strategies have been used for dissecting the metabolic activation of the egg. These are (1) activator candidate approaches, (2) *in vivo* approaches, (3) *in vitro* approaches, and (4) use of permeabilized cells as *in situ* models.

1. Activator Candidate Approaches

By "activator candidate" I refer to studies that test whether specific enzymatic or structural changes are potential candidates affecting metabolic change. For example, one might posit that protein phosphorylation activity is the master switch at fertilization. To test this, one could examine the pattern and amount of protein phosphorylation before and after insemination. Alternatively, one might posit that protease activity is the switch and test this by examining proteolytic activity directly.

a. Phosphorylation. Studies on changes in the phosphorylation status of proteins in sea urchin eggs are not simple, since the unfertilized egg is relatively impermeable to inorganic phosphate. Keller *et al.* (1980) circumvented this problem by analyzing the ability to phosphorylate endogenous substrate proteins in homogenates prepared before and after fertilization. This study showed that there were differences, suggestive of both dephosphorylation and phosphorylation, at fertilization.

Another approach to bypass the permeability problem is to incubate eggs in very high levels of ^{32}P; eventually enough will enter to assess the phosphorylation status of proteins. Using this approach with eggs of *Arbacia punctulata* showed that shortly after fertilization there was increased phosphorylation of the ribosomal protein S6 (Ballinger and Hunt, 1981; Ballinger *et al.*, 1984). This finding raised the exciting possibility that this change might be related to the large turn on of protein synthesis at fertilization. However, a similar study in eggs from two other species showed that although there were comparable increases in protein synthesis, there was no change in the phosphorylation status of this protein (Ward *et al.*, 1983). Thus, there is a change in some cases, but its role is not clear.

Another approach has been to examine the types of protein kinases and protein phosphatases present in eggs. These studies reveal the presence of tyrosine protein kinases, which become activated 5–10 min after fertilization (Kinsey, 1984; Satoh and Garbers, 1985). The role of this kinase(s) and the nature of the substrates phosphorylated is not known. The ubiquity of such kinase activity in terms of growth factor action in somatic cells makes studies of these enzymes in eggs of particular interest.

The immature eggs of organisms that are stimulated by either hormone or fertilization to undergo germinal vesicle breakdown as a part of the egg maturation process exhibit large changes in protein phosphorylation. Part of this appears to be a consequence of the so-called "maturation promoting factor" activity (see, e.g., review by Lohka and Maller, 1987). In eggs of *Urechis*, which are stimulated to begin maturation by fertilization, there is a large step-up in phosphorylation and also in the activity of protein kinases (Meijer *et al.*, 1982). The phosphorylation changes include an early period of both increases and decreases in phosphorylation of specific proteins and then a later generalized increase in the phosphorylation status of many cell proteins. The latter change might be related to phosphorylation of nuclear membrane proteins. The role of the early changes, however, is not known and could be of great importance.

b. Protease Activity. Historically, one of the earliest enzymatic changes described after fertilization was an increase in the activity of various proteases. This was studied intensively by Lundblad and Runnstrom in the late 1940s and 1950s (see, e.g., Lundblad and Runnstrom, 1962). Subsequent work showed that there were at least two proteases released from the cortical granules at fertilization (Vacquier *et al.*, 1972, Carroll and Epel, 1975). It is still unclear as to whether any of the protease activities first described by Lundblad are cytoplasmic and, therefore, involved in egg activation; the possibility is intriguing and merits further work.

c. Other Covalent Modifications. Other types of covalent modification could be associated with fertilization. Examples of these, such as protein myristylation, ADP ribosylation, acetylation, or methylation have not yet been examined, perhaps because of the difficulty of labeling unfertilized eggs; again, these possibilities should be explored.

2. In Vivo *Approaches*

In vivo approaches require first describing *what* occurs and *when* it occurs. Such descriptive and temporal listings can be an especially helpful tool in causal analysis (see, e.g., the article by Epel *et al.*, 1974). Referring to the master switch model in Fig. 2, for example, one might assume that change A which occurs early could be involved in controlling change B, but that change B could not be involved in controlling change A; determining the sequence of events clearly has heuristic value.

Once the sequence of events has been described, one can examine the consequences of interfering with specific events or of inducing specific changes. This approach has the advantage of working with the intact cell, but the disadvantage of often depending upon pharmacological agents which of course are rarely specific.

Another disadvantage, which is becoming apparent from recent experimental dissections of control mechanisms, is that many cells do not use just one regulatory or effector mechanism to control a specific cell function; the consequence is that if the experimenter interferes with one mechanism, an alternative one might take over. One example has come about from studies on secretion in various cell types. Gomperts (1986) has noted that exocytosis can be mediated either by phospholipase C action, utilizing IP_3 to control Ca^{2+}, by diacylglycerol-mediated protein kinase action or by phosphorylation via cyclic AMP-dependent kinases. Inhibiting one mechanism need not prevent the cell response since the alternative regulatory mechanism might compensate. If cells can compensate and use alternate mechanisms, then a danger in pharmacological studies is that one might infer that a particular mechanism is not important when in fact it is the normal *in vivo* regulator.

In spite of the above caveats, we have learned much about regulation of fertilization activities through such *in vivo* approaches. One example is the question of whether the increased protein synthesis that begins 5–10 min after fertilization is required for initiation of DNA synthesis that begins at 30 min after fertilization. This question can be approached by preventing protein synthesis, as through fertilizing eggs in the presence of emetine. Under these conditions, even when the protein synthesis is almost 99% inhibited, eggs will still go through the first cycle of DNA synthesis (Wagenaar and Mazia, 1978). This means that the increased protein synthesis that occurs after fertilization is not required or involved in the later onset of DNA synthesis.

Another example of inhibiting one process and assessing the effects on a later event comes from asking whether the cortical granule exocytosis, which begins at 30 sec after fertilization and is completed by 60 sec, is required for the insertion of the $Na^+–H^+$ exchanger into the plasma membrane. One can approach this by inhibiting the cortical reaction through placing the eggs under high hydrostatic pressure immediately after insemination. When this is done, one finds that the major portion of the fertilization acid release associated with $Na^+–H^+$ exchange still occurs, even though the cortical granule exocytosis has been prevented (Schmidt and Epel, 1983; Swezey *et al.*, 1987). This removes from consideration the possibility that the $Na^+–H^+$ exchanger is inserted into the plasma membrane via cortical granule exocytosis, as occurs in some systems (Gluck *et al.*, 1982).

A third example is the dissection of the role of pH_i and Ca_i in egg activation. As was noted earlier, and will be described in more detail below, the two major ionic changes with consequences for development are a transient rise in Ca_i and a permanent change in pH_i. One can induce the pH_i change directly by incubation of the eggs in ammonia or other weak bases (Shen and Steinhardt, 1978; Johnson and Epel, 1981). This also induces certain postfertilization events, such as increased protein synthesis and initiation of DNA synthesis

(Epel et al., 1974; Mazia and Ruby, 1974), suggesting a direct role of the pH_i change in these synthetic events. When pH_i is increased with ammonia, the calcium increase does not occur (Poenie et al., 1985) and such calcium-dependent changes as activation of NAD kinase or the induction of the cortical granule exocytosis do not take place (see, e.g., Epel et al., 1974, 1981). Thus, using ammonia, one can induce pH_i changes but not calcium changes and can use this intervention to assess which events are Ca^{2+} or pH dependent.

Conversely, allowing the Ca^{2+} change to occur while preventing the pH_i changes can be achieved by activating eggs with the calcium ionophore A23187 in sodium-free seawater. When this is done, NAD kinase activation and cortical granule exocytosis occur (Epel et al., 1981), but the pH_i changes do not take place (Shen and Steinhardt, 1979; Johnson and Epel, 1981). There is also no initiation of DNA synthesis and no increase in rate of protein synthesis (Winkler et al., 1980; Dubé et al., 1985).

3. In Vitro Approaches

Typical in vitro approaches involve breaking up the cell by homogenization as a prelude to assaying activities in the extract or for isolating organelles or regions of the cell. These approaches provide the advantage of a defined or well-controlled system; variables can be limited and indeed purification of individual components can be achieved and reconstitution attempted. A major disadvantage is that the cell integrity is destroyed.

Another problem is the replacement or substitution of the intracellular milieu. Should one utilize a medium that mimics the intracellular one? Or should one utilize a medium that provides optimum activity for the particular process being studied?

A problem with optimization is that such optimum conditions may not be present in vivo. For example, studies on protein synthesis in sea urchin lysates indicate that the optimum Mg^{2+} concentration is 2–4 mM (Winkler and Steinhardt, 1981). Yet the free magnesium concentration in the cell is 5 mM (Sui and Shen, 1986) which is quite inhibitory for protein synthesis, at least in vitro (Winkler and Steinhardt, 1981). This suggests that using the best medium for studying a particular process may be misleading since cells most likely proceed at some "best fit" for the manifold processes occurring in vivo.

Another aspect of choosing the "right" media is the choice of the proper anion. Cells do not have high chloride concentrations, but rather utilize proteins, amino acids, bicarbonate and other small molecules as the major intracellular anions. Indeed, inclusion of chloride in the media can have chaotropic affects and release proteins from cell structures (Sasaki and Epel, 1983). One approach to circumvent such problems, is to substitute organic anions in the media, such as gluconate or glutamate (Mazia et al., 1981; Baker et *al.*, 1980). There are surely better media than these and better mimics of the intracellular milieu will be eagerly awaited.

4. Permeabilized Cell Studies

A new approach, which lies in between the *in vitro* cell homogenization approach and an *in vivo* approach is to use permeabilized cells. These have the advantage of retaining cell structure while having an open plasma membrane so that one can discern the effects of changing various parameters, especially small ions (see, e.g., Baker *et al.*, 1980; Suprynowicz and Mazia, 1985). The disadvantage of this system is that some proteins are generally lost from these leaky cells. Another problem is the aforementioned lack of a good intracellular milieu. One can simulate the major ionic constituents but ultimately missing are the small molecular weight "somethings" which could be important.

5. Concentrated Cell Lysates

A fourth approach, which is proving particularly useful in the analysis of meiosis, is to use extremely concentrated cell lysates, typically with no or little extracellular milieu added (see, e.g., Masui and Shibuya, 1987). The system is ideal for frog oocytes which are fragile and can be disrupted by gentle centrifugation. These lysates have the advantage of retaining proteins in as concentrated a form as possible. As there is no plasma membrane, one can easily ascertain the effects of different media, ions, etc. (see, e.g., Lohka and Maller, 1987).

IV. EXPERIMENTAL ANALYSIS OF EGG ACTIVATION

In the following sections I will go through the changes known to occur after fertilization, and describe their role (if any) in subsequent steps of egg activation. Are there causal–linear chains? Branched chains? Master switches or parallel pathways?

A. Ionic Changes—Calcium

1. The Calcium Increase

As noted, one of the early (i.e., less than 30 sec) postfertilization changes in the sea urchin egg is a large and transient increase in calcium ion, which increases from a resting level of approximately 100 nm to about 2 μM (Poenie *et al.*, 1985; Swann *et al.*, 1987). Pharmacological studies had initially suggested that this calcium increase might be the major and primary trigger for egg activation. For example, eggs will initiate many of the synthetic sequellae of fertilization if Ca^{2+}_i is raised by incubation of eggs in the calcium ionophore A23187 (Steinhardt and Epel, 1974). Conversely, one can inject eggs with EGTA—which will prevent the Ca^{2+} rise—and prevent the activation of the egg by sperm (Zucker and Steinhardt, 1978).

Although this suggests that calcium is critical, the recent analysis of egg activation in terms of PIP_2 metabolism suggests that the role of calcium may be more complicated. First, as noted, there is a calcium-activated phospholipase C (Whitaker and Aitchison, 1985); increasing calcium levels as would occur in ionophore-treated eggs could activate this lipase which would produce both IP_3 and DAG. Many of the subsequent consequences of fertilization might then be mediated through the DAG-related regulation of protein kinase C. Similarly, the effects of EGTA injection could result from preventing phospholipase C activation and the subsequent production of DAG.

2. Calcium Targets

a. Cortical Granule Exocytosis. Irrespective of these caveats, we know a number of events which are calcium-mediated as shown by direct *in vitro* analysis. The most dramatic of these is the induction of cortical granule exocytosis by micromolar amounts of calcium (Vacquier, 1975; Baker *et al.*, 1980). The best description of this has come from studies with isolated cortices, and a number of proteins and factors involved in this calcium-mediated exocytosis are beginning to be identified (Baker *et al.*, 1980; Sasaki, 1984; Zimmerberg *et al.*, 1985).

At present, the only known roles of this exocytosis are for elevation of the fertilization membrane and formation of the hyaline layer. The fertilization membrane has several roles, including a physical block to polyspermy as well as protection for the early embryo (see, e.g., Kay and Shapiro, 1985). The exocytosis is not a prerequisite for development; development proceeds normally if the secretion is blocked by high hydrostatic pressure (Schmidt and Epel, 1983).

b. NAD Kinase. Calcium is also an activator for the calmodulin-regulated enzyme, NAD kinase (Epel *et al.*, 1981). The activity of this enzyme can be assessed *in vivo* by monitoring the postfertilization conversion of NAD to NADP–NADPH (Epel, 1964). This change occurs during the period when the calcium level is high (Swann *et al.*, 1987), and *in vitro* studies show that it is regulated by calcium via calmodulin (Epel *et al.*, 1981). The concentrations of calcium governing this reaction are 0.1–1 μM, which is in the *in vivo* range (Swann *et al.*, 1987).

The targets of increased NADP–NADPH have not been explicitly identified. Clearly the redox potential of the cell is altered, and one can imagine a myriad of resultant effects on biosynthesis. Also, experiments with partially activated eggs suggest a role of reduced pyridine nucleotides in initiation of DNA synthesis (Whitaker and Steinhardt, 1981).

c. Lipoxygenase. A third role for the calcium is the activation of a lipoxygenase. *In vitro,* the sea urchin enzyme converts free arachidonic acid to hydroxy fatty acids, primarily 11-HETE and 12-HETE (Perry and Epel, 1985a; Hawkins and Brasch, 1987). A similar activity can be seen *in vivo;* if one loads eggs with [³H]arachidonic acid and then fertilizes them, the free arachidonic acid is converted into products similar to those produced *in vitro* (Perry and Epel, 1985a, b).

A problem with interpreting this observation, however, is that only the *free* [³H]arachidonic acid added during the experiment is converted to HETE. Most of the added [³H]arachidonate is linked into phospholipids, but this arachidonate is not released at fertilization through the action of phospholipase. Does this mean that there is no increased phospholipase A_2? Or does this mean that there is a large reservoir of phospholipids in the egg, making it difficult to detect release and conversion of this radioactively labeled arachidonate? Clearly, it is important to see whether 11-HETE and 12-HETE are produced *in vivo* and if so to discern their role in egg activation.

d. Protein Kinase C. Presumably this enzyme is active after fertilization. The circumstantial evidence is that diacylglycerol increases after fertilization (Ciapa and Whitaker, 1986) and that phorbol ester and other diacylglycerol analogs will induce Na^+–H^+ exchange (Swann and Whitaker, 1985; Shen and Burgart, 1986; Lau *et al.,* 1986). This suggests that the Na^+–H^+ exchanger is controlled by protein kinase C, but there is yet no direct evidence for this. If protein kinase C is activated, however, then the temporally coincident rise in calcium would augment the activity of this enzyme. An array of protein phosphorylations might then ensue, in addition to the putative phosphorylation of the Na^+–H^+ exchanger.

e. Other Roles for Elevated Calcium. Undoubtedly the calcium increase has other roles in egg activation in addition to those listed above. *In vivo* studies, for example, suggest some role in the large increase in protein synthesis after fertilization (Winkler *et al.,* 1980). There is also the aforementioned role for calcium in the augmentation–propagation of PIP_2 hydrolysis via its effect on phospholipase C (Whitaker and Aitchison, 1985). We recently have described a Ca^{2+}-sensitive PIP kinase, whose action would produce more PIP_2 for this phospholipase (Oberdorf *et al.,* 1988). Clearly there are additional calcium targets and their further description will be eagerly awaited.

3. Calcium and Egg Activation?

The above summary makes clear that elevated cytosolic calcium can have many effects. Are any of these Ca^{2+} targets specifically involved in egg activation leading to cell division? Or are these calcium-mediated actions on sidetracks not necessarily involved in forwarding cell activation?

As noted above, there are no calcium targets *clearly* involved in promoting the subsequent later events of development. The Ca^{2+}-mediated cortical reaction is not necessary for the arousal of cell activity since development proceeds if this Ca^{2+}-mediated exocytosis is prevented by high pressure. The action of Ca^{2+}-mediated NAD kinase, lipoxygenase, and protein kinase would seem important, but specific roles for these enzymes have not yet been shown.

I noted earlier roles for calcium in the initial signal transduction process of fertilization and perhaps this is the major function for the calcium rise in the activation process. Calcium-mediated PIP phosphorylation (Oberdorf *et al.*, 1988) and PIP_2 hydrolysis (Whitaker and Aitchison, 1985) would yield more Ca^{2+} and DAG, and the synergism of Ca^{2+} and DAG on protein kinase C might be the major consequence leading to cell activation. If so, the action of protein kinase C would then be the critical one in forwarding egg activation. So far, the only suggested role for this enzyme is activating $Na^+–H^+$ exchange. The resultant pH_i change may be the critical consequence of the activation sequence, and this possibility is discussed below.

B. Ionic Changes—Intracellular pH

1. The pH_i Increase Is Important

A large number of *in vivo* studies indicate that the postfertilization pH_i increase of about 0.4 units is critical for egg activation. Since the pH_i increase is sodium-dependent and affected by amiloride or amiloride derivatives, one can assess the effects on activation when eggs are fertilized in sodium-free seawater or in the presence of the amiloride drugs (Johnson *et al.*, 1976; Shen and Steinhardt, 1979; Swann and Whitaker, 1985). Such studies show that egg activation is severely repressed when the pH_i increase is prevented. The movements of the sperm and egg pronucleus are prevented and the sequence of postfertilization synthetic changes, such as increased protein synthesis, do not take place (Dubé *et al.*, 1985; Schatten *et al.*, 1985). Exceptions are those events directly mediated by calcium, such as the cortical granule exocytosis and activation of NAD kinase; these occur when $Na^+–H^+$ exchange is inhibited during the activation process, as when eggs are activated in sodium-free seawater with calcium ionophore (Epel *et al.*, 1981).

One can similarly assess the role of pH_i changes by inducing the pH_i increase directly, as by incubating eggs in weak bases such as ammonia. When this is done the pH_i rises to levels even greater than the postfertilization increase (Shen and Steinhardt, 1978, 1979; Johnson and Epel, 1981). The consequences of ammonia incubation on egg activation include such postfertilization events as centration of the egg nucleus and activation of protein and DNA synthesis (Epel *et al.*, 1974; Mazia and Ruby, 1974; Mar, 1980).

In vitro studies on protein synthesis also support the idea that pH_i changes are a major factor leading to increased protein synthesis. Winkler and Lopo have developed excellent *in vitro* protein synthesis systems, which exhibit marked pH sensitivity (see review by Winkler and Grainger, 1987). In these systems there is little protein synthesis below pH 7.0 and a marked acceleration above this pH (see Fig. 3).

2. Problems with the Simple Interpretation that the pH_i Rise Is Important

Although the above results strongly support the idea that the pH_i change is important, there are alternative explanations for these results. For example, part of the evidence rests on experiments where eggs are fertilized in Na^+-free seawater. Such a medium could also have effects on intracellular calcium levels. When free Ca^{2+} increases in cells, the extra Ca^{2+} is often removed from the cell by $Na^+–Ca^{2+}$ exchange (Baker *et al.*, 1980). If this exchange is responsible for decreasing the Ca^{2+} level after fertilization, then the inhibitory

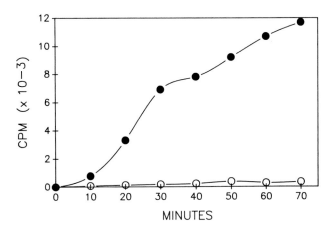

Fig. 3. The pH dependence of protein synthesis in a cell-free system prepared from a sea urchin egg lysate. pH 7.4 (●); pH 6.9 (○). [Redrawn from data of Winkler and Steinhardt (1981).]

effects of low-Na$^+$ media might result not from its effects on pH$_i$ but from the retention of a high and potentially toxic calcium level.

The interpretation of the effects of amiloride are similarly complicated since this drug has other nonspecific effects (see discussion in Epel and Dubé, 1987). Even the amiloride derivatives might be problematic, since fairly high concentrations still must be used (Swann and Whitaker, 1985).

The effects of ammonia similarly may not be as direct and simple as was previously assumed. For example, one can raise pH$_i$ directly by simply raising the extracellular pH (pH$_o$) (Johnson and Epel, 1981). When this is done, protein synthesis is partially activated but there is no initiation of DNA synthesis such as would occur during incubation in ammonia (Dubé and Epel, 1986). Also, the stimulation of protein synthesis at a particular pH$_i$ caused by a high pH$_o$ is not as great as that stimulated by ammonia at a similar pH$_i$ (i.e., when the pH$_i$ is raised to a similar intracellular level by ammonia). This suggests that ammonia is having effects in addition to those of simply raising pH$_i$ (Dubé and Epel, 1986).

Finally, one must be cautious about extrapolating *in vitro* findings to the *in vivo* case. A pertinent example is comparing the pH effects from *in vitro* studies on protein synthesis to the *in vivo* situation. Both *in vivo* and *in vitro* studies are consistent as regard the importance of pH$_i$ in turning on protein synthesis *immediately* after fertilization. One can manipulate the pH$_i$ of embryos by placing them in weak acids to return the pH$_i$ of the embryos to that of the unfertilized egg. If this is done during the first 20 min after insemination, there is inhibition of protein synthesis as expected if it is pH$_i$ sensitive (Dubé *et al.*, 1985). However, if the pH$_i$ is dropped to the unfertilized level after the eggs have been fertilized for 20 min or longer, there is little effect on protein synthesis (Dubé *et al.*, 1985). Thus, there is a discrepancy between the effects of pH$_i$ in vivo and *in vitro*.

The *in vitro* studies show high pH sensitivity (Fig. 3), but this may be a case of overdesigning the media for optimal conditions. Using the permeabilized cell system, we find that the pH sensitivity of protein synthesis is markedly dependent on the medium. If a gluconate–glycine medium is used, which is optimized for protein synthesis, then protein synthesis is indeed very pH sensitive (Winkler and Steinhardt, 1981). However, if one uses a glucamine–gluconate media optimized for studying calcium pumping (Suprynowicz and Mazia, 1985) in permeabilized cells (not necessarily optimum for protein synthesis), then protein synthesis does not show such a marked pH sensitivity (Fig. 4). To be sure, the glycine–gluconate media is undoubtedly more "physiological" (eggs contain glycine as opposed to glucamine), but I make this comparison to illustrate the problem of media, optimization, and pH sensitivity. This problem of optimization was also referred to earlier in terms of the finding that cells contain "inhibitory" levels of magnesium, at least as far as protein synthesis is concerned.

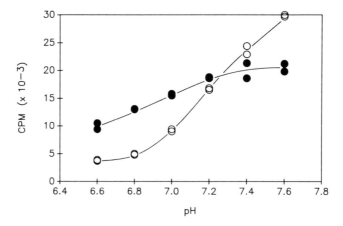

Fig. 4. The pH dependence of protein synthesis in a permeabilized cell system prepared from 20-min old sea urchin embryos. Cells were prepared as described by Swezey and Epel (1988) and then suspended in a glycine–gluconate media (○) similar to that used by Winkler and Steinhardt (1981) or a glucamine–glucanate media (●) similar to that of Suprynowicz and Mazia (1985). The media was supplemented with an amino-acid incorporating system similar to that of Winkler and Steinhardt (1981). Incorporation is expressed as counts per minute (cpm) incorporated into protein after 20-min incubation at the indicated pH.

3. How Might pH Changes Activate the Egg?

In spite of the above caveats in interpretation of experiments where pH$_i$ is altered both *in vivo* and *in vitro*, the simplest interpretation of the *in vivo* studies still is that the pH$_i$ increase is indeed necessary for activation of the egg. An importance is also suggested by the ubiquity of the change, which is seen in so many cases where cell metabolism and activity are increased, be it by fertilization in the case of eggs or by growth factors in the case of differentiated cells (Epel and Dubé, 1987).

The aforementioned study of Dubé *et al.* (1985) indicates that the major effect of the pH$_i$ change may occur in the first few minutes after insemination. Lowering pH$_i$ during this period prevents activation of the egg; however, when the pH$_i$ was reduced to the prefertilization level at later times after fertilization, i.e., later than 10–20 min after insemination, the cells now progress through DNA synthesis and mitosis albeit at a considerably reduced rate (and cytokinesis is impaired). These studies suggest that the pH$_i$ increase is especially critical for 10–20 min postfertilization and that the later role of elevated pH$_i$ is for accelerated rate.

The nature of this initial pH-sensitive process apparently necessary for activation is important to know. Given the wide range of cell activities that are affected (microtubule events involved in male and female pronucleus migration, protein synthesis, DNA synthesis), it is probable that this pH-sensitive target is some global factor in the egg. An intriguing candidate is cell structure. This could be through some direct pH effect on structural macromolecules, or could be an indirect effect via a pH-sensitive enzyme, such as protein kinase or protein phosphatase.

An alternative role for the pH_i change, which could also be related to cell structure affects, is that the Na^+-H^+ exchange activity is critical both to raise pH_i and also to bring in cations to balance charge on critical proteins and other cell buffers (Epel and Dubé, 1987; Epel, 1988). Protons are lost from the cell during the bout of exchange, and the charges on the cell buffers are initially balanced by Na^+ and later by K^+ [the Na^+ that is brought into the cell is subsequently changed for K^+ via action of the Na^+,K^+-ATPase (Payan et al., 1983)]. This view does not mitigate the importance of pH_i changes, but implies that *both* the Na^+ and H^+ components of the exchange are important. The mechanism of ammonia action could then be to both raise pH_i and also to balance charge.

C. Permeabilized Cell Approach

The above analysis of cell activation has emphasized the first three study approaches—activator candidate, *in vivo,* and *in vitro* approaches. We have recently begun using permeabilized cells as a means to probe activation at fertilization and the findings are truly unexpected (Swezey and Epel, 1987, 1988). Our results indicate a change in activity in five out of the six enzymes we have assayed so far. The change appears to be calcium mediated and may represent a global change in the cell, but one which is *only* apparent in the permeabilized cells. (It is not apparent in homogenates.)

Most of this work has focused on the enzyme glucose-6-phosphate dehydrogenase (G6PD). This enzyme was long ago described as undergoing a change from an insoluble to a soluble form following fertilization (Isono, 1963). However, this change was only apparent in low-ionic-strength media; in high-salt media, a milieu more characteristic of the *in vivo* state of the cell, the enzyme was always soluble, and there was no change in activity at fertilization (Isono, 1963). This could mean that the postfertilization change was an artifact.

The reality of this phenomena was supported when we found that this post-fertilization shift in locale could be attained in the high-ionic-strength media characteristic of the cell if the protein concentration was kept high, as could be achieved by adding high concentrations of ovalbumin to the homogenate (Swezey and Epel, 1986). We wondered whether permeabilized cells, which retain the high protein content characteristic of the *in vivo* state, would also show a change in G6PD activity. To our surprise we found that the activity in the permeabilized, unfertilized egg was barely measurable—only about 7% of the total available activity was apparent. If the eggs were fertilized and then permeabilized, we observed a 15 to 20-fold increase in activity to about 15–20% of the available activity. A similar behavior was seen for a number of other enzymes assayed, suggesting that this is a pervasive or global change affecting many enzymes all at once (Swezey and Epel, 1987, 1988).

The nature of the G6PD change after fertilization is not clear but appears to be related to both the Ca^{2+} and pH_i increase. Calcium appears to be indirectly required for the change, and the shift in activity is not induced in eggs by simply increasing pH_i (as by incubation in ammonia). However, the pH_i increase accentuates the extent of the activation of the enzyme; if one prevents the pH_i increase as by activating eggs in sodium-free media, G6PD is activated to about only two-thirds of the normal level. So, although the pH_i increase by itself will not activate this enzymatic change, it augments the response mediated by calcium (Swezey and Epel, 1988).

This new experimental approach using permeabilized cells is a potentially exciting one as it reveals large changes in enzymic activity which are not apparent from standard *in vitro* studies. We are now carrying out *in vivo* measurements to see if these enzymic changes also occur in the intact cell. If so, the study of fertilization with these permeabilized cells could throw much light on the activation process.

Our current speculations are focused on changes in intracellular structure–environment, since the pH–activity profile of the enzyme in the permeabilized cell differs radically from the similar profile in solution (Swezey and Epel, 1988). If so, this could suggest a Ca^{2+}-mediated, pH-requiring alteration of cell structure as a primary consequence of the fertilization response.

V. SUMMARY AND OVERVIEW

This review of the egg activation process has summarized the evidence that changes in Ca^{2+} and pH_i appear to be the primary effectors of structural and metabolic change after fertilization. A major question I have raised about the initial signal transduction event is whether it is a typical sperm receptor-mediated event coupled to PIP_2 hydrolysis, or whether the fertilization process is unique and involves the injection of an "activator" by the sperm.

I have also raised some concerns regarding the Ca^{2+}_i and pH_i changes. If Ca^{2+} is important in activation, what are the targets for elevated Ca^{2+} that act to effect metabolic change? These may be some of the documented Ca^{2+}-sensitive changes of fertilization, such as NAD kinase; if so, however, it is still unclear how these act to forward new cell activity. A potentially exciting target is the apparent Ca^{2+}-mediated shift in G6PD activity seen in permeabilized cells. However, our analysis indicates that Ca^{2+} does not *directly* effect the enzyme in the permeabilized cell, suggesting it may be a secondary consequence of the Ca^{2+} rise. As a final consideration, it may be that the critical Ca^{2+}-dependent change(s) has not yet been described.

Alternatively, is it possible that the sole role of the Ca^{2+} increase is to propagate the cortical reaction around the egg to effect fertilization membrane elevation? Or is its role to simultaneously produce DAG via the Ca^{2+}-sensitive lipase? The production of this protein kinase C activator, in concert with the Ca^{2+} rise, would activate the kinase and this might ultimately be the main effector of cell change.

The protein kinase C-mediated pH_i increase could well be the major effector of new cell activity, but as noted there are alternative interpretations of the experiments supporting the idea that a pH_i change is critical. A major problem is that simply raising the pH_i by raising pH_o is not as effective an activator as when the pH_i is raised to the same level by ammonia. Perhaps ammonia has additional effects, such as acting—in addition to its affect on pH_i—as a cation to balance protein and buffer charge (the Na^+–H^+ exchange activity alters both Na^+ and H^+).

Alternatively, the major effect of ammonia (and pH_i?) may be to initiate a "Ca^{2+} clock" in the cell. When eggs are activated with ammonia a program of cytoplasmic Ca^{2+} changes is also initiated which may be critical for the later mitotic stages (Poenie *et al.*, 1985). [It should be emphasized, however, that the initiation of DNA synthesis and the increased rate of protein synthesis *precede* the ammonia-induced Ca^{2+} rise, indicating that the regulators of these synthetic events are apparently calcium independent (or, alternatively, could elevated pH_i make some processes sensitive to the resting Ca^{2+}_i level?).]

Assuming that pH_i is important, it would appear that the crucial pH-dependent event is occurring early, during the first 10–20 min after fertilization. It will be critical to understand the nature of this event and its relationships, if any, to the calcium rise.

I close on the question of whether we have really described the major concomitants of fertilization. Until recently, it was assumed that the major consequences had been described, such as those depicted in Table I and Fig. 1. But the permeabilized cell data suggests that there may be global changes in cell activity at fertilization. At the moment, these are seen in permeabilized cells, and it is unclear whether a similar modulation of cell activity occurs in the intact cell. If the permeabilized cell is an accurate reflector of the *in vivo* situation, then understanding this global change will represent a major new insight into cell activation.

REFERENCES

Baker, P. F., Knight, D. E., and Whitaker, M. J. (1980). The relation between ionized calcium and cortical exocytosis in eggs of the sea urchin, *Echinus esculentus*. *Proc. R. Soc. London. Ser. B* **207**, 149–161.

Ballinger, D. G., and Hunt, T. (1981). Fertilization in sea urchin eggs is accompanied by 40S ribosomal subunit phosphorylation. *Dev. Biol.* **87**, 277–285.

Ballinger, D. G., Bray, S. J., and Hunt, T. (1984). Studies of the kinetics and ionic requirements for the phosphorylation of ribosomal protein S6 after fertilization of *Arbacia punctulata* eggs. *Dev. Biol.* **101**, 192–209.

Carroll, E. J., Jr., and Epel, D. (1975). Isolation and biological activity of the proteases released by sea urchin eggs following fertilization *Dev. Biol.* **44**, 22–32.

Ciapa, B., and Whitaker, M. (1986). Two phases of inoistol polyphosphate and diacylglycerol production at fertilization. *FEBS Lett.* **195**, 347–351.

Clapper, D., and Lee, H. C. (1985). Inositol trisphosphate induces calcium release from non-mitochondrial stores in sea urchin egg homogenates. *J. Biol. Chem.* **260**, 13947–13954.

Dale, B., De Felice, L. J., and Ehrenstein, G. (1985). Injection of a soluble sperm fraction into sea urchin eggs triggers the cortical reaction. *Experientia* **41**, 1068–1069.

Dubé, F., and Epel, D. (1986). The relation between intracellular pH and rate of protein synthesis in sea urchin eggs and the existence of a pH-independent event triggered by ammonia. *Exp. Cell Res* **162**, 191–204.

Dubé, F., Schmidt, T., Johnson, C. H. and Epel, D. (1985). The hierarchy of requirements for an elevated intracellular pH during early development of sea urchin embryos. *Cell (Cambridge Mass.)* **40**, 657–666.

Epel, D. (1964). A primary metabolic change of fertilization: Interconversion of pyridine nucleotides. *Biochem. Biophys. Res. Commun.* **17**, 62–69.

Epel, D., and Dubé, F. (1987). Intracellular pH and cell proliferation. *In* "Control of Animal Cell Proliferation" (A. L. Boynton and H. L. Leffert, eds.), Vol 2, pp. 364–394. Academic Press, Orlando, Florida.

Epel, D., Steinhardt, R. A., Humphreys, T., and Mazia, D. (1974). An analysis of the partial metabolic derepression of sea urchin eggs by ammonia: The existence of independent pathways. *Dev. Biol.* **40**, 245–255.

Epel, D., Patton, C., Wallace, R. W., and Cheung, W. Y. (1981). Calmodulin activates NAD kinase of sea urchin eggs: An early event of fertilization. *Cell (Cambridge, Mass.)* **23**, 543–549.

Gluck, S., Cannon, C., and Al-Awquati, Q. (1982). Exocytosis regulates urinary acidification in turtle bladder by rapid insertion of H^+ pumps into the laminal membrane. *Proc. Natl. Acad. Sci. U.S.A.* **79**, 4327–4331.

Gomperts, B. D. (1986). Calcium shares the limelight in stimulus coupling. *Trends Biochem. Sci.* **11**, 290–292.

Hawkins, D. J., and Brasch, A. R. (1987). Eggs of the sea urchin, *Strongylocentrotus purpuratus*, contain a prominent (11R) and (12R) lipoxygenase activity. *J. Biol. Chem.* **262**, 7629–7634.

Isono, N. (1963). Carbohydrate metabolism in sea urchin eggs. IV. Intracellular localization of enzymes of the pentose phosphate cycle in unfertilized and fertilized eggs. *J. Fac. Sci., Univ. Tokyo* **10**, 37–53.

Johnson C. H., and Epel, D. (1981). Intracellular pH of sea urchin eggs measured by the dimethyloxazolidinedione (DMO) method. *J. Cell Biol.* **89**, 284–291.

Johnson, J. D., Paul, M., and Epel, D. (1976). Intracellular pH and activation of sea urchin eggs after fertilization. *Nature (London)* **262**, 661–664.

Kamel, C. C., Bailey, J., Schoenbaum, L., and Kinsey, W. (1985). Phosphatidylinositol metabolism during fertilization in the sea urchin egg. *Lipids* **20**, 350–356.

Kay, E. S., and Shapiro, B. M. (1985). The formation of the fertilization membrane of the sea urchin egg. *In* "Biology of Fertilization" (C. B. Metz and A. Monroy, eds.), Vol. 3, pp. 45–81. Academic Press, Orlando, Florida.

Keller, C., Gunderson, G., and Shapiro, B. M. (1980). Altered *in vitro* phosphorylation of specific proteins accompanies fertilization of *Strongylocentrotus purpuratus* eggs. *Dev. Biol.* **74**, 86–100.

Kinsey, W. H. (1984). Regulation of tyrosine-specific kinase activity at fertilization. *Dev. Biol.* **105**, 137–143.

Lau, A. F., Royson, R. C., and Humphreys, T. (1986). Tumor promoters and diacylglycerol activate the Na^+–H^+ antiporter of sea urchin eggs. *Exp. Cell Res* **106**, 23–30.

Loeb, J. (1911). How does the act of fertilization save the life of the egg? *Harvey Lect.* 1910–1911, 22–30.

Lohka, M. J., and Maller, J. L. (1987). Regulation of nuclear formation and breakdown in cell-free extracts of amphibian eggs. *In* "Molecular Regulation of Nuclear Events in Mitosis and Meiosis" (R. A. Schlegal, M. S. Halleck, and P. N. Rao, eds.), pp. 67–110. Academic Press, Orlando, Florida.

Longo, F. J., Lynn, J. W., McCulloh, D. H. and Chambers, E. L. (1986). Correlative ultrastructural and electrophysiological studies of sperm–egg interactions of the sea urchin, *Lytechinus variegatus*. *Dev. Biol.* **118**, 155–166.

Lundblad, G., and Runnstrom, J. (1962). Distribution of proteolytic enzymes in protein fractions from non-fertilized eggs of the sea urchin *Paracentrotus lividus*. *Exp. Cell Res.* **27**, 328.

McCulloh, D. H., and Chambers, E. L., (1986). When does the sperm fuse with the egg? *J. Gen. Physiol.* **88**, 38a–39a.

Mar, H. (1980). Radial cortical fibers and pronuclear migration in fertilized and artificially activated eggs of *Lytechinus pictus*. *Dev. Biol.* **78**, 1–13.

Masui, Y., and Shibuya, E. K. (1987). Development of cytoplasmic activities that control chromosome cycles during maturation of amphibian oocytes. *In* "Molecular Regulation of Nuclear Events in Mitosis and Meiosis" (R. A. Schlegal, M. S. Halleck, and P. N. Rao, eds.), pp. 1–42. Academic Press, Orlando, Florida.

Mazia, D., and Ruby, A. (1974). DNA synthesis turned on in unfertilized sea urchin eggs by treatment with NH_4OH. *Exp. Cell Res.* **85**, 167–172.

Mazia, D., Paweletz, N., Sluder, G., and Fenze, E. M. (1981). Cooperation of kinetochores and pole in the establishment of monopolar mitotic apparatus. *Proc. Natl. Acad. Sci. U.S.A.* **78**, 377–381.

Meijer, L., Paul, M., and Epel, D. (1982). Stimulation of protein phosphorylation during fertilization-induced maturation of *Urechis caupo* oocytes. *Devel. Biol.* **94**, 62–70.

Oberdorf, J., Vilar-Rojas, C., and Epel, D. (1988). Characterization of PIP kinase of sea urchin eggs, *Devel Biol.*, in press.

Oberdorf, J. A. Head, J. F., and Kaminer, B. (1986). Calcium uptake and release by isolated cortices and microsomes from the unfertilized egg of the sea urchin *Strongylocentrotus purpuratus*. *J. Cell. Biol.* **102**, 2205–2210.

Payan, P., Girard, J. -P., and Ciapa, B. (1983). Mechanisms regulating intracellular pH in sea urchin eggs. *Dev. Biol.* **100**, 29–38.

Perry, G., and Epel, D. (1985a). Characterization of a Ca^{+2}-stimulated lipid peroxidation system in the sea urchin egg. *Dev. Biol.* **107**, 47–57.

Perry, G., and Epel, D. (1985b). Fertilization stimulates lipid peroxidation in the sea urchin egg. *Dev. Biol.* **107**, 58–65.

Poenie, M., and Epel, D. (1987). Ultrastructural localization of intracellular calcium stores by a new cytochemical method. *J. Histochem. Cytochem.* **35**, 939–956.

Poenie, M., Alderton, J., Tsien, R., and Steinhardt, R. (1985). Changes of free calcium levels with stages of the cell division cycle. *Nature (London)* **325**, 147–149.

Sasaki, H. (1984). Modulation of calcium sensitivity by a specific cortical protein during sea urchin egg cortical vesicle exocytosis. *Dev. Biol.* **101**, 125–135.

Sasaki, H., and Epel, D. (1983). Cortical vesicle exocytosis in isolated cortices of sea urchin eggs: Description of a turbidometric assay and its utilization in studying effects of different media on discharge. *Dev. Biol.* **98**, 327–337.

Satoh, N., and Garbers, D. L. (1985). Protein tyrosine kinase activity of eggs of the sea urchin *Strongylocentrotus purpuratus:* The regulation of its increase after fertilization. *Dev. Biol.* **111**, 515–519.

Schatten, G., Bestor, T., Balczon, R., Henson, J., and Schatten, H. (1985). Intracellular pH shift leads to microtubule assembly and microtubule-mediated motility during sea urchin fertilization. *Eur. J. Cell Biol.* **36**, 116–127.

Schmidt, T., and Epel, D. (1983). High hydrostatic pressure and the dissection of the fertilization responses. I. The relationship between cortical granule exocytosis and proton efflux during fertilization of the sea urchin egg. *Exp. Cell Res.* **146**, 235–248.

Shapiro, B. M., Schackmann, R. W., and Gabel, C. A. (1981). Molecular approaches to the study of fertilization. *Annu. Rev. Biochem.* **50**, 815–843.

Shen, S. S., and Burgart, L. J (1986). 1,2-diacylglycerols mimic phorbol 12-myristate 13-acetate activation of the sea urchin egg. *J. Cell. Physiol.* **127**, 330–340.

Shen, S. S., and Steinhardt, R. A. (1978). Direct measurement of intracellular pH during metabolic derepression of the sea urchin egg. *Nature (London)* **272**, 253–254.

Shen, S. S., and Steinhardt, R. A. (1979). Intracellular pH and the sodium requirement at fertilization. *Nature (London)* **282**, 87–89.

Steinhardt, R. A., and Epel, D. (1974). Activation of sea urchin eggs by a calcium ionophore. *Proc. Nat. Acad. Sci. USA* **71**, 1915–1919.

Sui, A. -L., and Shen, S. S. (1986). Intracellular free magnesium concentration in the sea urchin egg during fertilization. *Dev. Biol.* **114**, 208–213.

Suprynowicz, F. A., and Mazia, D. (1985). Fluctuation of the Ca^{2+}-sequestering activity of permeabilized sea urchin embryos during the cell cycle. *Proc. Nat. Acad. Sci. USA,* **82**, 2389–2393.

Swann, K., and Whitaker, M. (1985). Stimulation of the $Na^+–H^+$ exchanger of sea urchin eggs by phorbol ester. *Nature (London)* **314**, 274–275.

Swann, K., and Whitaker, M. (1986). The part played by inositol trisphosphate and calcium in the propagation of the fertilization wave in sea urchin eggs. *J. Cell Biol.* **103**, 2333–2342.

Swann, K., Ciapa, B., and Whitaker, M. (1987). Cellular messengers and sea urchin egg activation. *In* "Molecular Biology of Invertebrate Development" (J. D. O'Connor, ed.), pp. 45–69. Alan R. Liss, New York.

Swezey, R. R., and Epel, D. (1986). Regulation of glucose-6-phosphate dehydrogenase activity in sea urchin eggs by a reversible association with cell structural elements. *J. Cell Biol.* **103**, 1509–1515.

Swezey, R. R., and Epel, D. (1987). Regulation of egg metabolism at fertilization. *In* "Molecular Biology of Invertebrate Development" (J. D. O'Connor, ed.), pp. 71–86. A. R. Liss, New York.

Swezey, R. R., and Epel, D. (1988). Enzyme activity revealed in electrically permeabilized sea urchin eggs. *Proc. Natl. Acad. Sci. U.S.A.* **85**, 812–816.

Swezey, R. R., Schmidt, T., and Epel, D. (1987). Effects of hydrostatic pressure on actin assembly and initiation of amino acid transport upon fertilization of sea urchin eggs. *In* "Current Perspectives in High Pressure Biology" (H. W. Jannasch, R. E. Marquis, and A. M. Zimmerman, eds.), pp. 95–110. Academic Press, London.

Turner, P. R., Sheetz, M. P., and Jaffe, L. A. (1984). Fertilization increases the polyphosphoinositide content of sea urchin eggs. *Nature (London)* **310**, 414–415.

Turner, P. R., Jaffe, L. A., and Fein, A. (1986). Regulation of cortical vesicle exocytosis by inositol 1,4,5-trisphosphate and GTP binding protein. *J. Cell Biol.* **102**, 70–76.

Turner, P. R., Jaffe, L. A., and Primakoff, P. (1987). A cholera toxin-sensitive G protein stimulates exocytosis in sea urchin eggs. *Dev. Biol.* **120**, 577–583.

Trimmer, J. S., and Vacquier, V. D. (1986). Activation of sea urchin gametes *Annu. Rev. Cell Biol.* **2**, 1–26.

Vacquier, V. D. (1975). The isolation of intact cortical granules from sea urchin eggs: Calcium ions trigger granule discharge. *Dev. Biol.* **43**, 62–74.

Vacquier, V. D., Epel, D., and Douglas, L. A. (1972). Sea urchin eggs release protease activity at fertilization. *Nature (London)* **237**, 34–36.

Wagenaar, E. B., and Mazia, D. (1978). The effect of emetine on first cleavage division in the sea urchin, *Strongylocentrotus purpuratus*. *In* "Cell Reproduction: In Honor of Daniel Mazia" (E. R. Dirksen, D. M. Prescott, and C. F. Fox, eds.), pp. 539–546. Academic Press, New York.

Ward, G. E., Vacquier, V. D., and Michel, S. (1983). The increased phosphorylation of ribosomal protein S6 in *Arbacia punctulata* is not a universal event in the activation of the sea urchin egg. *Dev. Biol.* **95**, 360–371.

Whitaker, M. J., and Aitchison, M. (1985). Calcium-dependent polyphosphoinositide hydrolysis is associated with exocytosis in vitro. *FEBS Lett.* **182**, 119–124.

Whitaker, M. J., and Irvine, R. F. (1984). Inositol 1,4,5-trisphosphate microinjection activates sea urchin eggs. *Nature (London)* **312**, 636–639.

Whitaker, M. J., and Steinhardt, R. A. (1981). The relation between the increase in reduced nicotinamide nucleotides and the initiation of DNA synthesis in sea urchin eggs. *Cell (Cambridge, Mass.)* **25**, 95–103.

Whitaker, M. J., and Steinhardt, R. A. (1985). Ionic signaling in the sea urchin egg at fertilization. *In* "Biology of Fertilization" (C. B. Metz and A. Monroy, eds.), Vol. 3, pp. 168–222. Academic Press, Orlando, Florida.

Winkler, M. M., and Grainger, J. L. (1987). Regulation of protein synthesis during early development in sea urchin eggs. *In* "Molecular Biology of Invertebrate Development" (J. D. O'Connor, ed.), pp. 95–116. A. R. Liss, New York.

Winkler, M. M., and Steinhardt, R. A. (1981). Activation of protein synthesis in a sea urchin cell-free system. *Dev. Biol.* **84**, 432–439.

Winkler, M. M., Steinhardt, R. A., Grainger, J. L., and Minning, L. (1980). Dual ionic controls for the activation of protein synthesis at fertilization *Nature (London)* **287**, 558–560.

Zimmerberg, J., Sardet, C., and Epel, D. (1985). Exocytosis of sea urchin egg cortical vesicles *in vitro* is retarded by hyperosmotic sucrose: Kinetics of fusion monitored by quantitative light microscopy. *J. Cell Biol.* **101**, 2398–2410.

Zucker, R., and Steinhardt, R. A. (1978). Prevention of the cortical reactions in fertilized sea urchin eggs by injection of calcium-chelating ligands. *Biochim. Biophys. Acta* **541**, 459–466.

Index

A